Trace Element Speciation for Environment, Food and Health

Trace Element Speciation for Environment, Food and Health

Edited by

Les Ebdon, Les Pitts
University of Plymouth, UK

Rita Cornelis
University of Gent, Belgium

Helen Crews
Central Science Laboratory, Sand Hutton, York, UK

O.F.X. Donard
Université de Pau et des Pays de l'Adour, France

Philippe Quevauviller
European Commission, Brussels, Belgium

RS•C
ROYAL SOCIETY OF CHEMISTRY

ISBN 0-85404-459-0

A catalogue record for this book is available from the British Library

Published by The Royal Society of Chemistry,
Thomas Graham House, Science Park, Milton Road, Cambridge CB4 0WF, UK

Registered Charity Number 207890

For further information see our web site at www.rsc.org

Typeset in Great Britain by Vision Typesetting, Manchester
Printed by MPG Books Ltd, Bodmin, Cornwall, UK

Foreword

The growing awareness of speciation is reflected by the increasing number of analyses performed in research and routine laboratories in various fields *e.g.* food and agriculture, environment, medicine, industry, in which species are determined. The determination of the 'chemical forms of elements' forms the basis on which to understand the bio-geochemical cycle of contaminants in terrestrial and aquatic ecosystems and for detecting possible harmful substances which might be toxic to biota and humans. Besides the importance of this tool for risk assessment studies, speciation is also highly relevant for testing the quality of products, *e.g.* the amount of essential elements in food products, impurities in pharmaceutical products or chemical substances *etc.* and is, therefore, of potential interest to industry.

The evolution of this awareness is, however, quite paradoxical. Dramas were necessary to alert the public to toxic 'forms' of elements, *e.g.* the high toxicity of methylmercury identified in Minamata (Japan) in the 1950s, or organotin impurities in medicaments (the 'Stalinon' problem) in the 1960s. Studies of trace elements partitioning in sediments and soils were initiated in the 1970s, *e.g.* to evaluate the mobility and/or bio-availablility of heavy metals, but speciation was still considered to be an 'academic exercise' at that time, rather than a regulatory tool.

At the beginning of the 1980s, high mortality of oysters in the Arcachon Bay (France) due to tributyltin leached from antifouling paints, and subsequent economic problems in this area, justified strong research efforts which were crowned with success and opened the way to a larger field of investigations. Results were reflected in an increasing awareness from the legislative point of view, since national regulations were implemented (*e.g.* the banning of TBT-based antifouling paints, systematic monitoring of methylmercury in fish by control laboratories, *etc.*); furthermore, the term 'elements and their compounds' appeared in EC Directives related to environmental protection. This decade was also marked by a flourishing development of new instruments and methods for the determination of a wide variety of chemical species.

One could have thought that the 1990s would have been a key decade for speciation with an expanding market for new instruments and a large range of regulations. This trend was, however, not so marked as was anticipated. Huge efforts were actually made to improve the state-of-the-art of speciation analysis within Europe, *e.g.* projects funded by the European Commission (BCR programme and successors) which enabled cross-checking of techiques developed by expert EU laboratories and the demonstration of the comparability of data. Six to seven years were necessary to obtain a good picture of the state-of-the-art for the speciation of some elements, *e.g.* As, Hg, Pb and Sn, and to certify suitable reference materials for quality control purposes. While these efforts were highly necessary and permitted the identification of robust methodologies and the rejection of unqualified techiques, a gap existed in the transfer of knowledge from expert laboratories to routine laboratories. This was recognised by some instrument manufacturers at the beginning of the 1990s, but efforts were limited and still did not result in the creation of new markets. Moreover, much remains to be done to disseminate information on the importance of pursuing research and development efforts in the area of speciation, in particular for decision-makers, *e.g.* regulators and industrialists.

The start of the 21st century will represent a cornerstone for speciation. Some projects have recently started, aiming to develop simple methodologies which are readily marketable and usable by routine control laboratories (provided along with Standard Operating Procedures). Certified reference materials are now increasingly available for the quality control of speciation analysis and all the requirements are met for speciation to be considered at the same level as trace organic analysis.

There is still a clear need, however, to enhance multi-disciplinary collaborations and boost communication of research results to decision makers and end-users (*e.g.* legislators, industrialists, routine control laboratories). The best strategy for a way forward will be to share views and expertise among experts from different disciplines with complementary experience, to communicate with decision-makers and end-users and to discuss efforts necessary for the transfer of knowledge to routine laboratories. This has been well understood by the European scientific communty which decided to act through the development of a Thematic Network entitled 'Speciation 21' funded by the European Commission's Standards, Measurements and Testing programme, and run for two years (1998–2000).

The 'Speciation 21' network aims were to tackle problems related to the lack of communication between scientists, industry representatives and legislators for the possible improvement of written standards and EC regulations. As stressed before, legislation at present mainly concerns total element concentrations, which in many cases is insufficient for an accurate risk evaluation (*e.g.* for environment contamination, food quality, health risks *etc.*). The main objective of the network was to bring together scientists with a background in analytical chemistry interested in speciation method development, with potential users from industry and representatives from legislative bodies. The network organized a series of meetings to debate all the important questions related to environmental, food and occuaptional health aspects of speciation. This book gives a detailed review of the state-of-the-art of speciation issues in the occupational health, food and environment sectors, along with the main conclusions of round-table discussions held during expert meetings; it provides an insight into applied research in the speciation field, which was recognised to be in support of Community policies and industrial competitiveness. As such, this document opens new areas of investigation, which should enable speciation to progress and find its rightful place in the analytical world of the 21st century.

Philippe Quevauviller
Brussels, June 1999

IUPAC Definitions for Terms Related to Chemical Speciation and Fractionation of Elements

In order to circumscribe the domain of element speciation, it is mandatory to quote the IUPAC (International Union for Pure and Applied Chemistry) recommendations for the definition of terms related to the chemical speciation of elements: 'chemical species', 'speciation analysis', 'speciation of an element', 'speciation and fractionation'.[1]

Chemical species. <chemical elements> specific form of an element defined as to isotopic composition, electronic or oxidation state, and/or complex or molecular structure.

Speciation analysis. <analytical chemistry> analytical activities of identifying and/or measuring the quantities of one or more individual chemical species in a sample.

Speciation of an element; speciation. Distribution of an element amongst defined chemical species in a system.

It is, however, often not possible to determine the concentration of the different individual chemical species that sum up the total concentration of an element in a given matrix (*e.g.* in metal–humic complexes or metal complexes in biological fluids), which means that it is impossible to determine the speciation. The practice is then to identify various classes of species of an element and to determine the sum of its concentrations in each class. This practice is useful and will continue. Such fractionations can be based on many different properties of the chemical species, such as size, solubility, affinity, charge and hydrophobicity. Consistent with restriciton of the use of the term speciation to defined chemical species, fractionation has been defined as follows:

Fractionation. Process of classification of an analyte or a group of analytes from a certain sample according to physical (*e.g.* size, solubility) or chemical (*e.g.* bonding, reactivity) properties.

The terms are described in detail in the original manuscript.[1] Methodological approaches to achieve speciation and fractionation analysis are also reported.

Reference

1. D.M. Templeton, F. Ariese, R. Cornelis, L.-G. Danielsson, H. Muntau, H.P. van Leeuwen and R. Lobinski, IUPAC recommendations, *Pure Appl. Chem.*, 2000, **72**, 143.

Rita Cornelis, Ghent, Belgium

Contents

Environment

Food

Health

Overview

List of Abbreviations

Abbreviations, in particular those for chemical species, are generally defined at their point of use. A list of commonly used abbreviations follows.

AAS   atomic absorption spectrometry
AES   atomic emission spectrometry
AFS   atomic fluorescence spectrometry
APCI   atmospheric pressure chemical ionisation
API   atmospheric pressure ionisation
AQA   analytical quality assurance
ASE   accelerated solvent extraction
BCR   Community Bureau of Reference
CAC   Codex Alimentarius Commission
CE   capillary electrophoresis
CRM   certified reference material
CSV   cathodic stripping voltammetry
CT   cold trapping
CV   cold vapour
CZE   capillary zone electrophoresis
ECD   electron capture detector
EQT   environmental quality target
ES   electrospray
FAFS   flame atomic fluorescence spectrometry
FID   flame ionisation detector
FPD   flame photometric detector
GC   gas chromatography
HG   hydride generation
HPLC   high performance liquid chromatography
IC   ion chromatography
ICP   inductively coupled plasma
LC   liquid chromatography
LCM   laboratory control material
MAE   microwave assisted extraction
ML   maximum level
MS   mass spectrometry
NIES   National Institute of Environmental Studies (Japan)
NIST   National Institute of Standards and Technology (USA)

NRCC   National Reasearch Council of Canada
OEL   occupational exposure limit
PEL   permissible exposure limit
PNEC   predicted no effect concentration
RM   reference material
SFE   supercritical fluid extraction
SPE   solid phase extraction
SPME   solid phase microextraction
TBC   toxicological benchmark concentration
TDI   tolerable daily intake
TDS   total diet study
TLR   threshold limit range
TLV   threshold limit value
TWI   tolerable weekly intake

Contributors

D. Amouroux, *Laboratoire de Chimie Analytique Bioinorganique et Environnement, UMR CNRS 5034, Université de Pau et des Pays de l'Adour, 64000 Pau, France*

F. Ariese, *Department of Analytical Chemistry and Applied Spectroscopy, Vrije Universiteit, Amsterdam, The Netherlands*

T. Berg, *Danish Veterinary and Food Administration, Institute of Food Chemistry and Nutrition, Mørkhøj Bygade 19, DK-2860 Søborg, Denmark*

S. Branch, *RHM Technology, The Lord Rank Centre, Lincoln Road, High Wycombe, Bucks. HP12 3QR, UK*

C. Cámara, *Departamento de Química Analitica, Universidad Complutense de Madrid, Avenida Complutense S/N, 28040 Madrid, Spain*

R. Cornelis, *Laboratory for Analytical Chemistry, University of Gent, Proeftuinstraat 86, B-9000 Gent, Belgium*

H.M. Crews, *Central Science Laboratory, Sand Hutton, York YO41 1LZ, UK*

G. Darrie, *Elementis-Chromium, Eaglescliffe, Stockton-on-Tees, ST16 0QG, UK*

O.F.X. Donard, *Laboratoire de Chimie Analytique Bioinorganique et Environnement, UMR CNRS 5034, Université de Pau et des Pays de l'Adour, 64000 Pau, France*

J.H. Duffus, *The Edinburgh Centre for Toxicology, 43 Mansionhouse Road, Edinburgh EH9 2JD, UK*

L. Ebdon, *Plymouth Environmental Research Centre, University of Plymouth, Drake Circus, Plymouth PL4 8AA, UK*

H. Emons, *Research Center Jülich, Environmental Specimen Bank, D-52425 Jülich, Germany*

P. Fodor, *Department of Chemistry and Biochemistry, University of Horticulture and Biochemistry, Villányi u. 29–32, H-1114 Budapest, Hungary*

I. Giráldez, *Departamento de Química y Ciencia de los Materiales, Escuela Politécnica Superior, Universidad de Huelva, Campus de La Rábida, 21819-Palos de la Frontera (Huelva), Spain*

J.L. Gómez-Ariza, *Departamento de Química y Ciencia de los Materiales, Escuela Politécnica Superior, Universidad de Huelva, Campus de La Rábida, 21819-Palos de la Frontera (Huelva), Spain*

M. Horvat, *Department of Environmental Sciences, Jožef Stefan Institute, Jamova 39, 1000 Ljubljana, Slovenia*

M. Huvinen, *Occupational Health Service, Outokumpu, PO Box 27, Riihitontuntie 7D, FIN-02201 Espoo, Finland*

M. Jalón, *Gobierno Vasco, Departamento de Sanidad, María Díaz de Haro 60, 48010 Bilbao, Spain*

E.H. Larsen, *Danish Veterinary and Food Administration, Institute of Food Chemistry and Nutrition, Mørkhøj Bygade 19, DK-2860 Søborg, Denmark*

G. Lespes, *Laboratoire de Chimie Analytique Bioinorganique et Environnement, UMR CNRS 5034, Université de Pau et des Pays de l'Adour, 64000 Pau, France*

M.L. Macho, *Gobierno Vasco, Departamento de Sanidad, María Díaz de Haro 60, 48010 Bilbao, Spain*

S. Moesgaard, *Pharma Nord, Sadelmagervej 30–32, DK-7100 Vejle, Denmark*

R. Morabito, *ENEA – TEIN/CHIM, SP Anguillarese 301, I-00060 Rome, Italy*

E. Morales, *Departamento de Química y Ciencia de los Materiales, Escuela Politécnica Superior, Universidad de Huelva, Campus de La Rábida, 21819-Palos de la Frontera (Huelva), Spain*

R. Morrill, *Pharma Nord, Sadelmagervej 30–32, DK-7100 Vejle, Denmark*

G.M. Morrison, *Department of Water, Environment, Transport, Chalmers University of Technology, SE-41296 Göteborg, Sweden*

R. Muñoz-Olivas, *Departamento de Química Analitica, Universidad Complutense de Madrid, Avenida Complutense S/N, 28040 Madrid, Spain*

B. Neidhart, *GKSS Research Centre, Institute of Physical and Chemical Analysis, Max-Planck Straße, 21502 Geesthacht, Germany*

L. Pitts, *Plymouth Environmental Research Centre, University of Plymouth, Drake Circus, Plymouth PL4 8AA, UK*

S. Pugh Williams, *Occupational Health Services, INCO Europe Ltd, Clydach, Swansea SA6 5QR, UK*

Ph. Quevauviller, *European Commission, Competitive and Sustainable Growth Programme, rue de la Loi 200 (MO75 3/9), B-1049 Brussels, Belgium*

S. Rauch, *Department of Water, Environment, Transport, Chalmers University of Technology, SE-41296 Göteborg, Sweden*

E. Rosenberg, *Institut für Analytische Chemie, Technische Universität, Wien, Austria*

D. Sánchez-Rodas, *Departamento de Química y Ciencia de los Materiales, Escuela Politécnica Superior, Universidad de Huelva, Campus de La Rábida, 21819-Palos de la Frontera (Huelva), Spain*

M. Unger-Heumann, *Merck KGaA, Dep. SLP PM SPA, D-64271 Darmstadt, Germany*

I. Urieta, *Gobierno Vasco, Departamento de Sanidad, María Díaz de Haro 60, 48010 Bilbao, Spain*

P. Van Dael, *Nestlé Research Centre, PO Box 44, Vers-chez-les-Blanc, CH-1000 Lausanne, Switzerland*

R. Wood, *Food Standards Agency, c/o Institute of Food Research, Norwich Research Park, Colney, Norwich NR4 7UA, UK*

General Aspects

CHAPTER 1

Accuracy and Traceability in Speciation Analysis

PHILIPPE QUEVAUVILLER

1 Introduction

Traceability issues are of increasing concern in all fields where chemical measurements form the basis for decisions. The concepts of accuracy and traceability as applied to environmental analysis are, however, still prone to misunderstanding. Some years ago, Horwitz stated that 'considerable evidence exists in the literature that few analytical chemists pay attention to the question of the reliability of the analytical results they produce. These chemists believe that a natural law exists in measurement science, that if the directions for conducting a measurement are followed, the true value necessarily results'.[1] This corresponds to the long-term debate about precision or reproducibility over accuracy,[2] which is now relayed by on-going discussions on accuracy and traceability: while accuracy refers to the closeness of analytical values to 'true values' (trueness) and among various repetitions (precision), the term traceability implies a link between the data obtained and established references, through an unbroken chain of comparisons, all with stated uncertainties. Recent controversial discussions have illustrated the misunderstanding, which may occur among the analytical community with respect to accuracy and traceability issues in the area of speciation analysis, with possible consequences on environmental data interpretation.

Speciation analysis is no longer a new feature. IUPAC defines this term as 'the analytical activity of identifying and measuring the quantity of one or more individual chemical species in a sample'.[3] The speciation of an element is defined as 'the distribution of defined chemical species of an element in a system'.[3] Chemical species of some elements (*e.g.* organotins, organomercury compounds) are now included in the list of substances to be determined in the frame of international environmental programmes, requiring an increasing knowledge and care with respect to the quality control (including traceability issues) of all monitoring steps (from sampling to reporting data).[4]

Analytical techniques used for the determination of chemical species are generally based on a succession of steps (*e.g.* extraction, derivatisation, separation, detection) all of which are prone to various sources of systematic errors. Within the last decade, international collaborative efforts (through inter-laboratory studies and certification of reference materials) have enabled the systematic study of hyphenated techniques used for the determination of chemical species of, *e.g.*, arsenic, chromium, mercury, lead, tin and selenium in environmental matrices (water, fish or mussel tissues, sediments).[5-7] The determination of operationally-defined element fractions (extractable forms of trace elements) in sediment and soil matrices has also been collaboratively studied, mainly to harmonise and standardise extraction schemes, in order to improve the comparability of data (stressing that, strictly speaking, this type of determination should not be covered by the term 'speciation').[8] In this context, all of these collaborative efforts have been understood as being directed towards the drive for accuracy (trueness and precision as defined below). It has been recognised recently that these achievements have actually led primarily to the establishment of reference points (*e.g.* certified values in reference materials), which do not necessarily correspond to 'true values', but offer a mean with which laboratories may compare their data internationally and, hence, achieve traceability. This ambiguity still generates confusion and misunderstanding among the scientific community. This chapter discusses this issue, focusing on analytical measurements only. Extending discussions on general traceability issues would imply an examination of steps prior to laboratory work (sampling, storage, *etc.*) which is beyond the scope of this contribution.

2   Accuracy

The accuracy concept covers the terms *trueness* and *precision*.[9] Trueness is defined as 'the closeness of agreement between the "true value" and the measured value', whereas precision is 'the closeness of agreement between the results obtained by applying the same experimental procedure several times under prescribed conditions'. Trueness relies on the *true value* of the substance to be measured, which is defined as 'a value, which would be obtained by measurement, if the quantity could be completely defined and if all measurement imperfections could be eliminated'. We will discuss later in this chapter what the practical implications for speciation analysis are.

There are three recognised ways to evaluate accuracy:[2]

1   comparison with an 'independent' method (*i.e.* with different measurement principles and different sources of errors),
2   comparison with other laboratories, and
3   use of Certified Reference Materials.

The principle of establishing certified values of reference materials on the basis of comparisons of independent methods used by different (independent) laboratories has been followed by the European Commission (BCR and its successors).[10] The certified values were long considered to reflect the best estimates of the true values of the certified substances. As discussed below, some may argue that certified values actually represent reference points to achieve *traceability*, but that they are not necessarily to be considered as true values for the verification of *accuracy*. Strictly speaking, measurement results are accurate when both the result and its uncertainty are described in units from the 'Système International' (SI units),[11] *i.e.* the kg or the mole for analytical measurements. Previous discussions have underlined that the use of SI units in chemical measurements is quite unrealistic and is only applicable if primary methods ('method having the highest metrological quality, for which a complete uncertainty statement can be written down in terms of SI units', *e.g.* gravimetry, titrimetry, coulometry, IDMS) are involved.[12,13] This will be discussed later in this chapter.

3   The Traceability Concept as Applied to Speciation

Traceability is defined as 'the property of the result of a measurement or the value of a standard whereby it can be related to stated references, usually national or international standards, through an unbroken chain of comparisons all having stated uncertainties'.[9] The application of this concept to chemical measurements has been extensively discussed over the past ten years.[11,14] The practice often differs from theory: indeed, analytical chemists usually describe their results (amounts of chemical compounds) in terms of weight or mass, whereas metrologists underline that weighing ignores the chemical nature of the measurand and the fact that particles interact with each other in chemical reactions and not masses of matter.[15] It is recognised that the actual measurement of the 'amount of substances' correspond to approximations, consisting of measuring ratios of 'weight' and converting them into ratios of amount by means of 'atomic weights' or 'molecular weights', which is considered accurate enough for most chemical purposes.[15] In the strict metrological sense, the 'approximations' do not allow one to demonstrate that the measurements are traceable to the relevant SI unit, *i.e.* the mole. The arguments developed by metrologists are that, in any chemical reaction, the masses of reacting compounds change (even if the effect is extremely small) since there is energy uptake or production. Furthermore, mass is a property of matter with is basically inert, which is not the case for (amount of) particles, which explains why the SI system distinguishes 'mass' from 'amount of substance'. As stressed above, the practice is far removed from theory, and a recent discussion of the key elements of traceability was conducted between analytical chemists and metrologists, to understand the concept and make it applicable to routine chemical measurements.[16] The three key elements concern (1) the link to stated references, (2) the unbroken chain of comparison and (3) the stated uncertainties. The following paragraphs examine how these concepts apply to speciation analysis.

3.1   Stated References for Speciation Analysis

3.1.1   Generalities

The 'stated references' may be reference methods, reference materials or, as already said above, the units of the *Système International* (SI).[17] The mole and the kg are the SI units which underpin chemical measurements; the mole relates atomic/molecular entities to a macro-scale *via* classical chemical reactions and provides the basis for analytical techniques such as, for example, gravimetry, titrimetry and electrochemical measurements (keeping in mind that the kg is necessary to define the mole).[16] In theory, perfect traceability could be established if each atom/molecule of a certain substance could be counted one by one on a microscopic scale. In practice, the measurements correspond to approximations, *e.g. via* comparisons of amounts, of instrumental response generated by a number of particles, *etc.* Basically, establishing SI traceability nowadays implies the ability to demonstrate to what extent the approximations made are clearly related to the stated references.[16] Typically, many chemical measurements are actually traceable to either a reference material or to a (reference) method. In the field of speciation analysis, the 'stated references' can hardly be established to the mole given the present state of knowledge. Indeed, as discussed below, the techniques used involve a series of analytical steps, which multiplies approximations that are still not under control. The references, in this case, are either pure substances or Certified Reference Materials (when they exist). At present, there are no real 'reference methods' in speciation to which results can be traceable, with the exception of operationally-defined parameters (*i.e.* 'forms' of elements defined according to an extraction protocol,[8] *e.g.* a single or a sequential extraction protocol, which represents in this case the reference. As stressed above, however, this type of measurement is not really considered as being covered by the term 'speciation').

3.1.2   CRMs as Stated References

It has been emphasised that the 'reference' represented by a CRM is not always reliable since, in many cases, the RM does not have the 'same' matrix as the unknown sample.[11] In most cases, CRMs represent a compromise with respect to the matrix of the unknown sample, which will be useful as a quality control tool. Analytical chemists should not expect more from CRMs than they can offer, *i.e.* they are to be considered as useful tools for validation and not calibration tools for applying 'correction factors' to measurement results. In other words, if an error is detected in a method when analysing a CRM, this error has to be removed *before* the analysis of the unknown sample, and not corrected for on the basis of the deviation observed for the CRM results. This may sound trivial, but the validation process of a method, involving the use of CRM(s), is required prior to measuring unknown samples! Furthermore, it should be stressed that a correct result obtained with a CRM does not give a full assurance that 'correct results' will be achieved when analysing unknown samples, due to differences in matrix composition.

In addition, the question of traceability of CRMs, representing complex chemical systems, to SI units is still an open debate. The new ISO definition of a CRM implies 'traceability to an accurate realisation of the unit in which the property values are expressed'.[18] As with the traceability to SI units, this 'accurate realisation' is often difficult to demonstrate in the field of speciation analysis. In practice, the approximations made at different analytical steps do not give proof that a 100% recovery has been obtained (*e.g.* extraction recovery, derivatisation yield). The approximations are actually more valid when the certified values have been obtained in the frame of inter-laboratory studies involving several (independent) laboratories using a variety of different techniques. Even then, in the absence of a 'definitive method', the collaboratively obtained value is considered to reflect the 'state-of-the-art' of a given method (hence a good reference point), but not necessarily the 'true value' of the measured chemical compound.

As stressed before,[19] traceability of chemical measurements to the SI unit (*i.e.* to the mole) is achievable for relatively 'simple' determinations such as trace elements in seawater. This has been made possible through measurement rounds between a few metrological laboratories using a so-called 'primary reference method' (in this case isotope dilution mass spectrometry).[11] The traceability of certified values of a RM to the mole is, therefore, theoretically achievable. What is demonstrated for water analysis, however, is far from being achievable for complex matrices requiring chemical pre-treatment, extraction, clean-up, separations, *etc.* In this case, the 'chain' will be broken at several stages and the traceability will rely on approximations, *i.e.* recovery estimates. The better these estimates, the closer is the achievement of traceability to the relevant SI unit, the mole. This goes hand-in-hand with the possible achievement of accuracy, *i.e.* the closeness to the 'true value', which is intimately linked with the possibility of achieving, and demonstrating, a 100% recovery of the measurand at each analytical step, where loss or contamination might occur.

Examples of CRMs developed over the past few years to serve as 'references' for speciation analysis are given in Table 1.[20]

3.1.3 Reference to Well-defined Species

As a general comment, one should consider that a given compound may have different species and/or different ways of being reported. An example is tributyltin, $(C_4H_9)_3Sn$, which may be reported as the cation TBT^+, or with its respective anion (*e.g.* chloride, acetate, oxide), or even as Sn, depending on the analytical techniques used.[7] Therefore, besides the definition of 'stated references' as physical entities (*e.g.* pure substances or CRMs), the results of speciation analysis require to be traceable to well-defined units and chemical forms. Potentially serious errors are made because results, submitted according to one chemical form, are compared with results reported in a different unit (*e.g.* comparison of TBT results reported as TBTCl with results reported as Sn). Strictly speaking the

Table 1 *Examples of Reference Materials certified for their contents in chemical forms of elements. The table lists CRMs available at BCR (European Commission), IAEA (Austria), NIES (Japan), NIST (USA) and NRCC (Canada). This list is not exhaustive*

CRM	Certified parameters & matrices	Producer
SRM 2108	Cr(III) and Cr(VI) in solution	NIST
SRM 2109	Cr(III) and Cr(VI) in solution	NIST
CRM 544	Cr(III) and Cr(VI) in lyophilised solution	BCR
CRM 545	Cr(VI) in welding dust (loaded on a filter)	BCR
SRM 1974a	Total mercury and methylmercury in mussel tissue	NIST
SRM 2974	Total mercury and methylmercury in mussel tissue	NIST
SRM 2976	Total mercury and methylmercury in mussel tissue	NIST
DORM-1	Methylmercury in fish muscle (dogfish)	NRCC
CRM 463	Total mercury and methylmercury in fish muscle (tuna)	BCR
CRM 464	Total mercury and methylmercury in fish muscle (tuna)	BCR
LUTS-1	Trace elements and methylmercury in lobster tissue	NRCC
CRM 580	Total mercury and methylmercury in sediment	BCR
IAEA 356	Methylmercury in sediment	IAEA
IAEA 086	Total mercury and methylmercury in human hair	IAEA
PACS-1	Butyltin compounds in marine sediment	NRCC
CRM 462	Butyltin compounds in coastal sediment	BCR
CRM 646	Butyl- and phenyltins in freshwater sediment	BCR
CRM 477	Butyltin compounds in mussel tissue	BCR
NIES 11	Total tin and tributyltin compounds in fish tissue	NIES
CRM 627	Organoarsenic compounds in fish muscle	BCR
CRM 605	Trimethyllead in urban dust	BCR

results of chemical determinations should be reported as 'amount (of substance) measurements' to comply with the requirements of the SI system.[11]

3.1.4   Reference Methods

The question has been raised as to whether the development of 'reference measurement' procedures, adequately applicable to real sample matrices, would not be a more desirable trend in the coming years, rather than trying to develop thousands of different matrix RMs.[11] 'Reference measurement methods', however, also need to be validated, so the necessity to develop suitable CRMs cannot be ignored. As stressed in the following section, the development of reference methods will depend upon the availability of adequate quality control tools, *e.g.*

reference materials naturally enriched with isotopically-labelled compounds, to be used for the verification of extraction recoveries.

3.2 Significance of the Unbroken Chain of Comparisons for Speciation

3.2.1 Traceability Links

As already mentioned, the determinations of chemical forms of elements are in most cases based on multi-step methods, representing multiple risks of contamination, or losses of analytes, during the measurement process.[7] Strictly (metrologically) speaking, measurements are, by definition, *quantifying* something, and chemical separations (*e.g.* extraction, clean-up, chromatographic separation) do not quantify. However, the separation of chemical species before measuring them is a very important step on the road of quantification.[11] For reliable and accurate measurements, a 100% complete separation is an essential condition for accurate quantification, but in itself it is not a measuring process.

3.2.2 Extraction Recoveries

The extraction of chemical forms of elements from various matrices is a complex matter in which two conflicting issues need to be combined, obtaining an adequate recovery on one side and preventing losses, especially destruction of the compound(s), on the other.[7] Basically, the extraction should be done in such a way that the analyte is separated from the interfering matrix without loss, contamination, change of the speciation, and with the minimum of interferences. Ensuring a good traceability of measurements implies that the extraction recoveries are verified. This can be done in several ways, as extensively discussed in the literature for speciation analysis.[21] Although extraction methodologies have been systematically studied for various organometallic compounds, *e.g.* organotins,[22] no general consensus has been reached so far on recovery experiments and corrections, which are described elsewhere.[21] The main limitations are due to the lack of CRMs (and their related drawbacks discussed in Section 3.1.2). Much work remains to be carried out to validate this step, which certainly represents the weakest part of the traceability chain in speciation analysis at the present. Advances could be made by the use of materials containing incurred chemical species (bound to the matrix in the same way as the samples to be analysed), preferably isotopically-labelled, enabling an accurate evaluation of the recovery, using isotope dilution techniques. Since such materials are not available at present, the extraction methods should be validated by comparison with independent methods (which gives good confidence on data comparability, but not necessarily on accuracy).

3.2.3 A Particular Case – Distillation

In speciation analysis, distillation is used mainly for the determination of methylmercury (MeHg) in environmental matrices. This procedure is often based on the addition of H_2SO_4 and KCl at a temperature of 145 °C, with verification of the distillation recovery by standard additions.[23] Recent discussions have illustrated the controversies that may occur between environmental chemists with respect to accuracy and traceability issues.[24-27] Artificial MeHg formation was noted by different groups, especially when water vapour distillation was used.[23] The effect was mainly noted in sediment samples and was not observed in fish samples. The phenomenon was observed by recording both the normal measurement and the formation of MeHg due to an enriched Hg^{2+} isotope spike. These observations threw doubts on the certified values of sediment CRMs, which justified the organisation of a workshop to discuss this matter.[24] Further discussions argued that artifacts observed by three laboratories were not proof that the certified values are biased (considering the good agreement obtained by various laboratories using independent techniques), because the free (and very reactive) Hg^{2+} used in spiking experiments, which cast doubt on the materials, can be a precursor in transformation mechanisms, including methylation. Therefore, the ionic mercury could have generated the artifacts.[25] It was stressed that the demonstration of artifact formation should rely on experiments with enriched isotopes of Hg^{2+} incorporated into the test material by a natural process.[27] A recent experiment has been performed,[28] consisting of the addition of $^{199}Hg^{2+}$ to different sediment samples, which resulted in an artifact formation demonstrated by an increase in $CH_3^{199}Hg^+$ as compared with $CH_3^{200}Hg^+$. However, this approach did not reflect real conditions. First, the spiking procedure did not mimic a natural enrichment process and second, the increase in $CH_3^{199}Hg^+$ simply confirmed that the artifact probably only arose from the addition of the Hg^{2+} spike. The best strategy to demonstrate effects on MeHg in certified samples would be to spike the material with methylated-enriched mercury isotopes. This would distinguish between 'real' artifacts related to the naturally-bound MeHg and 'artificial' artifacts coming from the MeHg spike used to check recovery.[27] This short account illustrates the difficulty in demonstrating traceability in multi-step techniques, as used in speciation analysis, and stresses that CRMs are good 'reference points' for achieving comparability of data (traceability to a common reference). In this case, the overestimation of the methylmercury content in sediments, due to possible artifact formation, cannot be firmly 'demonstrated' on the basis of the described experiments, and doubts about the validity of the certified values should not be generalised solely on this basis. These observations only showed that the distillation method should be carried out under strict quality control.

3.2.4 Derivatisation Yields

Many techniques used for speciation analysis rely on a derivatisation step, which may also be used to separate elements from their matrices and interferences, and to concentrate the species. The principle is to transform the actual compounds into their derivatised forms, which are more easily separated from each other by chromatography. Reactions currently employed in speciation analysis are hydride generation (formation of volatile hydride forms based on the addition of sodium borohydride), ethylation (using sodium tetraethylborate) and Grignard derivatisation (with Grignard reagents to achieve butylation, pentylation *etc.*). These methods have been reviewed recently.[29,30] Despite many publications describing these reactions, derivatisation procedures are far from being well controlled, due to an insufficient knowledge of the reaction mechanisms and lack of high purity reagents. In particular, the evaluation of derivatisation yields (recovery estimate) is hampered by a lack of primary standards ('standard having the highest metrological qualities and whose value is accepted without reference to other standards of the same quantity'[18]) of suitable purity and stoichiometry. This evaluation should indeed be carried out on the basis of pure derivatised compounds (*e.g.* the yield of an ethylation reaction used for the determination of methylmercury chloride should be evaluated with an ethyl-methylmercury standard). Recent progress has been made in a collaborative project on tin speciation, in which pure alkylated derivatives (ethylated and pentylated forms) of mono-, di- and tributyltin were prepared for the purpose of checking the yield of Grignard derivatisation.[31] This is certainly a promising approach, which should be extended to the determination of other chemical forms of elements, thus strengthening this link in the traceability chain. It has to be noted however, that derivatisation may be prone to uncontrolled sources of error, and that if this step can be avoided it is worthwhile considering the alternatives.[7]

3.2.5 Separation

The separation of chemical species of elements can only be performed by techniques that do not destroy the chemical forms (*e.g.* by heat-induced decay). Two basic methods are generally in use for speciation analysis – chromatography (liquid or gas) and cold trapping – coupled with various types of detectors.[32,33]

Cold trapping techniques have achieved much success thanks to their potential to determine traces of organotins in waters. The separation technique relies on prior derivatisation of the compounds (hydride generation) and separation in a simple chromatographic device (U-tube filled with chromatographic material) after cryogenic trapping. The separation is based on the sublimation points of the different (hydride) species. Identified drawbacks are related to the low resolution of the 'packed column', the limitation of the technique to species generating volatile hydrides, and the possible mis-identification of species of similar volatility.[7]

Chromatographic separation is certainly the most popular technique used in speciation analysis. Whilst the use of packed columns is decreasing (although still used in conjunction with high sensitivity detectors), due to the many pitfalls observed, *e.g.* for methylmercury analysis,[34] capillary GC is now very popular and widely used in speciation studies.[33] Liquid separation techniques (*e.g.* HPLC) offer more potential than gaseous separation methods,[35] and present the advantages of not requiring a prior derivatisation step. However, stationary phases in HPLC (*e.g.* ion exchangers, ion pairing) are still not widely available for speciation analysis.

The weak links of this analytical step are mainly related to the risks of losses (*e.g.* heat-induced decay at the injection, or instability of the species on the column) or misidentification. This part of the traceability chain is intimately linked to the pre-treatment steps, *i.e.* proper derivatisation in the case of GC and clean-up of the extracts to eliminate interfering substances. It also relies on the availability of proper chromatographic phases and calibrants.

3.2.6   Detection

In all analytical measurements, the determinand should, in principle, arrive alone at the detector, to avoid interferences. The choice of the detector in speciation analysis will depend on the chemical forms to be determined and the mode of separation used. Detectors are either element specific (*e.g.* AAS) or non-specific (*e.g.* FID, FPD, ECD). This step is the last one of the measurement chain where traceability can be lost. This break can be due to, for example, uncontrolled interferences (*e.g.* from the matrix), lack of optimisation of the temperature programme, gas flow rates *etc.* (*e.g.* for AAS techniques), *etc.* The on-line coupling of classical detectors (*e.g.* ICP-AES or ICP-MS, FPD, FID) to GC or HPLC has led to the elimination of the risks related to the off-line character of detectors, such as ETAAS (which cannot be applied in a continuous mode), and this has represented considerable progress for speciation studies. MS detection offers an additional advantage of allowing an on-line QA in the isotope dilution mode. This technique is certainly the only one that would enable one to firmly trace back isotopic forms of elements from extraction to detection. As stated in the previous sections, the possible demonstration of traceability to SI units for speciation studies would rely on the availability of samples containing incurred isotopically-labelled compounds, which would be detected (after extraction, derivatisation, if relevant, and separation) by isotope dilution MS or ICP-MS. Recent advances have been made in this field with the development of coupled HPLC–ID-ICPMS techniques for lead and tin speciation.[36]

3.2.7   Calibration

The importance of calibration is obvious, since all efforts to obtain a good sample and carry out the analysis under the best quality control possible, are spoiled if the calibration is wrong. In theory, firmly establishing traceability in analytical chemical measurements means that several 'primary' chemical reference materials in the form of (ultra-)pure substances are interlinked by well known, quantitative high precision–high accuracy chemical reactions (*e.g.* as available in titrations or precipitation processes).[11] In the case of speciation analysis, the 'links' which should constitute the traceability system are made of various analytical steps, which should be traceable to a limited set of primary RMs.

Calibration is perhaps one of the weakest links of the measurement traceability chain, since the risks of errors related to calibration are often not properly evaluated. Besides the choice of a suitable calibration procedure (*e.g.* standard additions, matrix-matching, *etc.*) and their respective advantages and draw-backs,[7] this step relies on calibrants of suitable purity and stoichiometry. This is certainly one of the major difficulties nowadays in speciation analysis, which may prevent proper traceability from being achieved, *i.e.* the insufficient availability of pure substances ('primary standards') in many instances (*e.g.* organic forms of arsenic, selenium *etc.*).

As mentioned above, techniques involving a derivatisation step present the particular case that they should, in principle, be calibrated with derivatives (since it is the derivatised forms that are separated and detected, not the species in their original anionic form) to preserve the traceability link. This implies that 'second-ary standards' ('standards whose value is assigned by comparison with a primary standard of the same quantity'[18]) should be available. This principle has been followed in recent inter-laboratory and certification studies on organotins, where alkylated derivatives of butyltin species have been prepared and distrib-uted to the laboratories taking part.[31,37] This example is, however, the only attempt, to the author's knowledge, to improve the traceability of speciation analysis. Such efforts are actually hampered by the lack of commercially avail-able 'primary' and 'secondary' standards.

3.3 Measurement Uncertainties

The uncertainty is defined as 'a parameter, associated with the result of a measurement, that characterises the dispersion of the values that could reason-ably be attributed to the measurand'.[9] Traceability and uncertainty are closely interconnected (see definitions given above). Various approaches exist for calcu-lating uncertainty (*e.g.* ISO, Analytical Methods Committee), and these have been reviewed recently and discussed in view of their application to routine analysis.[38] The purpose of the present chapter is not to discuss uncertainty in detail, but rather to provide some elements for discussion, with respect to its meaning for speciation analysis. In principle, speciation measurements would typically require that uncertainties from individual sources of variations (*i.e.* from each analytical step) be combined to produce an overall uncertainty for the measurement. The simplest way to calculate uncertainty for speciation measure-ments would be to consider the errors estimated from the different analytical steps, *i.e.*

$$y = y^1 \text{extraction} + y^2 \text{derivatisation} + y^3 \text{separation} + y^4 \text{detection}$$

The combined standard uncertainty ('uncertainty budget') would then be given by:

$$u(y) = \{u(y1)^2 + u(y2)^2 + u(y3)^2 + u(y4)^2\}^{1/2}$$

The theory, as applied to speciation is, here again, far from the current practice, owing to a lack of appropriate tools, and because real uncertainty measurements would be very time-consuming. To achieve the best state-of-the-art, one would need to employ a CRM with a matrix *identical* to that of the sample, containing known (incurred) amount(s) of the compound(s) to be measured, providing that the enrichment has been performed by a 'natural' process. The only actual proof that a full (extraction or distillation) recovery could be obtained would be *via* the use of a CRM isotopically-enriched with the compounds to be determined, and to determine the isotopes by IDMS or ID-ICPMS. The recovery experiments could then accurately determine the percentage recovered and its associated uncertainty. For techniques involving a derivatisation step, the derivatisation yields could be accurately determined, again with their associated uncertainties. The same principle applies to the separation (checking that a full separation is achieved and estimating related uncertainties) and to the detection (estimation of uncertainty linked to possible interfering substances). As stressed above, the use of matrix RMs is, however, not recommended for the evaluation of the uncertainty of analytical measurements owing to the differences in matrix composition.[11] This aspect of speciation analysis therefore remains an open research area, which will certainly generate fruitful (and controversial!) discussions in the years to come.

3.4   Conclusions

In conclusion to the above discussions, we may ask ourselves 'is accuracy achievable in speciation analysis?'. To respond to this question, we have to return to the basic principle of traceability, *i.e.* 'the ability to trace the result of a measurement back, with an acceptable uncertainty, to the closest form of an accurate realisation of the relevant SI unit (basic or derived) for the quantity measured or – failing these – of a relevant internationally recognised empirical unit for that quantity'.[11] The definition concerns a specific, defined entity, and applies to a quantitative amount of substance measured with a verifiable total uncertainty, involving a measurement conducted using the best possible method. Basically, the accuracy ('closeness to the true value') is achieved if the traceability of the measurement to the SI unit (the mole) is perfectly demonstrated. As claimed by metrologists, the most logical metrological way would then be to report the results as mole (of defined entity) per kilogram of material. It has been underlined in the various sections that the demonstration of this traceability is far from achievable in speciation analysis. Therefore, it is currently hardly possible to pretend that we may perform accurate measurements of the chemical species of elements.

The metrological principle implies that 'if the traceability of measurements is claimed to be other than the mole unit itself, but rather through a procedure, material or standard, then these must be credibly described and their relation to the mole clearly established'.[11] With respect to speciation analysis, the first clause can be easily fulfilled, *i.e.* traceability can be demonstrated to a pure substance, a CRM or a written procedure. The second part of the principle can hardly ever be achieved for the reasons discussed above.

All these considerations oblige us to admit that, to date, it is not possible to demonstrate accuracy in most speciation measurements, owing to the multiplicity of unknown variables. The only cases where accuracy can be achieved, concern the analysis of 'simple' matrices (*i.e.* not requiring chemical pre-treatment), and the determination of species using primary methods, one of the few examples at present being the determination of Cr-species in water by IDMS.[39] A promising advance is the recent development of HPLC isotope dilution ICPMS[40] applied to Sn- and Pb-speciation analysis, for which the only 'weak part of the chain' is the difficulty in accurately determining the extraction recovery. Analytical advances for improving the traceability (and the accuracy) of speciation measurements will certainly be an interesting area of research during the next two decades.

4 References

1. W. Horwitz, *AOAC Internat.*, 1992, 368.
2. B. Griepink, *Fresenius', J. Anal. Chem.*, 1990, **338**, 360.
3. D.M. Templeton, F. Ariese, R. Cornelis, L.-G. Danielsson, H. Muntau and H.P. van Leeuwen, *IUPAC Guidelines for Terms Related to Chemical Speciation and Fractionation of Trace Elements*, Pure Appl. Chem., 2000, **72**, 1453.
4. Ph. Quevauviller and O.F.X. Donard, *J. Environ. Monitor.*, 1999, **1**, 503.
5. S. Zhang, Y.K. Chau, W.C. Li and A.S.Y. Chau, *Appl. Organometal. Chem.*, 1991, **5**, 431.
6. M. Horvat, V. Manduc, L. Liang, N.S. Bloom, S. Padberg, Y.-L. Lee, H. Hintelmann and J. Benoît, *Appl. Organometal. Chem.*, 1994, **8**, 533.
7. Ph. Quevauviller, *Method Performance Studies for Speciation Analysis*, Royal Society of Chemistry, Cambridge, 1998 ISBN 0-85404-467-1.
8. Ph. Quevauviller, *Trends Anal. Chem.*, 1998, **17**, 289.
9. *International Vocabulary of Basic and General Terms in Metrology*, International Standardisation Organisation, Geneva, Switzerland, 2nd Edition, 1993.
10. Ph. Quevauviller and E.A. Maier, *Interlaboratory Studies and Certified Reference Materials for Environmental Analysis – The BCR Approach*, Elsevier, Amsterdam, 1999 ISBN 0-444-82389-1.
11. P. de Bièvre, in *Accreditation and Quality Assurance in Analytical Chemistry*, H. Günzler ed., Springer, Berlin, 1996.
12. M. Valcárcel and A. Rios, *Trends Anal. Chem.*, 1999, **18**, 570.
13. M. Thompson, *Analyst*, 1996, **121**, 285.
14. M. Válcarcel and A. Ríos, *Anal. Chem.*, 1993, **65**, 78A.
15. A. Marschal, *Bull. BNM*, 1980, **39**, 33.
16. M. Válcarcel, A. Rios, E. Maier, M. Grasserbauer, C. Nieto de Castro, M.C. Walsh, F.X. Rius, R. Niemelä, A. Voulgaropoulos, J. Vialle, R. Kaarls, F. Adams and H. Albus, *EUR Report*, European Commission, Brussels, 18405 EN, 1998.
17. B. King, *Analyst*, 1997, **112**, 197.
18. ISO, Guide 30: 1992, International Standardisation Organisation, Geneva, Switzerland, 1992.
19. M.C. Walsh, *Trends Anal. Chem.*, 1999, **18**, 616.

20. Ph. Quevauviller, in *Elemental Speciation – New Approaches for Trace Element Analysis*, K.L. Sutton and J.A. Caruso eds., Elsevier, Amsterdam, 1999.
21. Ph. Quevauviller and R. Morabito, *Trends Anal. Chem.*, 2000, **19**, 86.
22. C. Pellegrino, P. Massanisso and R. Morabito, *Trends Anal. Chem.*, 2000, **19**, 97.
23. H. Hintelmann, R. Falter, G. Ilgen and R.D. Evans, *Fresenius' J. Anal. Chem.*, 1997, **358**, 363.
24. R. Falter, H. Hintelmann and Ph. Quevauviller, *Chemosphere*, 1999, **39**, 1039.
25. Ph. Quevauviller and M. Horvat, *Anal. Chem.*, 1999, **71**, 155A.
26. N.S. Bloom, D. Evans, H. Hintelmann and R.-D. Wilken, *Anal. Chem.*, 1999, **71**, 575A.
27. Ph. Quevauviller, F. Adams, J. Caruso, M. Coquery, R. Cornelis, O.F.X. Donard, L. Ebdon, M. Horvat, R. Lobinski, R. Morabito, H. Muntau and M. Valcárcel, *Anal. Chem.*, 1999.
28. H. Hintelmann, *Can. J. Anal. Sci. Spectr.*, 1998, **43**, 182.
29. R. Ritsema, F.M. Martin and Ph. Quevauviller, in *Quality Assurance for Environmental Analysis*, Ph. Quevauviller, E.A. Maier and B. Griepink eds., Elsevier, Amsterdam, 1996.
30. R. Morabito, P. Massanisso and Ph. Quevauviller, *Trends Anal. Chem.*, 2000, **19**, 113.
31. F. Ariese, W. Cofino, J.L. Gomez-Ariza, G.N. Kramer and Ph. Quevauviller, *J. Environ. Monitor.*, 1999, **1**, 191.
32. O.F.X. Donard and F.M. Martin, *Trends Anal. Chem.*, 1992, **11**, 17.
33. R. Lobinski, *Appl. Spectrosc.*, 1997, **51**, 260A.
34. J.H. Petersen and I. Drabæk, *Mikrochim. Acta*, 1992, **109**, 125.
35. C. Harrington, *Trends Anal. Chem.*, 2000, **19**, 167.
36. S.J. Hill, L.J. Pitts and A.S. Fisher, *Trends Anal. Chem.*, 2000, **19**, 120.
37. R. Morabito, H. Muntau, W. Cofino and Ph. Quevauviller, *J. Environ. Monitor.*, 1999, **1**, 75.
38. A. Maroto, R. Boqué, J. Riu and F.X. Rius, *Trends Anal. Chem.*, 1999, **18**, 577.
39. R. Nusko and K.G. Heumann, *Anal. Chim. Acta*, 1994, **286**, 283.
40. A.A. Brown, L. Ebdon and S.J. Hill, *Anal. Chim. Acta*, 1994, **286**, 391.

CHAPTER 2

Quality Control in Speciation Analysis

ERWIN ROSENBERG AND FREEK ARIESE

1 Introduction

Chemical species of elements can be differentiated at different levels. At the nuclear level, isotope distributions can provide information on the environmental sources of certain elements, *e.g.* lead.[1] At the electronic level, the redox state of elements will have a strong influence on properties like solubility, binding and reactivity. A well-known example is the toxicity of Cr(VI), which is much higher than that of chromium in other redox states.[2] At the organometallic compound level, the nature and number of covalently bound ligands will also strongly influence the element's properties, such as the toxic properties of methylmercury in comparison with that of elemental or ionic mercury. Also the nature of more loosely bound counter-ions may influence the physico-chemical behaviour of elements. Finally, at an even higher level of complexity, binding to larger units such as proteins, soot particles or humic acids will have a profound influence on factors like mobility, stability and bioavailability.[3] These various types of species differences generally require dedicated analytical approaches. Naturally, different species will usually only be determined and analysed separately if this provides information essential for the understanding of the system under study.

An important parameter that determines the way in which we study and interpret species distributions is kinetic stability. Very labile species, or species undergoing rapid interchanges, may exist only on timescales much shorter than those typically used for off-line analysis. In such cases only fast, non-invasive methods are to be used, like ion-selective electrodes or real-time spectroscopic methods. More time-consuming methods can only be applied to sufficiently stable species. In this chapter we will focus on relatively stable organometallic species that will remain unaltered during sampling, storage, extraction from the matrix and analysis. It should be realised that information on loosely bound counter-ions or the type of binding to the matrix will usually be lost in the procedure.

Advances in analytical instrumentation have been significant over the past two decades, and in contrast to earlier methods being capable of determining only the total elemental composition (such as atomic absorption, AAS, or atomic emission spectroscopy, AES), analytical methods are nowadays capable of distinguishing and detecting various elemental species. In most cases, hyphenated methods are used[4-7] in which separation is achieved by a suitable chromatographic or electrophoretic technique and ideally element- or molecule-specific detection is carried out with spectroscopic or other types of selective detectors. In spite of these developments, organometallic analysis is still not carried out routinely in many analytical laboratories. Methods and instruments are still being developed, and typical intra- and inter-laboratory standard deviations tend to be substantial. In this chapter, several aspects of speciation analysis and a number of difficulties encountered in present-day organometallic quantification will be discussed.

As in other fields of analytical chemistry, speciation analysis requires quality assurance measures to be implemented in order to produce valid data. However, it appears that this aim is more difficult to reach than in other fields, such as inorganic or organic analysis. Various reasons can be given, and the following list is not exhaustive:

- elemental species may be inherently unstable – this may refer to *e.g.* the redox state of an element which under atmospheric conditions is oxidised quickly (*e.g.* $Fe^{2+} \rightarrow Fe^{3+}$) or they may be unstable upon extraction from the matrix, which in some cases has a stabilising effect on the species (for example, organotin species adsorbed to sediments are much more stable than in aqueous solution)
- elemental species may be changed during the extraction/digestion step. Speciation analysis often requires the extraction of the species from the matrix (*e.g.* sediments) or in some cases the total digestion of this matrix (*e.g.* biological tissue) in order to liberate the analytes completely, and extreme pH values or elevated temperatures often have to be employed
- elemental species may be changed or be lost during the sample clean-up (due to their sensitivity to oxidation, elevated temperature or irreversible interaction with the clean-up material)
- elemental species may be changed during the derivatisation: depending on what method is used for the actual analysis, species may have to be derivatised. This bears the risk of changing, obscuring or even losing the species information.

All these reasons may be responsible for altering the species information originally present in the sample. Since, however, the different species may largely differ in their physico-chemical properties, in their bioavailability, toxicity, environmental fate and transport, an essential part of the information intended to be gained through species-selective analysis is inevitably lost.

The implementation of quality assurance measures in a laboratory carrying out speciation analyses should ensure the following:

- accuracy of analytical data
- precision of analytical data
- comparability of analytical data (as a consequence of the two above features)
- stability over longer time frames.

Various aspects of quality assurance in speciation analysis have been dealt with in books,[8,9] proceedings of conferences[10-12] or individual publications.[13-17]

The obvious difficulty of carrying out speciation analysis under good quality control, gives rise to what we may call a 'Catch-22' situation: since there is no (generally accepted) method for the various tasks and problems of speciation analysis, there is no general knowledge on the distribution of elemental species in food and feed, the environment or the work place. As there is a lack of comparable, valid and complete data, there is a lack of legislation for determining and monitoring elemental species and, even if it existed, legislation could hardly be enforced at the present time. On the other hand, since there is only in exceptional cases legislation for the determination of elemental species, public awareness and interest in the development of valid methods or dedicated instrumentation is comparatively low.

One particular difficulty in the field of organometallic analysis is related to the fact that many different species are to be analysed in a single analytical run. This is done partly as a result of traditional analytical approaches. Formerly an analysis of tin, mercury, arsenic *etc.* meant a thorough acidic digestion followed by elemental analysis, typically by AAS. As the first methods were developed to determine the different organometallic species separately, it was, of course, expected that for a particular element the sum of its individual species would equal the total content as measured by AAS. A more complicated and more expensive method should not overlook certain species that would have been included in a standard, total element measurement. When legislative criteria are expressed in terms of total element contents, one can only use information on separate species if all major species have been included in the analysis.

This approach has led to the current situation that analytical laboratories are often expected to develop or optimise a method for *all* species of a given element. On the other hand, the various species of a given element often have very different physico-chemical properties, making this a very challenging task. For instance, extraction procedures optimised for the relatively non-polar tributyltin may not be very effective for the more polar species, monobutyltin.[18] Inorganic forms of arsenic require a totally different approach from arseno-sugars or arsenobetaine.[19,20] Apart from the physico-chemical properties, the concentrations of the various species (and thus the optimum sample intake or dilution/enrichment factor) can be very different. As a result, many methods existing today are carried out under what may be referred to as 'compromise optimal conditions', conditions developed in order to determine the full spectrum of species, but which may not be ideal for each individual analyte. When adopting such methods it should be confirmed whether it was optimised for the same species of interest. If necessary, separate approaches should be used, if certain important species cannot be adequately determined in a single analytical run.

It appears thus, that analytical chemists first have to demonstrate their ability to determine elemental species in the relevant (environmental, food and health related) matrices under good quality control and then to propose methods 'fit for purpose', in order that speciation analysis be generally accepted. In the following pages, we will discuss the difficulties that are encountered, and the measures that may be taken, for establishing speciation analysis under good quality control.

2   General Aspects in Quality Control of Speciation Analysis

In the following sections, we will briefly mention aspects and considerations of quality control for trace analysis in general, and particularly focus on the implications for speciation analysis. As mentioned previously, all considerations that apply to trace organic analysis, also apply to speciation analysis, but, in addition, particular precautions have to be taken in this field, due to the properties of organometallic species. An overview of the different steps in speciation analysis and the possible errors related to this each particular step is given in Figure 1.

3   Sampling, Conservation and Storage

At this point, it has to be emphasised that the final determination (separation and detection) of elemental species usually only accounts for the smallest fraction of error or uncertainty in the entire analytical process. Often the biggest source of errors in analytical chemistry is beyond the control of the analyst, since he/she is typically not involved in the sampling or all subsequent steps, such as conservation and storage of the samples, prior to the delivery of the sample to the laboratory. Thus, one has to assume that the sample that was delivered is representative of the analytical problem. In fact, any conclusions from the measurements are, strictly speaking, only valid for the sample that was analysed, and not necessarily for the population that may have been sampled in an inappropriate or non-representative way.

The, sometimes, limited stability poses additional problems in speciation analysis, when compared with trace organic analysis or the analysis of total metal concentrations. Although certain organometallic species appear to be remarkably stable (*e.g.* arsenobetaine – in contrast to the As(III)/As(V) ratio[21]) this is not generally the case, and may not be true for most relevant elemental species (*e.g.* the storage stability of organoselenium species has been found to be much lower than that of inorganic selenium species[22]). It is therefore recommended that biological or environmental samples be stored deep-frozen (at $-20\,^{\circ}\mathrm{C}$ or lower), but this may not always be possible in the field. Sending samples by mail or courier service at ambient temperatures should only be done if stability studies have shown the particular type of species/matrix combination to be sufficiently stable over several days. Aqueous samples may be stored at $4\,^{\circ}\mathrm{C}$, but should be processed as soon as possible after sampling. While the stability of

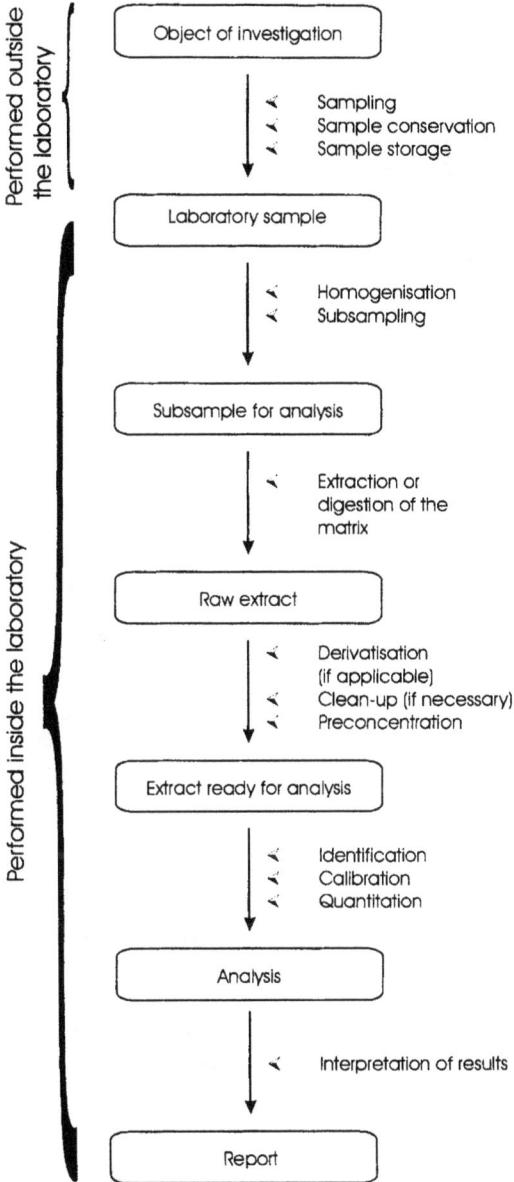

Figure 1 *Individual steps for speciation analysis with the typical tasks to be performed inside and outside the laboratory*

selenium compounds in solution has been investigated[22,23] there are only limited data available on organotin compounds in solution. However, the stability in freeze-dried sediments and mussel tissue has been investigated in detail, since this is one of the essential steps in the preparation of reference materials.[14]

When preparing (certified) reference materials, the stability should be guaranteed for several years. This often means that the samples should be air-dried, freeze-dried, and/or sterilised, before sealing into bottles or ampoules. During this treatment the species concentrations may be altered, but for (C)RMs this may be acceptable as long as no further changes occur after the certification.

It is evident that similar precautions have to be taken when storing calibrants. Pure calibrants are essential for calibration and recovery studies. It should be realised that for the calibrants the nature of any loosely bound counter-ions will depend on the synthetic route used in their production. These will generally be different from those encountered in natural samples. Some calibrant species feature only limited stability. For instance, mono-alkylated tin species are known to form oxide polymer networks upon contact with water or air moisture. Calibrant solutions of these species may therefore degrade upon repeated opening. Elements in different redox states [*e.g.* Cr(III)/Cr(VI) and As(III)/As(V)] tend to change their distribution when stored in solution, and organometallic species are degraded more or less quickly (depending on the degree of substitution) in aqueous solution. Even in organic solvents (such as methanol), degradation can occur remarkably fast, and this implies that the calibrants be prepared freshly at appropriate intervals (where, as a rule of thumb, the interval becomes shorter as the concentration of the calibrants decreases). However, small-scale preparations could lead to relatively large weighing errors. Whenever the solubility is not limited, it is advisable to prepare calibrant solutions at high concentrations in aprotic organic solvents, such as THF, and seal them under an inert atmosphere in a number of small ampoules for future use. Particular care has to be taken when mixed calibrant solutions are prepared, since there is the possibility of transalkylation reactions among organometallic species of different elements.

4   Subsampling

Since modern methods of speciation analysis are highly sensitive, it is, in most cases, sufficient to use only a small sample aliquot for analysis. A small sample intake per analysis is also desirable since it (i) minimises the consumption of (expensive) reagents and solvents, (ii) reduces the time needed for sample preparation due to the smaller sample/extract volumes to be handled and (iii) minimises the amount of matrix introduced into the analytical system and thus problems related to memory effects, contamination of the analytical system, column deterioration *etc.* On the other hand, problems with the homogeneity may arise, when the actual sample intake becomes smaller. In particular, it must be remembered that, due to the specific properties of elemental species, they may be homogenous on a larger scale, whilst being inhomogeneous on a microscopic scale. Different species may be enriched in, or adsorbed onto, distinct parts of the sample, *e.g.* fat-rich tissue or dissolved organic matter. If the (sub-)sampling process fails to be representative, the species distribution may be skewed. The question of homogeneity, and thus of minimum sample intake, is difficult to answer for new samples. Only for (certified) reference materials, which are sufficiently well characterised, is this value well defined, since it has been determined by extensive homogeneity testing. In practice, it is recommended to make a statistical analysis of the data to determine the contribution of sample in-homogeneity. If the analytical results of replicate analyses differ by significantly more than the standard deviation calculated from repeated analysis of the same sample, inhomogeneity of the sample may be suspected. If the scatter is reduced when the sample intake is larger, this is further evidence that the minimum sample intake was too small.

5 Extraction/Digestion

The extraction or digestion procedure in a speciation scheme aims to liberate the analytes from the matrix with minimal co-extraction of interfering compounds, and without changing the species information. Since metal species and particularly organometallic species tend to be bound rather strongly to the matrix (by adsorptive forces, ionic or van der Waals interactions), the extraction process must – from a physico-chemical point of view – apply energy to overcome the interaction of the analyte with the matrix. On the other hand, the energy applied (not necessarily thermal energy, but also chemical, electromagnetic or other forms of energy) should not be too large, since this may lead to decomposition of the analyte. Unfortunately, the range of acceptable working conditions where maximum recovery is achieved with minimum sample decomposition is often very narrow (as has been demonstrated for microwave assisted extraction (MAE) of organotin compounds from sediments[24]).

A suitable solvent for the extraction of metal species has two functions:

- it liberates the analyte from the matrix by weakening the bond to the matrix and/or by displacing it from the active sites of the matrix, and
- it stabilises the metal species by solvating or complexing it, thereby avoiding the re-equilibration and re-adsorption on to the matrix

The degree to which an extraction mixture succeeds in fulfilling both criteria determines the extraction yield for a particular analyte and a particular matrix.

The situation has probably been most extensively studied in the case of organotin species. A great number of more or less established extraction schemes is available in the literature (for a general assessment see refs. 25–27, for comparisons of methods applied to sediments see refs. 28–30, and for biota see refs. 31 and 32). Table 1 gives an overview of the different options that have been used for the extraction of organotin compounds from biota. The great variety of extraction conditions derives from the fact that the compounds to be analysed differ significantly in their polarity and in their physico-chemical properties. While monobutyltin is a highly polar compound that is adsorbed on to the matrix mainly through ionic interaction, triphenyltin is a highly lipophilic compound that has strong interaction with the matrix through van der Waals forces. It is thus almost impossible to optimise the extraction procedure to yield quantitative recovery for all analytes and any type of matrix. From the study quoted above,[25] the trends discussed in the following paragraph became evident.

Table 1 *Matrix of the different options for the extraction of ionic organometallic (organolead, organotin, organomercury) compounds (after ref. 25)*

Extraction solvent	Complexing agent	Acid	Method of extraction
• Methanol	• none	• none	• agitation
• Hexane	• tropolone	• HCl	• Soxhlet extraction
• Pentane	• diethyl-	• HAc	• refluxed
• Toluene	dithiocarbamate	• HBr	• ultrasonication +
• Methanol	• APDC		shaking
• Dichloromethane			• microwave-assisted
			extraction

For the trisubstituted organotin compounds the highest extraction yields can be achieved in the presence of acids in polar to medium-polarity solvents. The use of complexing agents such as tropolone does not improve the recovery since it does not form complexes with TPhT or TBT. This is in clear contrast to the di- and mono-substituted organotins, where the use of a complexing agent and/or of a polar solvent (in addition to acidic conditions) is favourable for obtaining good recoveries. The kind of acid used seems of less importance, although HCl appears to be somewhat more efficient than acetic acid. However, the concentration of acid used seems to be critical, since degradation of tri- and di-substituted organotin compounds was observed in extraction experiments with acid concentrations in excess of $1 \, \text{mol} \, l^{-1}$. The way in which the extraction is carried out plays a subordinate role, unless extraction conditions are so energetic that partial destruction of the analytes occurs (as has been demonstrated with microwave assisted extraction[24]).

Apart from the sample preparation schemes that provide *extraction* of the analytes from the sample matrix, schemes have been reported in which complete *solubilisation* (*digestion*) of the matrix is achieved. This can be achieved for biotic matrices by either an enzymatic approach, where an enzymatic digest of the tissue is performed using a mixture of lipases and proteases,[33,34] or by solubilisation of the tissue with tetramethylammonium hydroxide (TMAH).[34,35] Due to the complete digestion of the tissue, both methods provide comparatively clean extracts that can be injected into the GC after derivatisation, even without clean-up. Since the enzymatic method is carried out under very gentle conditions (37 °C and near-neutral pH for typically 12–24 hours) no problems of decomposition of the analytes are observed. It has, however, been suspected of giving rise to blank problems, due to metal traces introduced with the enzymes.[34] The TMAH solubilisation of the tissue is a much faster procedure, with complete solubilisation of the tissue being achieved within one to two hours at 60 °C.[36] Longer digestion times are not recommended since degradation of the organotin compounds may occur.

For aqueous samples, derivatisation can take place *in situ* and does not require a preceding extraction step if *in situ* aqueous-phase alkylation is chosen.

Of vital importance is the determination of the extraction yield of a given procedure. It is evident that, for a multi-species procedure, extraction yields will generally be different for individual compounds. In principle, three options exist which all have their merits, but in some cases also significant shortcomings.[37,38] These are:

- the use of spiking experiments,
- the repeated extraction of analytes from the same sample, and
- the use of certified reference materials.

The use of spiking experiments to determine extraction yields has to be considered a compromise, and is definitely not an ideal approach. It is known today that an analyte that is spiked into the matrix (usually in the form of a concentrated solution) is not necessarily bound to this matrix in the same way as would occur naturally. When the matrix has taken up the analytes over a longer period, it may have been introduced into tissue, or be bound in a stronger/different form to the matrix, than for the spiked analyte. As a consequence, the extraction of a spiked analyte will be easier to perform and produce higher recoveries, than that of a 'real' sample. If these recovery rates are used for correction, the results will be underestimated. When this method still needs to be used, it is recommended that the analytes are applied in a solvent that has the ability to wet the matrix, and that sufficient time is allowed for equilibration (*e.g.* overnight). However, the matrix and the analyte(s) need to be carefully considered in order to prevent, for example, decomposition of already unstable analytes during the spiking procedure. Following the fortification of the matrix with the calibrant solution, the matrix then needs to be dried to remove the residual solvent in a gentle stream of inert gas.[39,40] The determination of extraction recoveries by spiking is best performed with a blank matrix (one that does not contain the analyte under investigation). If an uncontaminated matrix is not available, the extraction efficiency can only be determined by adding a known amount of the analyte to the sample already containing the analyte, and by analysing the spiked sample together with the unspiked. The recovery is then calculated from the signal and concentration increments, instead of the absolute signal and concentration:

$$Y(\%) = \Delta m_{found}/\Delta m_{added} \times 100$$

where Y = extraction yield (in %); Δm_{found} = increment of analyte mass found; Δm_{added} = increment of analyte mass added. Alternatively, surrogates (which are similar to the analytes in their chemical structure and may be expected to show the same behaviour) or isotopically labelled compounds (which are chemically identical to the analytes) may be used for estimating the extraction yields. The latter technique certainly represents the best approach to determine recoveries from spiking experiments, provided that the above recommendations for spiking are adhered to. A similar approach, using isotopically labelled compounds, is the isotope dilution technique. In this, the alteration of the natural isotopic ratio of a certain element is achieved by the addition of an elemental species with a changed isotopic ratio. The quantitative determination of the species from the alteration of the isotopic ratio is used to compensate for matrix effects. However, the lack of suitable isotopically labelled compounds, and the necessity of mass spectrometric detection,[41,42] precludes the general applicability of this approach. Furthermore, it needs also to be mentioned that, in the case of isotope dilution, spiked calibrants will not bind to the matrix in the same way as naturally occurring analytes.

Experiments in which a naturally occurring sample is extracted repeatedly offer another method to determine the extraction yield where:

$$Y(\%) = m_1/\Sigma m_i \times 100$$

with: Y = extraction yield (in %); m_1 = mass of analyte extracted in the first extraction; m_i = mass of analyte extracted in the ith extraction step. In practice, such an experiment is carried out by extracting the sample repeatedly, under identical conditions, as long as the analytes can be detected in the extracts. Although this procedure does not suffer from the disadvantage that the analytes are spiked, it has to be considered that the first extraction step may irreversibly change the matrix, and that the extraction behaviour is thus altered significantly for the second, and all further extraction steps. In addition, this procedure is very tedious to perform and still does not guarantee that the analyte is quantitatively removed from the matrix. For example, if a second extraction yields no detectable analyte levels, it may be concluded that the first extraction was virtually complete, but it could also mean that only the loosely-bound fraction was extracted, and that other extraction conditions are required for the remaining fraction.

Where available, the use of a certified reference material (CRM) is the most desirable way for the determination of extraction yields. There are some restrictions that limit the usefulness of even this approach. The matrix of the CRM must match that of the sample (which underlines the importance of having a great variety of CRMs with different matrices). Even if the matrices of CRM and sample are the same, one must be aware of the differences between a processed matrix (*e.g.* ground, freeze-dried tissue) and fresh material. This can lead to significant differences in recoveries even when the matrices are, in principal, the same. Furthermore, the CRM should contain the analyte at a concentration that is similar to that of the sample. If the concentration levels vary widely, extraction recoveries may also differ in the same way. Finally, the analyte should not have been spiked to the CRM, but be present as a naturally occurring compound (to be consistent with what was said above for spiked samples). It is evident that these three conditions are impossible to satisfy, but the closer they are obeyed, the more realistic is the calculated recovery of the extraction procedure and the smaller is the uncertainty of the final result.

6 Derivatisation

Since many organometallic compounds are of a non-volatile nature, derivatisation of the analytes is required for those speciation techniques where separation is achieved by gas chromatography. Separation techniques such as liquid chromatography, ion chromatography, capillary electrophoresis and other liquid phase separation methods do not require derivatisation of the analytes with the associated possibilities of errors. Undoubtedly, the derivatisation step is one of the most critical in the sequence of steps in speciation analysis. Which derivatisation reaction can be applied depends largely on the species to be investigated. Again, the group of organometallic compounds with the general formula R_xM^{y+}, where R represents an alkyl or aryl group, and M can be *e.g.* Sn, Pb, or Hg, has been investigated in greatest detail.[43]

The most commonly applied derivatisation reactions are:

• hydride generation,
• alkylation with Grignard reagents, and
• alkylation with tetra-alkylborates

Hydridisation is usually performed with $NaBH_4$ directly in aqueous solution. An acidic medium (*ca.* pH 2) is required for hydride generation and the formation of volatile hydrides, as illustrated for alkyltin compounds:

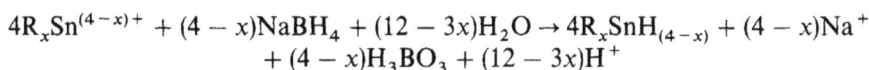

$$4R_xSn^{(4-x)+} + (4-x)NaBH_4 + (12-3x)H_2O \rightarrow 4R_xSnH_{(4-x)} + (4-x)Na^+ \\ + (4-x)H_3BO_3 + (12-3x)H^+$$

The volatile derivatives are stripped from the solution in a purge-and-trap system on to a cryogenic trap (a U-shaped tube filled with a chromatographic adsorbent and immersed in liquid nitrogen). Controlled heating of the cold trap allows separation of the derivatives according to their volatility.

Depending upon the element and the matrix, hydride generation (HG) has to be optimised with respect to the reagent concentration, the pH and the acid used.[44,45] The potential simplicity of the method (which can be used directly in aqueous solutions) and the high sensitivity, due to the achievable high preconcentration factors, are hampered by the fact that the presence of organic matter in high concentrations significantly reduces the derivatisation yield (as shown for butyltin[46] and for arsenic compounds[47]). High metal concentrations seem to cause an inhibition of the hydride formation.[44,45,48] In practice, the hydride generation technique is routinely applied to the determination of hydride-forming elements. In this case, different redox states may be distinguished under different reaction conditions [*e.g.* arsenic which can form arsine (AsH_3) from both As(III) and As(V) (arsenite and arsenate) at pH 1 while only arsenite is reduced at pH 5; or the differentiation of Se(IV) and Se(VI)]. $NaBH_4$ has also been used successfully for the derivatisation of butyltin compounds. Obviously, there is a steric influence on the derivatisation reaction. Methyltin compounds are readily derivatised with $NaBH_4$, whilst butyltin compounds require more powerful reducing conditions.[27,49] Phenyltin compounds are not well determined by hydride generation, since derivatisation yields are low and irreproducible, and the derivatives show poor chromatographic behaviour.[48-51]

In practice, the use of hydride generation for speciation analysis is dominant, when coupled to atomic absorption or atomic fluorescence systems as element-specific detectors. It may be coupled to both gas and liquid chromatographic separation techniques, potentially also with post-column oxidative degradation of the elemental species. For the speciation analysis of organometallic compounds, it shows severe limitations due to the limited separation power of the cryogenic trapping/U-tube chromatography set-up. On the other hand, the derivatives are not sufficiently thermally stable to allow high-resolution gas chromatography at elevated temperatures.

Traditionally, the second most commonly applied technique for speciation analysis is derivatisation by Grignard reagents. This technique has been used particularly for the speciation of organotin and organolead compounds and produces stable derivatives suitable for gas chromatography according to the relationship:

$$R_xM^{y+} + yR'MgX \rightarrow R_xR'_yM + yMg^{2+} + yX^-$$

Since the reagent quickly decomposes in aqueous solution, the reaction has to take place in an aprotic solvent. This requires extraction of the ionic compounds from the aqueous phase into a suitable organic solvent (*e.g.* hexane or isooctane), usually by the addition of a complexing agent such as diethyldithiocarbamate (DDTC), ammoniumpyrrolidine dithiocarbamate (APDC[52]) or tropolone.[53] After addition of the Grignard reagent and allowing the reaction to proceed for a few minutes, the excess of reagent has to be destroyed. This is normally done by the drop-wise addition of water. Since this reaction may be vigorous, efficient cooling of the reaction mixture is recommended and the addition of extra solvent (hexane or isooctane) as keeper helps to avoid analyte losses. Grignard reagents are very efficient tools for transforming ionic organometallic compounds into their fully alkylated derivatives. They are less sensitive to interferences than, *e.g.* hydridisation or alkylation with $NaBEt_4$, and the reaction generally has high yields. The influence of reaction parameters when using Grignard reagents has been studied extensively.[34,50,54-59] It has been demonstrated that the reaction time is not critical, since the reaction proceeds very quickly. In addition, the reaction temperature is not critical, but the concentration of the derivatisation reagent should be kept at 1 or $2 \, mol \, l^{-1}$ in order to ensure quantitative yields. The choice of Grignard reagent will mainly depend on the species to be determined. For steric reasons, derivatisation with short-chain Grignard reagents (*e.g.* MeMgCl) may be easier to perform than with those of longer chain length (*e.g.* PeMgBr), but this effect is only minor. For practical reasons, medium- to long-chain Grignard reagents are to be preferred, since they do not give rise to evaporation losses during subsequent pre-concentration steps.[56] In general, the Grignard derivatisation technique is excellently suited for the speciation analysis of organometallic compounds. It is, however, also the most tedious procedure, due to the solvent exchange and the multiple extraction steps, and bears the risk of contamination and sample losses when not performed appropriately.

Only a few interferences are observed when derivatising with Grignard reagents. One of these is the derivatisation of sulfur concurrently extracted from

Table 2 *Derivatisation reagents and conditions for organometallic speciation*

Reaction type and reagent	Typical experimental conditions
Hydride generation:	
• $NaBH_4$	• as 0.4–20% aqueous solution
• $LiB(C_2H_5)_3H$	• as 0.1% solution in THF
Grignard derivatisation:	
• methylation with MeMgCl	• as 1–2 mol L^{-1} solution in diethyl ether
• ethylation with EtMgCl	• as 1–2 mol L^{-1} solution in THF
• propylation with PrMgCl	
• pentylation with PeMgBr	
Alkylation/arylation with borate reagents:	
• ethylation with NaBEt$_4$	• as 0.5–10% aqueous solution
• propylation with NaBPr$_4$	• as 5–20% solution in THF
• butylation with (Bu$_4$N)BBu$_4$	• as 5–10% solution in acetone
• phenylation with NaBPh$_4$	

sulfur-rich sediments. During the derivatisation, dialkyl mono-, di- and tri-sulfides are formed which particularly interfere when flame photometric and mass spectrometric detection methods are used.[60] Therefore, different techniques have been designed to overcome this problem, either by oxidation of the sulfur compounds to the corresponding sulfones,[54] or by applying column chromatographic clean-up techniques.[61] However, with element-specific gas chromatographic detectors (such as the atomic emission detector or ICP-MS) these interferences are not observed.

The use of sodium tetraalkyl- (and aryl-)borates as derivatisation reagents for organometallic compounds bridges the gap between $NaBH_4$ and Grignard derivatisation reagents. It is particularly attractive, since it provides alkylated (and thus more stable) derivatives of the ionic organometallic compounds (as with Grignard reagents) and can be used in aqueous solution (under similar conditions as to $NaBH_4$). For the sample preparation procedure, the feasibility of *in situ* derivatisation means a significant reduction in the number of sample handling steps (and the potential risks for contamination or sample losses).[62,63] The derivatisation is usually carried out in buffered aqueous samples, or aqueous extracts, by the addition of the alkylborate solution. It is dexcribed by the reaction:

$$4R_xM^{y+} + yNaBR'_4 + 3yH_2O \rightarrow 4R_xR'_yM + yH_3BO_3 + 3yRH\uparrow + yNa^+$$

The resulting tetra-alkylated organometallic compounds are simultaneously extracted into a small volume of organic solvent (hexane or isooctane), which can be used for GC analysis after phase separation. Due to the fact that one to three alkyl groups have to be transferred to the mono- to tri-valent organometallic cations respectively, the derivatisation yields depend strongly on the degree of substitution and the nature of the organometallic moiety. Additional factors that influence the derivatisation yield are the pH of the sample extract and the reagent concentration, but the reaction time is not critical.[44] There is general agreement that for organotin compounds, the reaction produces the highest derivatisation yields at pH 4–5. At lower pH values, fast degradation of the alkylborate reagent takes place. At high pH, complexation reactions of the analytes and side reactions of the reagent presumably hamper the derivatisation. While the reaction time is not critical, which indicates that the derivatisation reaction and potential decomposition reactions of the reagent take place more or less instantaneously, the concentration of the reagent is critical. It has not only been shown that higher concentrations of NaBEt$_4$ produce higher derivatisation yields,[44,56] but it was also demonstrated that the gradual addition of the reagent in multiple portions increases the derivatisation efficiency.[64] This is in agreement with the assumption that the alkylborate reagent reacts quickly, but is also readily decomposed after addition to the sample extract. One also needs to consider that typical amounts or concentrations of the alkylborate reagents added to the extract (different volumes of 0.2–5% solutions) are on average about one order of magnitude lower than that of Grignard reagents used for the same purpose.

When samples with high metal or organic matter content are derivatised, the amount of reagent added should be increased, in order to compensate for additional reagent consumption through side reactions.[28,36,62,65-67] This effect can be minimised by moving away from the *in situ* procedure and separating the extract from the sample before derivatisation.[64,68]

In conclusion, *in situ* ethylation by sodium tetraethylborate may be considered an effective and very convenient technique for the derivatisation of organometallic (particularly tin, mercury and lead) compounds. It has to be kept in mind, however, that derivatisation yields are generally lower with alkylborates than with the corresponding Grignard reagent[56,64,68] and that the magnitude of the difference depends on the analyte, the matrix and the derivatisation conditions applied. As the derivatisation yield also tends to be more variable than with Grignard derivatisation, it is absolutely mandatory to monitor the derivatisaton yield with every batch of samples that is analysed. In this context it has to be mentioned that the stability of alkylborate reagents in aqueous solutions is very limited and they should thus be prepared fresh each day. Also, the degradation of the reagent during storage as the solid substance (*e.g.* by the reaction with moisture and oxygen) may occur, and is recognised by a change in colour of the white powder to yellow. As an alternative, it is recommended to dissolve the reagent in THF, in which it was found to be stable over a period of at least eight weeks.[69]

The disadvantage of the tetraethylborate reagent, that it cannot be used for the analysis of ethylated species of the relevant metals (particularly the environmentally relevant ethyllead compounds), has been overcome by the introduction of alternative alkylborate reagents. Recently, sodium tetrapropylborate and tetrabutylammonium tetrabutylborate reagents have become commercially available and have been used for speciation analysis. While the latter reagent is clearly not useful for speciation analysis of organotin compunds, it has been used successfully for the analysis of organolead compounds[70] and reduces the risk of sample losses during evaporative sample concentration steps, due to the lower volatility of the butyl in comparison with the ethyl derivatives. The propylborate reagent is, however, ideally suited for the simultaneous analysis of environment-ally relevant organolead and organotin compounds and has been optimised and used for their analysis with GC–AED[70] and GC–ICP-MS.[71] The optimum conditions for the use of the tetrapropylborate reagent are similar to those of the tetraethylborate reagent.

For the analysis of mercury and methylmercury, the use of sodium tetra-phenylborate as derivatisation reagent has been reported.[72,73] Due to the fact that sodium tetraphenylborate is a very mild derivatisation reagent, it can only be used for this task, but not for the derivatisation of organotin or organolead compounds. However, the lower reactivity of this reagent when compared with tetraethylborate or $NaBH_4$ minimises the risk of artifact formation, which has been strongly debated recently.[74,75]

In conclusion, it may be said that it is the responsibility of the analyst to decide which of the above discussed derivatisation techniques is most appropriate for the problem at hand. In the case of organometallic compound analysis, prefer-ence may either be given to a more robust but considerably more labour-intensive technique (Grignard derivatisation) or to a simple, potentially *in situ* derivatisation technique (alkylborate derivatisation) that has the disadvantage of lower derivatisation yields and robustness. In any case, the analyst has to prove the fitness of the applied method for the purpose, and from this point of view a derivatisation method that consistently produces lower (but still reasonably high) derivatisation yields that are monitored and verified by the use of control samples and reference materials may be acceptable.

Of course, a variety of other derivatisation methods exists for which, in principle, the same holds true: it does not matter whether the derivatisation reaction is, for example, a complex acylation/esterification procedure for the analysis of seleno-aminoacids[76] or a simple one-step reaction[77] – if the method produces reasonably high, stable derivatisation yields that are unaffected by interferences, it can be considered fit for purpose.

Unfortunately, as a result of derivatisation the detected compound is chemi-cally different from the original analyte. For each analyte, two different species – one underivatised and one derivatised – should thus be available for calibration and determination of recoveries. If these compounds are not commercially available they should be synthesised to a sufficient level of purity.[78]

In order to be able to monitor the derivatisation efficiency of organotin compounds with different degrees of substitution, the German standard for the determination of organotin compounds prescribes the use of differently al-kylated organotin calibrants (mono- and diheptyltin and tripropyltin chlor-ides).[79]

A particular point for consideration is the possibility of transformation of some species into others. Since organometallic species are often relatively un-stable, sample handling conditions, such as contact with air, extraction at higher temperatures or low pH, or derivatisation conditions, may sometimes lead to an alteration of the species information. For instance, dealkylation of TBT has been shown to occur during storage at higher temperature[14] or dur-

ing extraction or derivatisation.[25] In such cases the TBT measurements would
lead to low results, while those of dibutyltin (DBT) or monobutyltin (MBT)
could be too high as the result of this artifact. One should realise that such
transformations can lead to particularly large errors if the concentration of the
'starting compound' is much higher than that of the product species. For
instance, in the dried mussel powder of CRM 477 the triphenyltin (TPhT) level
is more than an order of magnitude higher than that of diphenyltin (DPhT). It
was observed that when the Grignard derivatisation was carried out at elevated
temperature, a minor (5%) transformation of TPhT occurred and that this
doubled the level of DPhT.[80] A similar artifact was reported for methylmercury determinations using derivatisation with tetraethylborate. A minor contamination level of the reagent with methyl groups could lead to elemental
mercury being derivatised into methylethylmercury, and quantified as
ethylated methylmercury. In environmental sediment samples, with elemental
mercury levels orders of magnitude higher than that of methylmercury, these
types of artifacts could lead to large errors in the quantification of the latter.[81]
In addition, the extraction/distillation procedure commonly applied in the determination of methylmercury was shown to produce significant artifacts
through methylation of inorganic mercury by the matrix constituents.[82,83]

In recent years, a number of methods have appeared in which separation is
based on liquid chromatography (LC).[6] Traditional LC detectors such as UV-
Vis absorption are not suitable for organometallic species at trace levels, but new
interfaces allow coupling to other types of detectors, such as ICP-MS[84,85] or
electrospray-MS/MS.[85] This would allow one to analyse underivatised organometallic species, thus significantly reducing the number of analytical steps
(and potential sources of error).

7   Separation

Since there are only a few detection techniques that can determine different
species in the presence of one another, a separation of the different species is
mandatory. The three principal techniques applied for this purpose are gas
chromatography (GC), liquid chromatography (LC), which includes ion
chromatography (IC), and electrophoretic methods, of which capillary (zone)
electrophoresis (CE or CZE) is the most important.

Gas chromatography is today the most commonly employed technique in
speciation analysis of organometallic compounds.[27,86] It is particularly attractive because of its high separation power (when used with capillary columns), and
due to the availability of highly sensitive and selective detectors such as mass
spectrometry (MS), atomic emission detection (AED), flame photometric detection (FPD) and electron capture detection (ECD). It requires, however, sufficient
volatility and thermal stability of the analytes, which in most instances is
achieved by derivatisation as outlined above. Once the analytes have been
extracted from the matrix and derivatised, the gas chromatographic separation
of elemental species poses no particular difficulties when compared with trace

organic analysis. Of course, chromatography-related problems are observed, such as split discrimination when split injection is used with compounds that span a wide volatility range, and peak tailing problems due to analytical columns starting to exhibit active sites or due to contamination of the precolumn (the use of which is generally recommended). It is worth mentioning that analysis sequences should be designed with care. Since calibrants are usually made up in clean solvents, they tend to 'flush' the chromatographic system, leaving behind active sites to which analyte molecules may adsorb. When a sample is injected, the response may be lower than expected due to adsorption of the analyte molecules onto the active sites of the liner, chromatographic column, transfer line or interface to the detector. After repeated injection of matrix-loaded samples, the matrix saturates all active sites in the chromatographic system, and the response increases and eventually reaches a stable value. When, after a number of sample injections a larger number of calibration solutions are injected successively, the chromatographic system will be flushed again and the response will decrease, due to the newly exposed active sites in the system. It is therefore recommended that samples and calibrants be measured alternately in a sequence, to keep the system conditioned in a constant state. Maximum care should be exerted to keep the gas chromatographic system as clean and as inert as possible. The liners used should be silanised and exchanged at intervals, particularly if extracts are injected without prior clean-up. Pieces of the polysiloxane rubber material in the liner, originating from septum damage by the syringe needle, are detrimental in obtaining reproducible results.

Peak tailing may also be an indication of thermal degradation of the organometallic compounds on the chromatographic column. The metal oxides formed (*e.g.* SnO_2, PbO) may cause adsorption and thus excessive peak tailing of the organometallic species. Depending on the type of detector used, the same phenomenon may also give rise to detector-based peak broadening, as has been observed for the AED.[87]

Cryo-trapping separation may be considered as a simplified form of gas chromatography. The short, packed column and the coarse temperature control of the system, however, only allow simple separations to be performed. Different chemical species that have similar volatility (*e.g.* $BuMeSnH_2$ and $PhSnH_3$) are not separated.[88]

Liquid chromatographic techniques are of great interest for speciation analysis since they do not require derivatisation of non-volatile or ionic organometallic compounds.[89-91] However, they are inferior to gas chromatography in terms of separation power and – probably with the exception of ICP-MS – available detectors are not sufficiently specific and sensitive at the same time. In the case of ionic organometallic compounds; ion chromatography or ion pairing chromatography offer the best separation. Reversed-phase chromatography can only be adapted, with difficulties, to the separation of ionic compounds, and does not provide adequate separation in most cases.

Capillary electrophoresis (CE) is one of the newer separation techniques. In the normal mode of operation, it is restricted to the separation of ionic compounds. Due to the small sample volume usually applied (in the low nL range),

CE requires coupling to the most sensitive detectors, to be useful for speciation analysis. The coupling to ICP-MS as a detector is promising, but still suffers from many difficulties in practice.[92,93] In particular, the question of robustness towards heavily matrix-loaded samples has not, so far, been addressed in a satisfactory way.

8   Detection

Almost any kind of available detector has been used for speciation analysis in combination with gas and liquid chromatographic separation or electromigration methods. From the point of view of quality assurance, compound- or element-specific detection is desirable and nonspecific detectors such as flame ionisation detection (FID) for gas chromatography or UV-absorbance detectors, although occasionally still applied, should be replaced by more selective detectors. Some detection principles require post-column derivatisation to establish selectivity, for instance with a fluorogenic label.[94] Even with element- or molecule-specific detectors, it has to be realised that under practical conditions they have a limited selectivity and may suffer from more or less serious interferences. This is most evident with optical detection such as flame photometric detection (FPD), atomic emission detection (AED) or atomic absorption spectrometric detection (AAS). In all three cases, background emission or absorption at the characteristic wavelength of the detected element can make the detection of (ultra)trace amounts, in the presence of a large amount of matrix, difficult. Typical selectivities that can be reached with these instruments are in the range of 10^3–10^4, and for most applications this is sufficient. However, particularly in cases of doubt (*e.g.* when peaks appear at unexpected retention times), one should try to ascertain the correctness of the element-specific response (*e.g.* by observing the characteristic emission spectrum in the case of atomic emission detection). Molecular mass spectrometry is equally suitable as a detector for speciation analysis. While GC–MS is well established as a workhorse technique in the analytical laboratory, this is not yet the case for liquid chromatography coupled to mass spectrometry (LC–MS). The recent improvement in LC–MS technology, brought about by the atmospheric pressure ionisation techniques [in the form of electrospray, ES, or atmospheric pressure chemical ionisation (APCI)], is soon expected to change this picture. Molecular mass spectrometry (coupled to both GC and LC) offers the advantage that not only the central metal atom is detected, but rather the whole molecule. This is a necessity in the identification of unknown compounds, but also helpful when, for example, assignments of retention times have to be made. Molecular information in the form of a mass spectrum is, of course, only possible when the MS is operated in scan mode. In order to obtain the sensitivity required for trace analytical determinations, the selected ion monitoring (SIM) mode is preferred. Since only two or three characteristic ions are used in this case, the detection is more susceptible to interferences. For quantitative analysis, acceptable ranges for the peak ratios of the characteristic ions (the quantifier and the qualifier ions) have to be defined. If the peak ratio is outside the acceptable range, an interference is likely to be present.

Elemental mass spectrometry, in most instances coupled to inductively-coupled-plasma atomisation and ionisation (ICP-MS), is presently one of the most sensitive element-specific detectors. It may be interfaced to both gas and liquid chromatography or to capillary electrophoresis. Since the chemical bonds are totally disrupted in the high-energy plasma, ICP-MS does not provide any information about the compound initially present. This information has to be deduced from the preceding separation step. In that respect, ICP-MS and molecular MS detection provide complementary information that can be used advantageously.

9 Calibration

A controversial issue in speciation analysis is how to obtain quantitative results and which quantification method is the most suitable. Various approaches have been suggested and applied, and each has both merits and shortcomings. The two principal possibilities are:

- quantification by external calibration and
- quantification by standard addition.

In the following, we will assume that the applied speciation scheme includes various sample preparation steps, such as extraction from the solid matrix, derivatisation, clean-up and determination by a suitable instrumental technique. In a case where the analysis does not require all of the above steps, the scheme can be simplified accordingly.

External calibration is certainly the simplest technique for quantification, but unfortunately it must be considered inappropriate in most cases, except perhaps for the simplest of matrices. External calibration can be performed at different levels of complexity. For analytical schemes that include derivatisation, the simplest way of performing an external calibration is by using derivatised calibrants. There are, though, a number of obstacles related to this. First, derivatised calibrants may be difficult to obtain, and even when they are available their stability and potential sources of losses (*e.g.* through degradation and/or solvent evaporation) have to be critically monitored when they are used for calibration. Examples include alkylated organotin compounds which can be synthesised then subjected to a series of distillation and/or recrystallisation steps before they are of satisfactory purity to be used as calibrants.[95] The disadvantage of this type of calibration is that it does not take into account the extraction, derivatisation, and clean-up steps, which usually have recoveries significantly less than 100%. Any loss factors occurring during these steps need to be quantified separately.

Carrying out an external calibration with underivatised calibrants is the next possibility, being more complex but at the same time more representative. Since the calibrants (*e.g.* ionic organometallic compounds) undergo the same derivatisation procedure as the samples (ideally, they should be derivatised on the same day and with the same batch of reagent under identical conditions as the samples), the day-to-day variability of the derivatisation efficiency should be compensated for. On the other hand, it has to be remembered that the derivatisation efficiency is often higher in pure solutions than in a matrix-loaded extract. Underivatised calibrants are also more widely available, but here again, the stability of the calibrants has to be scrutinised with extreme care – in particular, di- and mono-substituted alkyl-metal compounds are known to undergo degradation. Having to derivatise all calibrants at different levels, in order to establish a calibration curve, makes this method of calibration a tedious procedure.

The next stage of complexity for external calibration is to spike a blank matrix with the calibrants at different levels, and to subject the fortified material to the entire sample preparation procedure. This has to be considered the most representative way of (external) calibration, since the calibrants undergo all analytical steps in the same way as the analytes in the actual sample. Thus, all losses, be it during the extraction from the matrix, the derivatisation or the clean-up, are common to both the analytes and the calibrants and should thus cancel each other out. However, a disadvantage is that the analyst may not even be aware of the magnitude of these losses. The remarks made previously, concerning the difficulty of carrying out spiking experiments, apply here as well and may, in practice, impair the results. Blank matrices may not be available, or may not be very similar to the sample matrix. As discussed above, binding of the spiked calibrants to the matrix may not be as strong as in the case of naturally occurring species, thus leading to an overestimation of the recovery. On the other hand, the opposite has also been observed. Some organometallic compounds are more stable when adsorbed on to a sediment or other solid matrix. Since the decomposition rate is significantly reduced when compared with the species in the spiking solution (which is basically a consequence of the reduced accessibility and the changed chemical environment of the molecule, which may experience a different hydrophilicity or hydrophobicity than when in solution), recoveries may in such cases also be underestimated.

The second principal approach for determining analyte concentrations is to perform a calibration in the matrix through the 'method of standard addition'. The method of standard addition is based on adding defined increments of the analyte to the sample. After the analytes are allowed to equilibrate with the sample matrix, the sample is extracted and the extract is processed as a regular sample. From the known amounts added and the responses found for the fortified sample, it is possible to construct a calibration graph which has a non-zero intercept and meets the concentration axis at a value corresponding to the initial concentration of the sample (with a negative sign). From a statistical point of view, a number of conditions are to be fulfilled for a valid standard addition experiment. The analyte increments – usually three to four concentration increments are considered satisfactory – should be spaced in approximately equal distances. Furthermore, they should be chosen in such a way that the concentration is almost doubled at the highest concentration level. Under these conditions, the standard addition graph can be evaluated best and the confi-

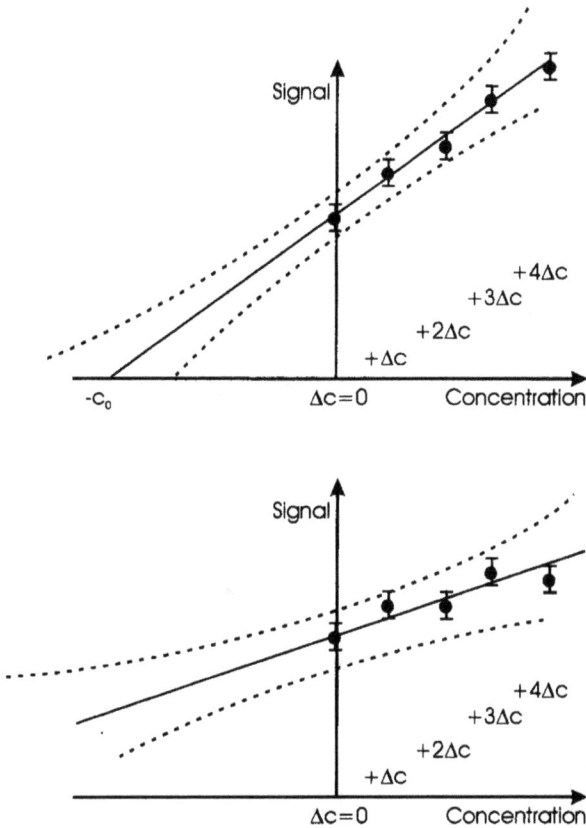

Figure 2 *Graphical representation of quantitative evaluation by standard addition with (a) suitable (top) and (b) unsuitable (bottom) concentration increments*

dence interval of the result (Figure 2, top) is minimised. If the concentrations with which the sample is fortified are too small in relation to the initial concentration, the slope of the calibration line becomes too low, and could even become negative as the result of random measurement errors. In other words, the confidence interval of the calibration can become too large to be practically useful (Figure 2, bottom). If, on the other hand, standard addition is carried out at a level much higher than that of the original analyte, the influence of matrix interferences may no longer be comparable (for instance, in the case of a limited number of active binding sites in the matrix).

It will be clear that the optimal spiking level will depend on the analyte level present in the sample. For analyte levels close to the LOD it will be advantageous to spike at more than twice the original level in order to decrease the influence of random analytical errors. Samples that already have high initial concentrations of the elemental species to be determined require particular attention. Spiking the sample to increase the analyte level to twice its initial concentration may already lead to levels exceeding the linear range of the

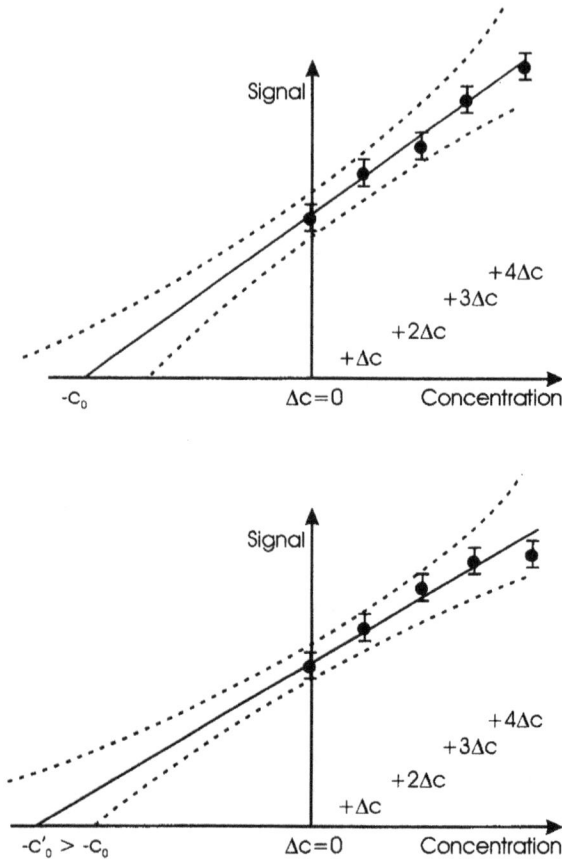

Figure 3 *Graphical representation of quantitative evaluation by standard addition with concentration increments that are (a) within the linear range (top) and (b) exceeding the linear range of response (bottom)*

detector. The consequence is that the slope of the calibration line is under-estimated and the analytical result is consequently overestimated (Figure 3). It is therefore essential to check that, after spiking, the concentrations lie within the linear range.

But even if these prerequisites are all fulfilled, there are some important limitations of the standard addition method which have to be kept in mind. Again, the question of how to spike correctly is raised. As discussed above, the added calibrants should ideally be incorporated into the matrix in the same way and of course in the same chemical form as the naturally incurred species. The standard addition approach requires that a series of measurements be performed for each sample that needs to be analysed, thus making the method much more time-consuming and costly. Also, some pre-knowledge of the analyte levels is needed in order to establish the proper addition levels. Furthermore, one has to be aware that, although the standard addition method is particularly attractive since it inherently corrects for proportional systematic errors, it does not correct for constant systematic errors (*e.g.* blank problems or constant sample losses).

Isotope dilution (ID) is a special form of standard addition that can be used to improve the accuracy of quantitative analysis. In contrast to inorganic analysis, however, suitable species with altered isotopic ratio are only rarely available commercially, since very few isotopically labelled organometallic compounds have been synthesised. The practical usefulness of ID techniques for speciation analysis is further limited due to the fact that – in order to compensate for all systematic errors during the sample preparation procedure – the species with altered isotopic ratio has to be added to the matrix. Thus, all the difficulties that apply to spiking a sample and that have been discussed above apply also to the isotopically labelled compounds.

10 Calculation of Results and Reporting

The final stage of the analysis is the calculation of the results and the reporting. In speciation analysis the same care has to be applied as in any other field of analytical chemistry, but the possibilities for introducing errors may be somewhat greater and are not always evident. In trace organic or inorganic analysis there is usually no doubt as to what species the final result refers to, either to the compound (in organic analysis) or to the metal (in inorganic analysis). Since speciation analysis lies between these two fields, results are reported both in terms of the central metal atom (*e.g.* tributyltin as tin) and of the analyte species itself (*e.g.* as tributyltin cation). Sometimes the listed concentrations refer to the complete compound, including the counter-ion that was used for calibration (*e.g.* as tributyltin chloride). In each case, the report has to state clearly to which species the results apply. This is particularly important, since the form in which the analyte is determined is usually not identical to the form in which the calibrants have been prepared, since a derivatisation step is usually involved (*e.g.* tributyltin chloride which, after ethylation, is detected as tributylethyltin). Conversion factors between the bare metal, the species to be detected and the derivative that is actually detected may significantly differ from unity. Finally, it should be realised that even when this is clear to the analyst it may still cause confusion to the end-users of the data.

A further issue of discussion is whether results should be corrected for recovery when this is known to be different from 100%. Although there is no general agreement on this topic in the scientific community, the authors are inclined to suggest the following procedure, which is adopted from trace organic analytical practice.[96]

Considering the concentration ranges and the difficulties of speciation analysis in complex matrices, a method that produces a recovery of 70–120% may be considered appropriate. Methods that produce recoveries higher than 120% (that is, methods that significantly overestimate the actual concentration) as verified by spiking experiments, or by the analysis of certified reference materials, may not be considered trustworthy. Methods that feature a recovery of less than 70% would seem to be unsatisfactory, but considering the difficulty of speciation analysis and the fact that we generally have to work with compromise methods which are designed to be applicable to a wider spectrum of analytes instead of a single species, we often have to accept methods that produce recoveries lower than 70% for some species. In this case, there is no doubt that a correction of the results for recovery has to be applied (when external calibration is used). Still, strong doubts will remain regarding a method that fails to produce quantitative recoveries, even if the recovery is fairly constant. If the matrix impedes the extraction/derivatisation procedure to such an extent that only, *e.g.*, 50% of the analyte is recovered, then it is likely that with a slightly different matrix or under marginally different conditions the recovery may vary significantly. Thus, if possible, the recovery should always be determined with the same matrix as the actual sample.

If the recovery is in the range of 70–120%, the correction for recovery is debatable.[37] One of the arguments against such corrections is that the recovery correction factors may be burdened by an uncertainty factor, which must be considered in the total uncertainty budget of the analysis. In contrast to this, uncorrected results include only the uncertainty related to the analysis itself. In addition, recovery factors are strictly speaking only valid for the matrix in which they have been determined. If they have to be determined in the actual sample, this can only be done by spiking experiments with all the associated (and previously discussed) shortcomings. This situation may be summarised as follows: 'Why should I correct data which I trust with a correction factor in which I have no confidence'?

There are, though, a number of arguments in favour of correcting data by empirically determined recovery factors. First, not correcting data from different laboratories whose methods have different recoveries would produce different results for the same sample. If the reproducibility of the measurement within the laboratories is sufficiently good, these results would be statistically different. In the case of organometallic compounds such differences are likely to occur when some laboratories use underivatised calibrants while others use alkylated compounds for external calibration. The former will typically have high recoveries, as only losses during the extraction step need to be accounted for, while with derivatised calibrants the recoveries tend to be much lower as they reflect the losses throughout the whole analytical procedure. Second, not correcting for recovery gives a negative bias to the results. This negative bias would diminish as the method and its recoveries improve, which means that the results (for instance the concentrations in a CRM) would not be consistent over time, but eventually shift to higher values. For these reasons, it is the authors' opinion that recovery corrections need to be applied based on the best available recovery determinations. The fact that the corrected results do not suffer from method-dependent biases outweighs the disadvantage related to the extra uncertainty. The correction factors applied should be clearly stated in the analytical reports.

11 Achieving Quality in Speciation Analysis on a Daily Basis

Achieving quality (and being able to demonstrate this) on a daily basis is one of the most difficult tasks for a routine laboratory. This holds true particularly for speciation analysis, due to the problems of low concentration levels, complex matrices, instability of analytes, complicated sample preparation schemes and risks of contamination.

A number of tools are at the disposal of analytical chemists to allow them to monitor (and to improve, if necessary) the quality of their analyses. These are, among others:

- use of validated protocols for sampling, sample storage and analysis
- monitoring of calibrant and (derivatisation) reagent stability
- monitoring of blanks
- monitoring of instrument performance
- use of control materials (reference materials, laboratory control materials, certified reference materials)
- participation in inter-laboratory comparisons.

It is generally recommended that only validated methods be used for speciation analysis. This requirement is, in practice, difficult to comply with, since the influence of the sample matrix on the analytical process is of such over-riding importance that, in principle, the application of a method to a new matrix requires its re-validation. If a complete validation cannot be done, at least the determination of the recovery has to be performed (even if the quantification is done by the standard addition technique). In exceptional cases it may be acceptable to use external calibration when the results are related to suitable internal standards. These internal standards are spiked in an appropriate way to the matrix (in which they must not be present initially) and undergo the entire sample preparation procedure. In organotin speciation, frequently used internal standards are tripropyltin and tricyclohexyltin, which are very similar to tributyltin and triphenyltin in their physico-chemical properties. They may be used either to monitor whether the recovery of the trisubstituted organotin compounds is within the normal range and thus does not need a recovery correction (see discussion above), or they may be used to correct for the recoveries of the trisubstituted organotin compounds, assuming that the recoveries of chemically similar species will also be similar. This assumption has been the basis of the suggested method of evaluation in the draft German standard DIN 38407-13.[79] Here, the response of the mono-, di- and tri-substituted organotin compounds is normalised to the response of monoheptyl-, diheptyl- and tripropyltin used as internal standards. This procedure is intended to make the data evaluation more robust against matrix influences.

For monitoring the stability of reagents or calibrants or for documenting blank values and instrument state, the use of control charts is highly recommended. Control charts plot the result of an analysis as a function of the date or series of analyses. Shewhart charts are the most frequently used control charts for this purpose.[97] They have a central line (equal to the mean of the analysis results, determined in a pre-period) and, at a distance of two standard deviations from this central line the warning limits and at $\bar{x} \pm 3s$, the control limits. An analytical process that is under control will mostly produce results that fall within the interval $\bar{x} \pm 2s$. If results repeatedly fall in the range between warning and control limits of the control chart, or if they exceed the control limits, there is reason to believe that the process is out of control (there are a number of exactly defined conditions when a situation is considered out of control, see, *e.g.*, ref. 97). In a case when an out-of-control situation has been encountered, routine analysis should be discontinued until the cause for the deviating results has been identified and eliminated.

Shewhart charts may be used for a number of purposes. The most common use is that of monitoring the analysis result (mean or single value) of a suitable control sample over time. Depending on the nature of the control sample, either the correct operation of the instrument can be demonstrated (*e.g.* if the control sample is a pure calibrant), or the success of the entire process, if the control sample is a real matrix sample. Shewhart charts may also be used to monitor procedure blanks or recoveries of the entire sample preparation procedure.

Whilst, in most cases, either single values or the mean from n repetitive measurements are plotted in a mean value chart, control charts can also be designed to monitor the range or the standard deviation of repetitive measurements. Out-of-control situations can thus not only be detected when there is a significant bias of the result, but also when the standard deviation becomes unacceptably large. Next to Shewhart control charts, the so-called Cusum control charts have gained some importance. In the Cusum chart, the *cumulative sum* of the parameter to be monitored is plotted over time. If the system is under control, the fluctuations around the target value are purely statistical and their sum should be close to zero. If there is a constant bias introduced into the measurement system, the mean would constantly shift to either higher or lower values. Although the Cusum chart has certain merits, it lacks the easy and intuitive appreciation that is offered by the Shewhart chart, since the value that is plotted is the cumulative sum, rather than the direct result of an analysis.

For the purpose of monitoring both the instrument and method performance, control samples are used. Suitable control samples may be certified reference materials (CRMs), reference materials (RMs), or laboratory control materials (LCMs). In order to be suitable, the used materials have to satisfy the following conditions:

- *stability*: the material must not change its composition over a long period of time under defined storage conditions (*e.g.* at $-20\,°C$), since the intention is to be able to monitor the method performance for an extended period and to ensure that results that have been measured with long time intervals are still comparable.
- *availability*: for the same reasons as given above, the control material should be available in sufficient quantity.

- *homogeneity*: the homogeneity, the concentration level and the minimum sample intake of a control sample are interrelated; the better the homogeneity of the material is, the smaller the recommended sample intake may be – which also depends of course on the concentration level of the analyte. Between-sample variations due to inhomogeneity should be insignificant in comparison with the measurement uncertainty.
- *assigned/certified value*: a control material may (but need not) have an assigned value for the analyte that is to be measured. If the assigned value has been obtained through the certification procedure of an internationally recognised body (*e.g.* BCR, NIST, NIES or NRCC),[98,99] the material is a certified reference material. When the certification has not been attempted, or has not been concluded successfully, the material is distributed as a (not certified) reference material. Since the stability and the homogeneity of this material have usually been demonstrated, it may still be of great value to analytical laboratories. The assigned value, however, is not given in the form of a certified value with its associated uncertainty, but instead as an indicative value or concentration range. Laboratory control materials may be considered the lowest level in this hierarchy of control materials. The most important require-ments for LCMs are that they be stable and sufficiently homogenous for the purpose they are used for. They do not have to have an assigned value, at least not through a formal procedure. Since LCMs may, for example, be the remainder of a previously analysed sample batch that is available in sufficient quantity and for which the homogeneity and the stability have been proven, they represent a low-cost alternative to reference materials for use in certain cases.

Control materials can be used for different purposes. First, they may serve to monitor the stability of the measurement system or the method. For this pur-pose, it is not necessary that the assigned value be known or even certified – in principle, the comparison of the results of different series of analyses may already give an indication whether the analytical system is under control. They can also be used to assess whether a new method yields results that are consistent with those obtained using a previous method. The second use of RMs is to assess the accuracy of the procedure. In this case, it is essential that the assigned concentra-tion value for the control material be precisely known with a statement of its uncertainty. Only then is it possible to assess recoveries for the method, which may, for example, be applied to the analysis of samples with a similar matrix. Although this approach is, in principle, acceptable, great care has to be exerted. This also holds true when certified reference materials are used for the calibra-tion of a method which is not generally recommended, but which, under certain conditions, is considered acceptable. Calibration by the use of a CRM means to calibrate a method whose systematic errors are either not known, or cannot be mastered, by the use of a matrix-matched calibrant of known concentration. Since certified reference materials are usually not available with analyte concen-trations at different levels, a one-point calibration is performed. However, par-ticularly in complex matrices, the assumption of a one-point calibration line going through the origin is not always justified. First, recoveries may be different at different concentration levels (which leads to a curvature of the calibration line) and also a one-point calibration is only possible in the absence of any constant, systematic errors.

A point of discussion is, at what frequency should control materials be used. It is a misconception from earlier times that certified reference materials are 'too valuable' (in terms of both availability and price) to use them on a regular basis. This is certainly not true. The costs of the use of reference materials will, in most cases, be negligible in comparison to the labour costs for the sample preparation and analysis, particularly when the analysis has to be repeated due to the analytical system being out of control, this not having been discovered early enough. Additionally, there is the potential for claims for compensation, due to incorrect analytical results.

Although the necessity of using control samples on a regular basis is well understood, the present situation is that we still face a great demand for suitable (certified) reference materials in speciation analysis.[100] As said before, only a few (C)RMs are available for metal (redox) and organometallic species in a few selected matrices. Usually, these RMs are only available with the species of interest at one concentration. This is unfortunate, since most analysts would like to have the species of interest at different concentration levels in various matrices. Furthermore, even if the matrix and the concentration level of the analyte are somehow matched to the actual analytical problem, the objection remains that reference materials and actual samples are vastly different in their physical composition. While reference materials are usually very fine freeze-dried powders and therefore very homogenous, actual samples may be inhomogeneous, of varying particle size and fresh (*i.e.* wet). Thus, they may behave differently in the sample preparation procedure, and recoveries determined from the analysis of the CRM may not be applicable to fresh samples.

An alternative – or rather a complement – to the use of CRMs is the participation in inter-comparison exercises for the purpose of quality assurance. Since the stability of the matrix materials used in inter-comparison exercises does not have to be guaranteed for years (in contrast to CRMs), fresh materials that are much closer to actual samples (*e.g.* tissue homogenates, wet sediments) can be used. Participating in such inter-comparisons may help identify potential problems associated with the analysis of fresh samples. Furthermore, inter-comparison exercises are able to react quickly, for example, to the public demand for quality control measures for a new analyte/matrix combination, since stability tests do not have to be performed for such extended periods as for the production of CRMs. A further benefit of the participation in inter-laboratory comparisons is the exchange of expertise in the technical discussion meetings which may help to overcome problems related to the analytical procedures in individual laboratories.

Table 3 *Uncertainties in certified levels for trace elements, organics and organometals in various CRMs.*

A: Trace elements in various matrices
Trace elements in soil CRM 141R
  Concentration range: $0.25\,mg\,kg^{-1}$ (Hg)–$683\,mg\,kg^{-1}$ (Mn)
  Average relative uncertainty ($n = 9$):   3.5%
Trace elements in plankton CRM 414
  Concentration range: $0.276\,mg\,kg^{-1}$ (Hg)–$299\,mg\,kg^{-1}$ (Mn)
  Average relative uncertainty ($n = 11$):   4.3%
Trace elements in lichen CRM 482
  Concentration range: $0.48\,mg\,kg^{-1}$ (Hg)–$1103\,mg\,kg^{-1}$ (Al)
  Average relative uncertainty ($n = 9$):   3.7%
Total average uncertainty trace elements ($n = 29$)   **3.9%**

B: Persistent organics in various matrices
Chlorinated biphenyls (CB) in milk CRM450 ($\mu g\,kg^{-1}$)
  Concentration range: $1.16\,\mu g\,kg^{-1}$ (CB 52)–$19\,\mu g\,kg^{-1}$ (CB 153)
  Average relative uncertainty ($n = 6$):   10.3%
Pesticides in animal fat CRM 430 ($mg\,kg^{-1}$)
  Concentration range: $0.2\,mg\,kg^{-1}$ (endrin)–$3.4\,mg\,kg^{-1}$ (p,p'-DDT)
  Average relative uncertainty ($n = 9$):   8.8%
Chlorinated biphenyls (CB) in cod liver oil CRM 349 ($\mu g\,kg^{-1}$)
  Concentration range: $68\,\mu g\,kg^{-1}$ (CB 28)–$938\,\mu g\,kg^{-1}$ (CB 153)
  Average relative uncertainty ($n = 7$):   7.6%
Polycyclic aromatic hydrocarbons (PAHs) in sewage sludge CRM 088
($mg\,kg^{-1}$)
  Concentration range: $0.42\,mg\,kg^{-1}$ (benzonaphthothiophene)
  –$2.16\,mg\,kg^{-1}$ (pyrene)
  Average relative uncertainty ($n = 8$):   8.2%
Total average uncertainty persistent organics ($n = 30$)   **8.6%**

C: Organometals in various matrices

compound	*concentration*	*uncertainty*	*rel. uncertainty*
Organotins in mussel powder CRM 477 ($mg\,kg^{-1}$ as cations)			
MBT	1.5	0.27	18.0%
DBT	1.54	0.12	7.8%
TBT	2.2	0.19	8.6%
Methylmercury in fish CRM 463 & 464 ($mg\,kg^{-1}$)			
MeHg	3.04	0.16	5.3%
MeHg	5.5	0.17	3.1%
Arsenic species in tuna fish CRM 627 ($mmol\,kg^{-1}$)			
Arsenobetaine	51.5	2.1	4.1%
DMA	2.04	0.27	13.2%
Trimethyllead in urban dust ($\mu g\,kg^{-1}$)			
TriML	7.9	1.2	15.2%
Methylmercury in sediment CRM 580 ($\mu g\,kg^{-1}$)			
MeHg	75.5	3.7	4.9%
Organotins in sediment candidate CRM 646 ($\mu g\,kg^{-1}$ as cations)			
TBT	480	50	10.4%
DBT	770	80	10.4%
MBT	610	110	18.0%
TPhT	29	9	31.0%
DPhT	36	7	19.4%
MPhT	69	17	24.6%
Total average uncertainty organometals ($n = 15$)			**12.9%**

12   Conclusions

In this chapter, we have discussed procedures for organometal analysis and pointed out some common pitfalls. We have focused on the sample analysis in the laboratory, but it should be realised that even larger errors can potentially result from improper sampling methods, instability during storage or transport, or sampling schemes that are not fit for purpose. More effective communication between the sampling crew, the analytical staff and the end-users could minimise such problems. In the laboratory, analytical problems can be related to the limited stability of the species and/or the calibration solutions, incomplete extraction, the necessity to carry out derivatisations, spiking problems, the fact that often a single, compromise method is used for a range of different species and the limited availability of suitable calibrants and CRMs. The consequence of these potential sources of error is that analytical uncertainties for organometallic species in real samples are typically larger than for other environmental pollutants such as heavy metals or persistent organic compounds. This is illustrated in Table 3 listing the relative uncertainties (95% confidence intervals) of a number of species compiled from recent certifications. For organometallic species these errors were still considered acceptable for certification purposes, reflecting the current state-of-the-art.

Regarding the instrumentation used for speciation analysis, currently there are no dedicated instruments on the market that would enable the extraction, derivatisation, separation and detection of organometallic species in a single, on-line procedure. Consequently, the various analytical steps will have to be performed in an off-line approach, with a risk of introducing manipulation errors, analyte degradation, evaporation losses *etc.* It is hoped that current efforts in the field of instrument development[101] will soon allow analytical laboratories to carry out organometallic species analysis more efficiently with a higher analytical precision. We would also like to stress that these improvements should be the result of continuous efforts in both research and routine laboratories. Rather than relying on rigidly specified standard methods, which carries the risk of all laboratories making the same systematic error and which tends to slow down development, laboratories and instrument manufacturers should be encouraged to develop new approaches. The comparability of both the new and established methods can be verified using CRMs.

13   References

1. R. Gwiazda, D. Woolard and D. Smith, *J. Anal. At. Spectrom.*, 1998, **13**, 1233.
2. M.J. Marques, A. Salvador, A.E. Morales-Rubio and M. de la Guardia, *Fresenius' J. Anal. Chem.*, 1998, **362**, 239.
3. R. Lobinski and J. Szpunar, *Anal. Chim. Acta*, 1999, **400**, 321.
4. R. Lobinski, *Appl. Spectrosc.*, 1997, **51**, 260A.
5. Ph. Quevauviller, *J. Chromatogr. A*, 1996, **750**, 25.
6. O.F.X. Donard and R. Ritsema, *Hyphenated Techniques Applied to the Speciation of Organometallic Compounds in the Environment*, in D. Barceló (ed.) *Environmental*

Analysis; Techniques, Applications and Quality Assurance, Elsevier, Amsterdam, 1993, Ch. 16.

7. Ph. Quevauviller (ed.), *Trends Anal. Chem.,* Special Issue on Speciation Analysis 2000, **19**(2 + 3).
8. Ph. Quevauviller, E.A. Maier and B. Griepink, *Quality Control of Results of Speciation Analysis,* in S. Caroli (ed.), *Element Speciation in Bioinorganic Chemistry,* Wiley, Chichester, 1996, Ch. 6, pp. 195–222.
9. Ph. Quevauviller, E.A. Maier and B. Griepink, *Quality Assurance for Environmental Analysis,* Elsevier, Amsterdam, 1995.
10. O.F.X. Donard and Ph. Quevauviller (eds.), Proceedings of the Workshop on Improvements in Speciation Analyses in Environmental Matrices, *Mikrochim. Acta,* 1992, **109**(1–4).
11. Ph. Quevauviller and R. Morabito (eds.), Proceedings of the Workshop on Trends in Speciation Analysis – Discussion on Organometallic Speciation, *Appl. Organomet. Chem.,* 1994, **8**(5).
12. Ph. Quevauviller (ed.), Proceedings of the Workshop on Trends in Speciation Analysis – Discussions on Inorganic Speciation, *Fresenius' J. Anal. Chem.,* 1995, **351**.
13. F. Adams and S. Slaets, *Trends Anal. Chem.,* 2000, **19**, 80.
14. J.L. Gómez-Ariza, I. Giráldez, E. Morales, F. Ariese, W. Cofino and Ph. Quevauviller, *J. Environ. Monit.,* 1999, **1**, 197.
15. B. Michalke, *Fresenius' J. Anal. Chem.,* 1999, **363**, 439.
16. B. Michalke, *Fresenius' J. Anal. Chem.,* 1994, **350**, 2.
17. E.H. Larsen, *Spectrochim. Acta Part B,* 1998, **53**, 253.
18. M. Ceulemans and F.C. Adams, *Anal. Chim. Acta,* 1995, **317**, 161.
19. T. Dagnac, A. Padro, R. Rubio and G. Rauret, *Talanta,* 1999, **48**, 763.
20. J.W. McKiernan, J.T. Creed, C.A. Brockhoff, J.A. Caruso and R.M. Lorenzana, *J. Anal. At. Spectrom.,* 1999, **14**, 607.
21. F. Lagarde, M.B. Amran, M.J.F. Leroy, C. Demesmay, M. Olle, A. Lamotte, H. Muntau, P. Michel, P. Thomas, S. Caroli, E. Larsen, P. Bonner, G. Rauret, M. Foulkes, A. Howard and E.A. Maier, *Fresenius' J. Anal. Chem.,* 1999, **363**, 18.
22. J.L. Gómez-Ariza, J.A. Pozas, I. Giráldez and E. Morales, *Int. J. Environ. Anal. Chem.,* 1999, **74**, 215.
23. C. Camara, P. Quevauviller, M.A. Palacios, M.G. Cobo and R. Munoz, *Analyst,* 1998, **123**, 947.
24. V. Schmitt, PhD. Thesis, University of Bordeaux, 1997.
25. C. Pellegrino, P. Massanisso and R. Morabito, *Trends Anal. Chem.,* 2000, **19**, 97.
26. R. Morabito, *Fresenius' J. Anal. Chem.,* 1995, **351**, 378.
27. M. Abalos, J.M. Bayona, R. Compaño, M. Granados, C. Leal and M.D. Prat, *J. Chromatogr. A,* 1997, **788**, 1.
28. M. Ceulemans and F. Adams, *Anal. Chim. Acta,* 1995, **317**, 161.
29. S. Zhang, Y.K. Chau, W.C. Li and A.S.Y. Chau, *Appl. Organomet. Chem.,* 1991, **5**, 431.
30. M. Abalos, J.M. Bayona and Ph. Quevauviller, *Appl. Organomet. Chem.,* 1998, **12**, 541.
31. F. Pannier, A. Astruc, M. Astruc and R. Morabito, *Appl. Organomet. Chem.,* 1996, **10**, 471.
32. J.L. Gómez-Ariza, E. Morales, R. Beltran, I. Giráldez and M. Ruiz-Benitez, *Analyst,* 1995, **120**, 1171.
33. D.S. Forsyth and W.D. Marshall, *Anal. Chem.,* 1983, **55**, 2132.
34. M. Ceulemans, C. Witte, R. Lobinski and F.C. Adams, *Appl. Organomet. Chem.,* 1994, **8**, 451.
35. Y.K. Chau, P.T.S. Wong, G.A. Bengert and J.L. Dunn, *Anal. Chem.,* 1984, **56**, 271.

36. J. Kuballa, R.-D. Wilken, E. Jantzen, K.K. Kwan and Y.K. Chau, *Analyst*, 1995, **120**, 667.

37. Ph. Quevauviller and R. Morabito, *Trends Anal. Chem.*, 2000, **19**, 86.

38. M. Thompson, S.L.R. Ellison, A. Fajgelj, P. Willetts and R. Wood, *Pure Appl. Chem.*, 1999, **71**, 337.

39. Ph. Quevauviller, *Appl. Organomet. Chem.*, 1994, **8**, 715.

40. R. Morabito, *Microchem. J.*, 1995, **51**, 198.

41. A.A. Brown, L. Ebdon and S.J. Hill, *Anal. Chim. Acta*, 1994, **286**, 391.

42. C.G. Arnold, M. Berg, S.R. Müller, U. Dommann and R.P. Schwarzenbach, *Anal. Chem.*, 1998, **70**, 3094.

43. R. Morabito, P. Massanisso and Ph. Quevauviller, *Trends Anal. Chem.*, 2000, **19**(2 + 3), 113.

44. F.M. Martin and O.F.X. Donard, *Fresenius' J. Anal. Chem.*, 1995, **351**, 230.

45. F.M. Martin, C.M. Tseng, C. Belin, Ph. Quevauviller and O.F.X. Donard, *Anal. Chim. Acta*, 1994, **286**, 343.

46. Ph. Quevauviller, F. Martin, C. Belin and O.F.X. Donard, *Appl. Organomet. Chem.*, 1993, **7**, 149.

47. A.G. Howard and S.D.W. Comber, *Mikrochim. Acta*, 1992, **109**, 27.

48. R.Ritsema, F.M. Martin and Ph. Quevauviller, in Ph. Quevauviller, E.A. Maier and B. Griepink (eds.), *Quality Assurance for Environmental Analysis*, Elsevier, Amsterdam, 1995, Ch. 19.

49. O.F.X. Donard and R. Pinel, in R.M. Harrison and S. Rapsomanikis (eds.) *Environmental Analysis Using Chromatography Interfaced with Atomic Spectroscopy*, Ellis Horwood, Chicester, 1989, Ch. 7.

50. M.D. Müller, *Anal. Chem.*, 1987, **59**, 617.

51. W.M.R. Dirkx, R. Lobinski and F.C. Adams, in Ph. Quevauviller, E.A. Maier and B. Griepink (eds.), *Quality Assurance for Environmental Analysis*, Elsevier, 1995, Ch. 15.

52. W.M.R. Dirkx, R.J.A. van Cleuvenbergen and F.C. Adams, *Mikrochim. Acta*, 1992, **109**, 133.

53. J. Szpunar-Lobinska, M. Ceulemans, W. Dirkx, C. Witte, R. Lobinski and F.C. Adams, *Mikrochim. Acta*, 1994, **113**, 287.

54. I. Fernández-Escobar, M. Gilbert, A. Messeguer and J.-M. Bayona, *Anal. Chem.*, 1998, **70**, 3703.

55. J.L. Gómez-Ariza, E. Morales, I. Giráldez and R. Beltran, *Int. J. Environ. Anal. Chem.*, 1997, **66**, 1.

56. M.B. de la Calle Guntinàs, R. Scerbo, S. Chiavarini, Ph. Quevauviller and R. Morabito, *Appl. Organomet. Chem.*, 1997, **11**, 693.

57. M. Harino, M. Fukushima and M. Tanaka, *Anal. Chim. Acta*, 1992, **264**, 91.

58. J.A. Stäb, U.A.Th. Brinkman and W.P Cofino, *Appl. Organomet. Chem.*, 1994, **8**, 577.

59. J.A. Stäb, W.P. Cofino, B. van Hattum and U.A.Th. Brinkman, *Fresenius' J. Anal. Chem.*, 1993, **347**, 247.

60. Y. Cai, R. Alzaga and J.-M. Bayona, *Anal. Chem.*, 1994, **66**, 1161.

61. P. Schubert, I. Fernández-Escobar, M. Gilbert, E. Rosenberg and J.-M. Bayona, *J. Chromatogr. A*, 1998, **810**, 245.

62. J.R. Ashby and P.J. Craig, *Appl. Organomet. Chem.*, 1991, **5**, 173.

63. S. Rapsomanikis, *Analyst*, 1994, **119**, 1429.

64. P. Schubert, E. Rosenberg and M. Grasserbauer, *Int. J. Environ. Anal. Chem.*, 2000, **78**, 185.

65. C. Carlier-Pinasseau, G. Lespes and M. Astruc, *Appl. Organomet. Chem.*, 1996, **10**, 505.

66. W.M.R. Dirkx, R. Lobinski and F.C. Adams, *Anal. Chim. Acta*, 1994, **286**, 309.
67. C. Carlier-Pinasseau, G. Lespes and M. Astruc, *Environ. Technol.*, 1997, **18**, 1179.
68. Y.K. Chau, F. Yang and M. Brown, *Anal. Chim. Acta*, 1997, **338**, 51.
69. P. Schubert, E. Rosenberg and M. Grasserbauer, *Fresenius' J. Anal. Chem.*, 2000, **366**, 356.
70. K. Bergmann and B. Neidhart, *Fresenius' J. Anal. Chem.*, 1996, **356**, 57.
71. T. De Smaele, L. Moens, R. Dams, P. Sandra, J. Van der Eycken and J. Vandyck, *J. Chromatogr. A*, 1998, **793**, 99.
72. V. Minganti, R. Capelli and R. De Pellegrini, *Fresenius' J. Anal. Chem.*, 1995, **351**, 471.
73. Y.-C. Sun, J. Mierzwa, Y.-T. Chung and M.-H. Yang, *Anal. Comm.*, 1997, **34**, 333.
74. R. Falter (ed.), Sources of error in methylmercury determination during sample preparation, derivatisation and detection, Special Issue, *Chemosphere*, 1999, **39**, 1037.
75. P. Quevauviller and M. Horvat, *Anal. Chem.*, 1999, **71**, 155A.
76. M.B. de la Calle-Guntinas, C. Brunori, R. Scerbo, S. Chiavarini, P. Quevauviller, F.C. Adams and R. Morabito, *J. Anal. At. Spectrom.*, 1997, **12**, 1041.
77. P.C. Uden, S.M. Bird, M. Kotrebai, P. Nolibos, J.F. Tyson, E. Block and E. Denoyer, *Fresenius' J. Anal. Chem.*, 1998, **362**, 447.
78. J.A. Stäb, B. Van Hattum, P. De Voogt and U.A.Th. Brinkman, *Mikrochim. Acta*, 1992, **109**, 101.
79. Deutsches Institut für Normung, DIN 38 407-13 (draft), *Gemeinsam erfaßbare Stoffgruppen (Gruppe F), Teil 13: Verfahren zur Bestimmung ausgewählter Organozinnverbindungen mittels Gaschromatographie (F 13)*, Beuth, Berlin, 1999.
80. F. Ariese, IVM, Free University Amsterdam, unpublished data.
81. C.M. Tseng, A. de Diego, J.C. Wasserman, D. Amouroux and O.F.X. Donard, *Chemosphere*, 1999, **39**, 1119.
82. H. Hintelmann, R. Falter, G. Ilgen and R. D. Evans, *Fresenius' J. Anal. Chem.*, 1997, **358**, 363.
83. N.S. Bloom, J.A. Colman and L. Barber, *Fresenius' J. Anal. Chem.*, 1997, **358**, 371.
84. G.K. Zoorob, J. W. McKiernan and J.A. Caruso, *Mikrochim. Acta*, 1998, **128**, 145.
85. J.W. Olesik, J.A. Kinzer, E.J. Grunwald, K.K. Thaxton and S.V. Olesik, *Spectrochim. Acta Part B*, 1998, **53**, 239.
86. J. Szpunar-Lobinska, C. Witte, R. Lobinski and F.C. Adams, *Fresenius' J. Anal. Chem.*, 1995, **351**, 351.
87. P. Schubert, PhD Thesis, Vienna University of Technology, 1999.
88. Ph. Quevauviller, R. Ritsema, R. Morabito, W.M.R. Dirkx, S. Chiavarini, J.-M. Bayona and O.F.X. Donard, *Appl. Organomet. Chem.*, 1994, **8**, 541.
89. C. Sarzanini and E. Mentasti, *J. Chromatogr. A*, 1997, **789**, 301.
90. C.F. Harrington, G.C. Eigendorf and W.R. Cullen, *Appl. Organomet. Chem.*, 1996, **10**, 339.
91. K.M. Attar, *Appl. Organomet. Chem.*, 1996, **10**, 317.
92. B. Michalke and P. Schramel, *Fresenius' J. Anal. Chem.*, 1997, **357**, 594.
93. J.W. Olesik, J.A. Kinzer and S.V. Olesik, *Anal. Chem.*, 1995, **67**, 1.
94. R. Compano, M. Granados, C. Leal and M.D. Prat, *Anal. Chim. Acta*, 1995, **314**, 175.
95. F. Ariese, W. Cofino, J.L. Gómez-Ariza, G. Kramer and Ph. Quevauviller, *J Environ. Monit.*, 1999, **1**, 191.
96. D.E. Wells, J. de Boer, L.G.M.Th. Tuinstra, L. Reuthegard and B. Griepink, *Fresenius' Z. Anal. Chem.*, 1988, **332**, 591.
97. K. Danzer, *Significance of Statistics in Quality Assurance*, in H. Günzler (ed.), *Accreditation and Quality Assurance in Analytical Chemistry*, Springer, Berlin, 1996, pp. 105–134.

98. Ph. Quevauviller and E. Maier, *Interlaboratory Studies and Certified Reference Materials for Environmental Analysis – The BCR Approach*, Elsevier, Amsterdam, 1999.

99. S.A. Wise, *Standard Reference Materials for the Determination of Trace Organic Constituents in Environmental Samples*, in *Environmental Analysis: Techniques, Applications and Quality Assurance*, D. Barceló (ed.), Elsevier, Amsterdam, 1993, pp. 403–446.

100. P. Quevauviller, *Trends Anal. Chem.*, 1999, **18**, 76.

101. B. Rosenkranz, Ph. Quevauviller and J. Bettmer, *Int. Lab.*, 2000, **30**(3), 20.

CHAPTER 3

Sample Treatment and Storage in Speciation Analysis

J.L. GÓMEZ-ARIZA, E. MORALES, I. GIRÁLDEZ AND
D. SÁNCHEZ-RODAS

1 Introduction

Speciation analysis is today performed routinely in many laboratories to control the quality of the environment, food and health. Chemical speciation analyses generally include the study of different oxidation states of elements (*e.g.* As, Cr, Se, Fe) – stable/non-inert species, or individual organometallic species (*e.g.* methylarsinic acid, methylmercury or tributyltin) – stable/inert compounds. Strictly speaking, the word speciation should not be applied to the determination of 'extractable forms' of metals,[1] which are operationally defined on the basis of single or sequential extractions.

The major target species therefore include well-known organometallic environmental pollutants which have created great environmental concern in the last few decades[2,3] (*e.g.* butyl- and phenyl-tin, alkyl-lead), products of transformation of toxic elements (*e.g.* methylmercury, organoarsenic) and complexes of essential and toxic metals and non-metals with biomolecules,[4] selenomethionine, selenocystine. The oxidation state assessment of some elements such as As(III) and (V), Se(IV) and (VI), or Cr(III) and (VI) is also of great interest.

The need for the determination of individual chemical species occurs especially when these species are known to have a different impact and behaviour, *e.g.* toxicity, mobility, or bioavailability. The oxidation state of the elements is one of the reasons for their differential toxicity: this is the case of Cr(VI) against Cr(III). The degree of alkylation is another important cause of toxicity, trialkyltins being more toxic than di- or mono-alkyltins, methylmercury (MeHg) more so than Hg(II) and so on. Sometimes the introduction of a toxic single ion into a complex organic molecule produces a clear reduction of the element's toxicity, such as As(III) against arsenobetaine (AsB). Generally, metal alkylation helps the molecule pass across the biological membrane and results therefore in accumulation in the food chain.[5-7] In other cases, toxicity is caused by the volatility of organometallic species and easy absorption through the lungs, as is the case of mercury.

The determination of chemical forms of elements is still an analytical challenge. The species are often unstable and concentrations found in the different matrices of interest are often in the $\mu g\,l^{-1}$ level or in some cases (*e.g.* alkylated forms of metals such as Sn or Pb) even in the $ng\,l^{-1}$ (*e.g.* estuarine waters) or $ng\,g^{-1}$ (sediments and biological tissues) range, whereas inorganic forms can be simultaneously present at thousand-fold higher levels. For this reason, sensitive and selective analytical atomic techniques (atomic absorption spectrometry, AAS; mass spectrometry, MS; flame photometric detector, FPD; inductively coupled plasma-atomic emission spectrometry, ICP-AES; inductively coupled plasma-mass spectrometry, ICP-MS; atomic fluorescence spectrometry, AFS, *etc.*) are being used as available detectors for speciation, generally coupled with some chromatographic step (GC, HPLC) to provide time-resolved introduction of analytes into the atomic spectrometer.[8]

The apparent feasibility of species-selective analysis for organometallic compounds and products of their degradation, down to the picogram level, has resulted in the increasing interest of national regulatory agencies, governmental and industrial quality control laboratories and consequently of manufacturers of analytical instrumentation.[9] Speciation analyses are, however, still developed by academic laboratories, because procedures usually involve many steps, which are frequently time-consuming and they generally involve a chromatographic step for species separation, which contributes to their unreliability. The complexity involved in the coupling between several instruments has a straightforward consequence in the duration of the analysis, as well as in the instrumentation costs. On the other hand, the sample preparation has to transfer the elemental species present in the sample into a solution suitable for separation by a chromatographic column. The preparation of such a solution requires the extraction and derivatization of analytes present in the sample and this involves numerous steps and manipulations that increase species losses and demand considerable operator skills and time. Therefore, in order to get these species-selective analyses to be performed in governmental and industrial quality control laboratories, the procedures have to be simplified and sample throughput increased, thereby reducing the use per unit time of expensive analytical instrumentation and the necessary, highly skilled analysts.[9] One of the most important factors is thus the improvement of sample preparation and treatment steps, *via* the introduction of a new generation of simple, rapid and reliable procedures for species extraction and derivatization, before their introduction to the tandem chromatograph/detector.

It is also true that analytical techniques for speciation analyses are sometimes not sensitive enough to control the effectiveness of regulations,[10] *e.g.* $0.2\,ng\,l^{-1}$ of tributyltin (TBT) established as Environmental Quality Target (EQT) in waters in the Netherlands.[11] This fact makes it essential to apply pre-concentration techniques suitable for both the different types of sample and species.

Simultaneous speciation of several elements is also an important task. Generally, problems arise during sample preparation (species extraction, enrichment and derivatization) due to the variety of conditions needed for each element and this has led to the introduction of new reagents such as sodium tetraethylborate ($NaBEt_4$) for the simultaneous derivatization of several elements[12] (*e.g.* tin, lead and mercury), as well as the implementation of techniques for efficient and accelerated isolation of the bulk of the analytes from the sample matrix, such as microwave assisted extraction.[9] Finally, the integration and automation of all steps between sampling and detection significantly reduces the time of analysis, increasing reproducibility and accuracy. For this reason, the development of new devices is needed for the full speciation of samples in a single instrumental device.[9,13]

2 Sample Treatment

The use of HPLC and especially GC, as time-resolved introduction techniques into the atomic spectrometer establishes some physico-chemical requirements for the analytes. This usually makes necessary a sample preparation procedure that includes the pre-treatment of the sample with some type of reagent to condition the matrix, or leach the species, for the extraction step, in which the species are completely isolated from the matrix. The species are then converted into suitable forms for the determination. Derivatization is a frequent approach for speciation and produces volatile and thermally stable species for GC, which, due to its good resolution and the availability of sensitive detectors, is still the most commonly used system, despite the increasing popularity of HPLC–ICP–MS coupling.[14] Finally, the extract is submitted to a clean-up step before its introduction into the chromatographic system. Sometimes, separation of the species from the matrix is combined with a pre-concentration step in order to improve the limit of detection of the method. Solid-phase extraction (SPE) and micro-extraction (SPME), immobilized chelating agents, chelating sorbents or ion-exchange are the most usual approaches for this purpose. These major steps are completed by a series of minor operations such as drying of the organic phase, dilution, pH adjustment, change of solvent, transfer from one vessel to another or phase separation.[9,15] As a general rule, sample preparation should lead to a shortening of the duration of the major steps, and reduction of the overall number of steps and manipulations and much research interest has been focused on these topics.

2.1 Species Pre-treatment and Extraction

Many methods have been proposed for species pre-treatment/extraction from environmental matrices: water, soils, sediments and biological materials. They can be categorized[16] according to solvent polarity, sample acidification to enhance the recoveries of the species, enzymatic hydrolysis for biotic samples and the use of chelating agents. Against these classical extraction techniques, other, more recent approaches such as supercritical fluid extraction (SFE), accelerated solvent extraction (ASE) or SPE, offer new possibilities in species treatment and offer such advantages as the drastic reduction of extraction time and incorporation into on-line flow analysis systems. In addition, although many of these extraction techniques offer good possibilities for species pre-concentration, SPE is the most versatile and powerful tool for this purpose.

2.2 Solvent Extraction Procedures

2.2.1   Organotin Species

Direct extraction of the sample with a non-polar solvent such as hexane, benzene, toluene, chloroform, pentane or dichloromethane separates low polarity species, *i.e.* multi-alkylated species, (*e.g.* tributyltin), but highly-polar mono- and dimethyl- or butyl-tin compounds including Sn^{4+} require complexation with a chelating agent such as fresh tropolone[16,17-20] or sodium diethyldithiocarbamate[21] solutions, to efficiently extract these species into the solvent. However, acidification of the sample with hydrochloric acid under stirring,[22-26] or sonication,[27-30] followed by sequential solvent extraction, enhances the recovery of organotin species, especially monoalkyl compounds (monobutyltin and monophenyltin). Salting out or ion-pairing with NaCl are used to increase the efficiency of the extraction of organotin compounds from the aqueous to the organic phase when hydrochloric acid is used, especially for biotic samples.[16] The addition of either hydrobromic[17,20,31-36] or acetic acid,[27-29] or a mixture, improves the extraction efficiency, possibly due to their ion-pair properties and the prevention of adsorption on to the walls of the container. These conditions are not directly applicable to methylated tin species which require a more drastic acidic treatment, using a combination of hydrobromic, hydrochloric, acetic and sulfuric acids at a high ionic strength.[17] There is no agreement regarding the polarity of the solvent used for extraction; medium polarity solvents are advisable to extract ionic organotin species, but these increase the number of substances that are co-extracted and impair the derivatization reactions.[16] Direct derivatization of sediment slurries and mussel tissues has been used by some workers to reduce the time of analysis. Sometimes the sample is homogenized in n-hexane for organotin pentylation by the Grignard reaction in diethyl ether.[37] However, Chau *et al.*[38] suggest the direct use of pentylmagnesium bromide with the sediment, to drastically improve the recovery of MBT species. This latter method can be applied to the simultaneous speciation of organotin and alkyllead in sediment.

When the analytical methods for organotin compounds are based on a separation by liquid chromatography, the species may be extracted from the sample (*e.g.* seawater) acidified with hydrochloric acid using a tropolone solution in chloroform, after which the extract is evaporated to dryness and re-dissolved in methanol for the chormatographic separation.[39]

Extraction with polar solvents such as aqueous hydrochloric acid,[40-44] hydrochloric or acetic acids in polar solvents (methanol, acetone),[43-56] acetic acid[26,44,57-63] or net polar organic solvents (methanol, butanol, methanol–ethyl acetate, methanol–dichloromethane)[30,40,64-66] or polar solvents in basic medium,[50,66] generally performed under sonication, has been used for the analysis of solid matrices for organotins. Generally, the extraction is followed by a non-polar solvent extraction (benzene, cyclohexane, toluene)[30,47,48,55,63] to recover the species for derivatization. Salting out reagents and tropolone have also been used to increase the species recovery.

2.2.2 Organolead Species

The extraction of lead compounds is very similar to that for organotins. Volatile non-polar tetra-alkylated compounds can be directly extracted from environmental samples with non-polar solvents such as benzene or hexane. Ionic alkyl compounds (R_2Pb^{2+} and R_3Pb^+) can be extracted and pre-concentrated from water samples in the presence of NaCl and sodium diethyldithiocarbamate in hexane,[67] benzene[68] or pentane.[69]

2.2.3 Organomercury Species

Sample preparation methods for the determination of $MeHg^+$ in sediments are based on the classic, but cumbersome, Westöö technique.[70] Most of the recent procedures, derived from this protocol, involve the following main steps: (i) organomercury compound liberation from the organic matrix (generally proteins) by displacing the mercapto-group with a halogen ion at low pH, (ii) selective extraction of organomercury species into toluene, (iii) purification of the organic extract by extraction with an aqueous solution of cysteine, (iv) redissociation of the organomercury–thiol complex in acid medium, (v) re-extraction in an organic solvent and pre-concentration by evaporation of the solvent. Other procedures are based on acidic leaching,[71–76] alkaline digestion[77–81] or steam distillation,[79–83] followed by one or two separation steps, *e.g.* solvent extraction, ion exchange, distillation or aqueous derivatization (hydride generation or ethylation). These sample preparation methods are not only laborious and time-consuming but also lack sufficient efficiency and reliability.[81,84,85] Hydrochloric acid leaching at room temperature does not quantitatively release methylmercury compounds from sediment samples.[79] Both alkaline digestion, by 25% KOH in methanol and distillation quantitatively release $MeHg^+$ from sediments, but it takes 1–6 h for complete extraction.[77–83] The diversity of leaching methods has made it necessary to make comparisons between them in order to achieve the optimum procedure. Huang[76] studied three procedures for sample preparation for the analysis of MeHg from environmental and biological matrices: (a) alkaline digestion of the biological sample followed by acidification with hydrochloric acid, (b) distillation of MeHg from sediment, soil and water samples in the presence of a L-cysteine/ascorbic acid/phosphoric acid/potassium bromide mixture and collection of the distillate in a $NH_2OH.HCl$ solution and (c) addition of EDTA to the sample solution to mask the interfering metal cations.

2.2.4 Organoarsenic and Organoantimony Species

Most living organisms reduce the toxicity of arsenic and antimony by incorporating them into organometallic molecules through metabolic pathways. Therefore, speciation methods have to be capable of extracting these compounds without structural modification. More ubiquitous organoarsenic environmental molecules are monomethylarsonic acid (MMA), dimethylarsinic acid (DMA), arsenobetaine $[(CH_3)_3As^+CH_2COO^-]$ (AsB) and arsenocholine $[(CH_3)_3As^+CH_2CH_2OH]$ (AsCh) and extraction procedures can be developed to isolate them from matrices. A complete inventory of arsenic- and antimony-containing biomolecules does not yet exist, although some progress has been made on the analytical characterization of different arseno-sugars. Testing the effectiveness of the procedures designed for this purpose is therefore difficult.

Due to the stability of methylated arsenic species, they are leached, together with total inorganic arsenic, using warm[86,87] or cold[88,89] concentrated hydrochloric acid from sediments and biological tissues. Crescelius et al.[90] recommend the use of acidic leaching (pH 2.3) for As(III) or basic leaching (pH 11.9) for As(V), MMA and DMA. Other weak leaching reagents such as acetate, citrate and oxalate buffers[91] selectively leach As(III) and phosphoric acid efficiently extracts total arsenic from soils.[92] Other experimental approaches for arsenic determination in aquatic sediments involve the use of extraction schemes to characterize the arsenic, bound to separate fractions of the sediments, such as exchangeable (adsorbed ions), carbonate, amorphous iron oxides, organics and residuals.[93] As most of the arsenic is usually associated with iron oxides,[94] a selective extraction method using hydroxylammonium hydrochloride as extractant, with special emphasis on the speciation of the different forms of arsenic [As(III), As(V), MMA and DMA] in the extract, has been performed.[95]

For antimony speciation in plants, Dodd et al.[96] homogenized frozen samples in a blender with acetic acid. Homogenates were sonicated for 1 h and left standing overnight; the extracts were analysed after filtration.

2.2.5 Organoselenium Species

Whilst there are a large number of extraction methods for organoarsenic and organotin compounds, the problem of leaching organoselenium compounds from environmental samples has not yet been completely solved.[97] A process, largely used for arsenic species (to extract inorganic and organic arsenic species) consisting of a mixture of water, chloroform and methanol (2 + 3 + 5) and evaporation to dryness, has been applied to selenium;[98] this method has been lately modified by Potin-Gautier et al.[99] and Emteborg et al.[100] the latter introducing oscillating magnetic stirring for 5 h. Other methods use a methanol–water mixture[101] (1 + 1) or the same mixture with 0.28 mol l^{-1} hydrochloric acid or 4% ammonia,[100] in all cases with two 30 min cycles of sonication. The samples were centrifuged, to collect the supernatant, filtered through a 0.45 μm pore size filter and brought to pH 9.5 by adding 5 ml of a 1% ammonia solution to ionize the selenium species present in the sample extract. The extracts were then passed twice through end-capped C_{18} cartridges and reduced in volume to 1–2 ml using rotary evaporation.

Pre-concentration of selenium (as organoselenium compounds) by 2–40-fold from wastewaters, plants and soils can be achieved by selective extraction of the element with 1-hexene from a perchlorate–bromide medium.[102] Living bacterial cells (*P. putida*) have been used to perform a separation and pre-concentration procedure for selenocystamine (Se-Cystm) and the determination of selenium in the slurry used ETAAS, after the addition of Pd as chemical modifier.[103] The procedure makes possible the determination of Se-Cystm at low levels, with detection limits of $0.01\,\mathrm{ng\,l^{-1}}$ for 1000 ml sample volumes and a cell growth period of 6 h, corresponding to an absolute limit of detection of 1 pg. The method was applied to the determination of Se-Cystm in natural waters, spiked with $5\,\mathrm{ng\,l^{-1}}$ of this compound, together with other selenium species.

Preparation of non-solid matrices, such as urine, is straightforward and only requires precise $(1 + 4)$ dilution with water prior to the analysis, although clean-up with C_{18} cartridges has been proposed for the determination of trimethylselenonium ion, selenomethionine, selenocysteine, selenite and selenate.[104]

2.3 Basic and Enzymatic Hydrolysis

Basic and enzymatic digestions are usually restricted to biotic samples. Enzymes, even non-specific protease enzymes, are capable of breaking down a wide range of proteins into their amino acid components, which makes the embedded metallic species more available to the extracting agents. Basic hydrolysis with reagents such as tetramethylammonium hydroxide (TMAH) is currently applied for organotin species, under warm temperature conditions (60 °C) for several hours.[105–107] Alternatively, ethanolic KOH at 60 °C for 90 min[108,109] or NaOH at 40 °C for 20 min[110] followed by pH adjustment, have also been proposed. Basic treatment is followed by liquid–liquid extraction with hexane or another non-polar solvent to isolate the species. Acid protein hydrolysis with HCl at 110 °C can also be used for selenium species leaching, but degradation of the selenium compounds has been observed during the treatment.[111]

Enzymatic hydrolysis is becoming a promising approach in metal speciation, especially for organolead and organotin compounds.[105–113] This method has been studied by Gilon *et al.*[114] for selenium applications, using a protease that breaks the peptide bonds of the proteins present in the material, at pH 7.5 (phosphate/citric buffer) and incubation at 37 °C for 24 h. A similar procedure can be used for organotins[107,115] or organoselenium[116] compounds, using a lipase-protease mixture under similar experimental conditions. Pannier *et al.*[112] also used a mixture of enzymes in aqueous phosphate buffer to release organotin compounds from marine biological samples and pronase, a protease with a broad specificity, isolated from *Streptomyses griseus*, has been used to monitor the levels of selenium in serum, which was digested with the enzyme at 37 °C overnight.[117]

2.4   Supercritical Fluid Extraction (SFE)

Interest in using supercritical fluids (SFs) for sample preparation originates in the common aim to reduce the consumption and disposal of and long-term exposure to, organic solvents. The exceptional characteristics of SFs as solvents are based on their gas-like and liquid-like properties.[118-120] Attractive characteristics of SFs are low viscosity and diffusion coefficients, much lower than those of liquids, that contribute to a rapid mass transfer of solutes and faster extractions than those in liquid extractions. On the other hand, densities are in the range of a liquid and this increases the solvating power of SFs at low temperature, which facilitates the extraction of thermally labile compounds. CO_2 is the most common SF used for extractions, owing to it being non-toxic, non-flammable, relatively cheap and easily obtained commercially. It has, however, low efficiency in the extraction of polar compounds and needs the addition of 'modifiers' and/or complexing agents for complete recovery of polar and non-polar species. Several studies consider the use of SFs for the extraction of arsenic, cadmium, copper, lead, manganese, mercury, selenium, tin and zinc,[121-134] although few papers consider the use of SFs as a sample preparation technique in metal speciation.[126-135]

This approach has been particularly employed for the extraction of organotin compounds. Several CO_2 modifiers have been tested, such as formic acid[128] and methanol.[131] Low recoveries have been obtained in the extraction of di- and mono-derivatives and two approaches have been made to improve the efficiency of the extractions: (i) the addition of complexing agents such as sodium diethyldithiocarbamate (DDC) and diethylammonium diethyldithiocarbamate (DEA-DDC)[131,132,136,137] and (ii) alkylation in the extraction cell prior to the extraction process.[138] SFE has been applied to different environmental matrices: seawater[129] using a combination of liquid–solid extraction on a disc followed by *in situ* Grignard ethylation and SFE, sediments,[130] soils,[131,132] potato and almond matrices[128] and biological samples.[133,135] The low recoveries for di- and, especially, mono-substituted compounds and the still lengthy (10–50 min) extraction of organotin species are drawbacks when using SFE, compared with the faster extraction procedures based on microwave-assisted methods.[9] Further developments are, therefore, needed to bring this methodology into routine analysis.

2.5   Accelerated Solvent Extraction (ASE)

Few automated extraction systems are currently available for efficiently processing organic and organometallic compounds in solid matrices. Several versions of automated Soxhlet extractors are offered, but all have limited capacity and long extraction times and require large solvent volumes. Supercritical fluid extractors were intended to overcome these problems, but present the problems outlined in the previous paragraph – matrix dependence of the extraction, time-consuming and incomplete recovery for polar organometallic species. Accelerated solvent extraction (ASE) constitutes an emerging and promising alternative, based on the use of organic solvents at high temperatures under pressure. Extraction time for ASE averages 11 min per soil sample. ASE has been used for organic compound extraction but only a few papers focus on metal speciation, and these consider the extraction of organotin compounds from sediments.[139,140] The extraction of organotin compounds in sediments can be achieved by treating 2.5 g of freeze-dried sample with 1 M sodium acetate and 1 M acetic acid in MeOH. The extraction cells, filled with solvent, have to be heated within 5 min to 100 °C, using three to five static cycles of 5 min. Between each static extraction cycle, 4 ml of solvent is renewed. The average recovery is 99%.

2.6 Solid Phase Extraction (SPE) for Species Separation and Pre-concentration

2.6.1 Chelating Sorbents

The use of chelating agents with a potential high selectivity, which can be modified by careful control of the experimental conditions, has allowed pre-concentration processes for metallic species. This pre-concentration process can be further improved by the chelating agent being rendered immobile by its retention on a solid support. This transforms the classical liquid–liquid extraction into a solid–liquid extraction, developed either in batch, in column or in micro-column systems, the last option making possible its incorporation into on-line flow analysis systems. Carrit[141] proposed the use of cellulose acetate loaded with dithizone for the separation and pre-concentration of trace elements. However, most of the chelating sorbents have been designed for mercury compounds, using sulfur reagents sorbed on to RP C_{18} material or other polymeric matrices. Several sorbents have been prepared by deposition of dithizone on silica gel,[142,143] polyurethane foam[144] and polystyrene beads[145,146] and used for the pre-concentration of trace metals. The recoveries are not good, however, because of the instability of the sorbent. Other chelating sorbents have been shown to avoid this drawback, such as dithizone-anchored poly(vinyl-pyridine),[147,148] dithizone immobilized on surfactant-coated alumina,[149] dithiocarbamates incorporated into mono-disperse polystyrene micro-spheres,[150] sodium diethyldithiocarbamate sorbed on to a C18 bonded silica column,[151,152] as sulfhydryl cotton fibre (SCF) packed in a screening column,[153] gel permeation[154] and pyrrolidinedithiocarbamate on a RP C18 column.[155-157]

The on-line determination of ultra-trace amounts of As(III) and As(V) in water can be achieved by the selective formation of the As(III)–pyrrolidine dithiocarbamate complex, over a sample acidity 0.01–0.7 mol l^{-1} of HNO_3, its adsorption on to the inner walls of the knotted reactor of the manifold made from 150 cm long, 0.5 mm i.d. PTFE tubing, elution with 1 mol l^{-1} HNO_3 and detection.[158] Total inorganic arsenic determination after reduction of As(V) to As(III) allows the calculation of As(V). The method can also be used for arsenic pre-concentration, obtaining enhancement factors of 22 if a pre-concentration time of 60 s and sample flow rate of 5 ml min^{-1} were used. The time required for a single determination is 200 s. Thionalide-loaded acrylic resin has also been used to separate and pre-concentrate As(III) and As(V) from natural waters.[159] Organotins can be pre-concentrated from marine water samples with tropolone impregnated on a macroporous polymer[160] (Amberlite XAD 2) and lead in natural waters by using chelating resin beads[161] – this method distinguishes between lead in stable complexes and the free lead ion.

2.6.2 Columns and Cartridges

The use of liquid–solid extraction in organometallic pre-concentration presents undoubted advantages, especially when the solid phase is housed in cartridges or disks that make possible its integration in on-line flow analysis systems. In addition, SPE has other advantages: (i) it requires less volume of solvent than traditional liquid–liquid extraction, (ii) it involves simple manipulations which are not time-consuming and makes possible *in-field* treatment of samples, (iii) the cartridges can be used for storage of the species and (iv) it provides high enhancement factors if great volumes of water are passed through the cartridge without breakthrough.

Cationic organometallic species can be pre-concentrated on reversed phase columns or cartridges. Adsorption of organotin compounds [monomethyltin (MMT), dimethyltin (DMT), trimethyltin (TMT), monobutyltin (MBT), dibutyltin (DBT), tributyltin (TBT), diphenyltin (DPT) and triphenyltin (TPT)] present in water samples can be achieved on a silica-gel C_{18} disposable column[162] and the species eluted later with a tetrahydrofuran–acetic acid mixture containing 0.5% (w/v) tropolone. This method pre-concentrates organotin species by a factor of 1000 (1 l to 1 ml) and the time required for water sample extraction is about 100 min. With simultaneous extractions, more than 20 samples could be analysed in a day. Organotin solid-phase extraction (SPE) can also be developed in other shaped-housings, disks and cartridges with different types of solid phase: Carbopack, C_{18}, C_8, C_2 and acid ethyl acetate, methanol, 1% (v/v) HBr solution in methanol and 1.0% (v/v) HBr and 0.1% (w/v) tropolone solution in methanol. With C_{18} columns or disks, only TBT and TPT are eluted, suggesting that SPE discs have advantages over SPE columns[163] due to the remarkable reduction of the background signal shown in the GC–ECD chromatograms. The use of C_{18} and Carbopack cartridges[164,165] gave quantitative recoveries for trialkylated tin compounds extracted from waters, if methanol was used as eluent. The complete elution of butyl- and phenyl-tin species, using 1.0% (w/v) tropolone solution in methanol, yielded enhancement factors of up to 5000 using 5000 ml water samples. Selective separation of the organotins from the C_{18} cartridge can be achieved by successive elution with solvents of different strength:[165] methanol for TBT and TPT, 0.4% (v/v) HBr solution in methanol for DBT and DPT and 0.4% (v/v) HBr and 0.1% (w/v) tropolone solution in methanol for MBT and MPT.

Anionic organometallic species can be pre-concentrated on anionic cartridges, which also avoid their interference and signal overlapping with cationic species, in complex species mixtures analysed by online, hyphenated instrumental systems. The overlapping of the peaks for As(III) and AsB, which occurs in the determination of As(III), As(V), MMA, DMA, AsB and AsCh by HPLC separation–atomic spectrometric detection, can be prevented by placing an anionic cartridge before the HPLC column.[166,167] Yalcin and Le[168] have studied the use of both anion exchange and reversed phase cartridges to separate As(III) and As(V) in water, which provided rapid separation, in 1.5 min and detection limits of 0.2 and 0.4 ng ml^{-1}, respectively, using hydride generation (HG)-AAS and HGAFS for detection.

Polar selenium compound pre-concentration can be performed by liquid–liquid extraction of piazselenols,[169] on-line co-precipitation[170] of selenium species with $La(OH)_3$, complexation and adsorption on activated carbon[171] and solid-phase extraction based on anionic sorbents.[172] Volatile selenides can be separated from water by helium stripping and are then swept into a cold trap containing solid sorbents such as activated carbon, glass-wool and GC stationary phases from where they may be thermally desorbed or extracted with organic solvents.[173,174] However, simultaneous pre-concentration of inorganic and volatile organic selenium species is a more difficult task. Gómez-Ariza *et al.*[175] proposed a SPE, based on the use of two different sorbent phases, octadecyl (C_{18}) and quaternary ammonium (SAX), in separate cartridges, in series, for the simultaneous pre-concentration of Se(IV), Se(VI), dimethylselenium (DMSe), dimethyl diselenide (DMDSe), diethylselenium (DESe) and diethyl diselenide (DEDSe) in water. Both cartridges have to be conditioned before the separation, the SAX cartridge is placed before the C_{18} cartridge and 1000 ml of water (at pH 7–8) passed through at 8 ml min^{-1}. The cartridges are then separated and the analytes eluted separately, Se(IV) with formic acid and Se(VI) with hydrochloric acid. The C_{18} cartridge is dried under a stream of nitrogen and the organic selenium species are eluted with CS_2.

2.7 Derivatization

Derivatization techniques in trace metal and organometallic compound speciation are usually related to GC analysis and involve volatile-species generation reactions based on hydride generation using sodium tetrahydroborate ($NaBH_4$) or alkylation by the Grignard reaction or sodium tetraethylborate ($NaBEt_4$).

2.7.1 Hydride Generation

Most of the hyphenated speciation systems involving hydride derivatization in a flow injection manifold complement the standard hydride generation methods and with the introduction of cryogenic and chromatographic separation steps provide the high sensitivity and speciation ability required for environmental applications.[8] The hydridization/cold trapping technique has a particular application for redox speciation of inorganic species of arsenic, antimony[176,177] and selenium,[178,179] although organometallic species of these elements, such as monomethylarsonate (MMA), dimethylarsinate[180] (DMA), dimethyl selenium

(DMSe), dimethyl diselenide (DMDSe) and diethyl selenium (DESe), can also be detected. Hydride generation allows the detection of As(III) and As(V), but does not distinguish between organic and inorganic species,[176] which is important in the analysis of biological samples. In addition, other naturally occurring higher organometallic forms of both arsenic and selenium compounds, such as arsenobetaine, arsenocholine, arseno-sugars, selenomethionine or selenocysteine, are not hydride-forming species and cannot be determined using this method. Hydride generation with $NaBH_4$ was seldom used in off-line determinations for tin compounds, due to the lack of hydride stability. However, hydridization/cold trapping allows the determination of butyltins[181] and highly volatile organotins (*e.g.* methyltin), which cannot be determined by most off-line methods.[16] Hydridization of alkyltins may be subject to interference in the presence of inorganic metals, often encountered in the analysis of contaminated environmental samples.[182] Phenyltins cannot be determined by this method. Finally, inorganic and organic species of mercury and lead can also be studied using a similar approach,[183,184] although the hydrides of organo-lead and -mercury compounds are not stable and prone to dismutation reactions.[185] When the sample is directly derivatized, the amount of derivatizing reagent required needs to be optimized according to the matrix characteristics,[44,46] since the matrix can inhibit the hydridization reaction.[186] In this regard, Ashby and Craig[25] reported that uncomplexed tropolone suppresses the hydride generation of organotin compounds.

Cryo-focusing (a packed chromatographic column immersed in liquid nitrogen) allows the simultaneous on-line pre-concentration and separation of the derivatized species. Pre-concentration factors in the range of 50–100-fold are possible and the introduction of a chromatographic packing allows the separation of most of the organometallic compounds of interest. Generally, non-polar stationary phases are used with higher loadings for the determination of methylated tin, alkylleads and methylated mercury species[165] and lower loadings for methylated arsenic[176] or selenium[178] species and butyltin compounds.[187] Cryogenic trap–hydride techniques are affected by interference problems arising in the generation stage, from the relatively high concentrations of metal ions and other hydride-forming elements present in the sample. The approach generally taken to overcome these interference problems has been the addition of masking agents such as cysteine. Pre-derivatization of the arsenic species by reaction with L-cysteine enhances the performance of hydride methods employed both for the determination of total arsenic and in cryogenic trapping of the hydrides of arsenic species.[188,189] The procedure has also been used in HPLC post-column hydride derivatization of arsenic species before their detection by AAS.[190] Three advantages can be outlined: (i) the addition of cysteine equalizes the sensitivities of the technique to arsenic, which is present as As(III), As(V), MMA and DMA,[191,192] (ii) interferences arising from transition elements and other hydride-forming elements are substantially reduced[192] and (iii) the optimum concentration of acid needed for the procedure is reduced. This is of particular importance for the determination of low levels of inorganic arsenic, as it results in a lowering of the acid-derived contribution to the blank, with its resulting influence on the improvement of detection limits. Other sulfur-containing complexants have been tested as pre-derivatization reagents:[193] 2,3-mercapto-1-propanol, 2-mercaptoethanol, 3-mercaptopropionic acid, thioglycollic acid and sodium diethyldithiocarbamate. Thioglycollic acid is more effective than cysteine in the equalization of arsine yields from different arsenic species and the pre-derivatization step is significantly faster. This reagent is therefore much better suited to closed, in-line, hydride generation systems.

In spite of the advantages of hydride generation–cold trapping techniques for arsenic speciation, some procedures yield good results from simplified, or screening, methods of speciation. Willie[194] proposed a 'rudimentary or 1st order' speciation scheme for arsenic in biological tissues, by partial solubilization of the sample in 50% HNO_3, conversion of non-hydride generating arsenic species into As(V) by microwave heating or on-line ultraviolet treatment under oxidizing conditions, hydride generation and detection for total arsenic. A second treatment of the sample without photo-oxidation or microwave treatments was performed to obtain the combined result for As(III), As(V), MMA and DMA. Nielsen and Hansen[195] evaluated total arsenic in a mixture of As(III) and As(V) by on-line reduction with 0.50% (w/v) ascorbic acid and 1.0% (w/v) potassium iodide in 4 M HCl at 140 °C. Using the same system without heating and employing dilute hydrochloric acid, As(V) was not converted into arsine, thereby allowing the selective determination of As(III).

2.7.2 Alkylation Reactions

Alkylation with Grignard reagents (*e.g.* methylation, ethylation, propylation and hexylation) has been widely used as a derivatization technique in the organometallic speciation of a wide range of environmental samples. However, the method is time-consuming, since the procedure includes solvent extraction of the organometallic species, complete alkylation with a Grignard reagent, reduction of interferents and concentration of analytes by gentle evaporation of the solvent prior to injection into the gas chromatograph. In addition, the use of a Grignard reagent requires the previous complexation of the organometallic compound to favour its transference to the non-polar solvent used for the Grignard reaction. The most common complexing reagents are tropolone for organotin compounds and diethyldithiocarbamate for the extraction of organo-lead, -tin and -mercury.[196,197] Further requirements in the Grignard reaction are strict anhydrous conditions for the reaction and the use of aprotic solvents, which implies a solvent exchange when polar solvents are used as extracting agents of species from the sample. Stäb *et al.*[198] tested the derivatization yield of Grignard reactions in the case of organotin compounds and found that it is not affected by the species counter-ion, but is strongly dependent of the elapsed time in solution for mono- and di-organotin compounds. When the derivatization reaction is performed directly on a sediment sample, a large excess of derivatizing reagent is necessary, due to the presence of side reactions, such as the formation of dialkyl mono- and di-sulfides.[138] Recently, Fernández-Escobar *et al.*[199] have performed a clean-up procedure to eliminate the alkyl-sulfur species that interfere in the organotin determination by GC–MS or GC–FPD and are not removed by the conventional desulfuration procedures. The method is based on the oxidation of all the sulfur species with dimethyldioxirane to sulfones or sulfur oxides. Sulfones are easily eliminated by alumina adsorption chromatography, due to their higher polarity than organotins and the sulfur oxides are spontaneously evaporated. The process occurs in a few minutes and the excess of dimethyldioxirane is easily removed by evaporation under a nitrogen stream before the Al_2O_3 clean-up step.

Another drawback of alkylation reactions is the loss of the most volatile species, especially with methyl and ethyl Grignard reagents.[25] In addition, methyl derivatives do not allow the determination of naturally occurring methyl compounds.[200]

The simultaneous derivatization of tin, lead and mercury species using a suitable Grignard reagent has been reported.[197] Sodium diethyldithiocarbamate (DDTC) is often used as a complexing agent for the extraction of these compounds, although optimum reaction conditions, such as pH and buffer medium, are different for all three elements and therefore make simultaneous sample preparation very difficult.[201–203] Ceulemans and Adams[12] recommended the use a pH between 4 and 5 with acetate buffer for this purpose.

2.7.3   Aqueous Alkylation

The possibility of aqueous alkylation reduces the number of treatment steps in water analysis for speciation and is a good alternative for this type of matrix.[12] A new alkylating reagent, sodium tetraethylborate ($NaBEt_4$), has been developed for this purpose, which performs alkylation reactions in aqueous media under buffered conditions.[25] This reagent allows the simultaneous extraction/derivatization of organometallic species from solid matrices in a buffered medium, the ethylated species being recovered with a non-polar solvent. The procedure may involve a single step[106,204–206] or use two steps: solid phase extraction/ethylation and separation in a non-polar solvent.[207] Carlier-Pinasseau et al.[208] proposed the acid leaching of organotin compounds from solid matrices followed by one-step simultaneous ethylation/extraction using $NaBEt_4$ and iso-octane. Other approaches include the use of $NaBEt_4$ in NaOH–MeOH,[209] the combination of *in situ* ethylation of mercury species (e.g. Hg^{2+} and $MeHg^+$) and volatilization of the species from the water sample by an argon stream. The volatilized species are trapped at ambient temperature in a Tenax trap column and desorbed by controlled heating (purge-and-trap).[210]

Other reagents have been studied for the aqueous alkylation of metals and organometallic compounds. De Smaele et al.[211] claim that the versatility of sodium tetra(n-propyl)borate, $NaBPr_4$, as an aqueous, *in situ* derivatization reagent for organometallic compounds makes possible the derivatization of important ethyl compounds of lead and mercury. The method can be applied to the simultaneous determination of tin, mercury and lead in environmental samples. Finally, tetrabutylammonium tetrabutylborate can be used for butylation of lead in waters.[212,213]

2.8 Fast Extraction Methods

2.8.1 *Microwave-assisted Sample Preparation*

The efficiency of microwave-assisted extraction (MAE) for sample preparation in environmental applications has been shown for different solid matrices (*e.g.* soils, sediments and biological tissues) and constitutes a valuable tool for the rapid treatment of solid samples in organometallic speciation analysis.[214] However, a careful optimization of the conditions of the microwave extraction procedure is required, in terms of stability of the target compounds in a microwave field, prior to speciation analysis. Essential parameters, such as the extraction medium, power applied and exposure time have to be optimized.[54] This approach has been successfully adapted to organotin,[14,53,215] organomercury[216,217] and organoarsenic[218] compound determination. Microwave-assisted extraction is considerably faster than other sample preparation and extraction procedures: a typical sample treatment may take only 3 min.

An important drawback of microwave-assisted acid extraction or leaching is that analytes in the polar leachate have to be derivatized and transferred into a non-polar solvent to produce a solution suitable for GC analysis and this increases the number of steps in the analytical procedure. This obstacle can be eliminated by reducing sample preparation for speciation analysis to only one step by (i) combining leaching, derivatization and liquid–liquid solvent extraction to produce directly a solution of the analytes capable of being analysed by GC (microwave-assisted derivatization solvent extraction, MADSE) or (ii) integrating leaching, derivatization and liquid–gas extraction (purge). This constitutes an alternative for more volatile derivatives, which can be recovered in a cryo-trapping device to avoid losses (microwave-assisted purge and trap, MAPT).

Microwave-assisted solvent extraction can be used to perform the separation of tin[219,220] and mercury[217,221] species from biological material, using tetramethylammonium hydroxide or methanolic KOH solutions for hydrolysis and leaching of mercury species from the biological tissues, assisted by microwaves, at a power of 40–60 W for 2–4 min. A mixture of methanol/water (1:1) has also been used for direct leaching of arsenic species[218] from this type of matrix. The determination of $MeHg^+$ in sediments can be a single-step, rapidly performed process using microwave-assisted leaching of the species with 10 ml of $2\,mol\,l^{-1}$ HNO_3 for 3–4 min at a power of 40–60 W.[216] In this case, under optimum conditions, sample throughput is restricted by the later derivatization, purge and trap and elution/detection steps (about 20 min per sample), rather than by the sample preparation (species lixiviation) step. A similar procedure has been proposed for the speciation of mercury in urine.[222] Simultaneous extraction of organo-tin and -mercury species present in sediments and biological tissues can be rapidly and simply achieved following a protocol also based on an open focused microwave system.[223] All these procedures could be considered as *fast sample leaching methods*. More frequently, a two-step approach (leaching plus derivatization/extraction) or, alternatively, a one-step approach (MADSE) is used for biological sample complete treatment.[224] For organotin speciation in sediments,[16] leaching was performed with acetic acid in 3 min, followed by extraction/derivatization (ethylation) in 5 min, the extract then being analysed by capillary GC with MIP detection, which limits the sample throughput due to the duration of the GC run (10 min). This drawback can, however, be overcome using the faster multi-capillary gas chromatography technique (3 min per chromatographic run).[225] This approach can also be used for HPLC organotin determination with microwave-assisted leaching of the species from sediments with acetic acid (3 min) in the presence of sodium 1-pentanesulfonate as ion pair agent, for later direct injection into the LC device.[226]

2.8.2 Solid Phase Micro-extraction (SPME)

This experimental approach offers a rapid method for organometallic species pre-treatment on the basis of simultaneous aqueous ethylation (NaBEt$_4$) and SPME, prior to analysis by GC. Although aqueous ethylation has considerably reduced sample preparation time, classical liquid–liquid extractions are still tedious and time consuming and sometimes need large volumes of highly pure organic solvents. Micro-extraction has therefore been applied to metal speciation for organo-tin, -mercury and -lead compounds.[227–232] The use of the SPME technique, in combination with multi-elemental detection using hyphenated techniques such as CGC–ICP-MS, makes possible the simultaneous determination of mercury, tin and lead organometallic compounds in aqueous samples, by sorption of ethylated derivatives on a poly(dimethylsiloxane)-coated fused silica fibre.[233] Ethylation/SPME has also been used for the speciation of mercury[234] and mixtures of Hg^{2+} and alkylated Hg, Pb and Sn species[235] in urine, using GC–MS. Limits of detection can be improved with this procedure, by the use of ion-trap GC–MS-MS, which makes possible DLs of between 7 and 22 ng l^{-1} for all the species studied. An alternative alkylation procedure based on methylation of mercury species with methylpentacyanocobaltate(III) or methyl-bis(dimethylglyoximato)pyridinecobalt(III) can also be used for the SPME of this element from water and water/soil slurries.[236]

Organoarsenic compounds can be determined in waters by SPME after dithio-derivatization to form stable five-membered ring structures. For cold vapour (CV)-AAS, calibration curves are linear over three orders of magnitude and limits of detection are less than 6×10^{-9} M, which represent an improvement of over 400 compared with conventional solvent extraction.[237] Other reagents such as thioglycollic acid methyl ester (TGM) can be used to form non-polar derivatives extractable into cyclohexane, for further chromatographic determination.[238] However, this derivatization reaction has not been used, so far as we are aware, for SPME with this element. Finally, triphenylarsine and triethylarsine can be extracted from water by SPME, using poly(dimethylsiloxane)-modified fused-silica fibres.[239]

2.8.3 Fully Automated Systems for Speciation

Manual handling of samples for organometallic speciation leads to long analysis times, low efficiency, poor reproducibility, multiplication of analytical errors and changes in the conditions of the analysis. Automation of hyphenated techniques should be the approach taken to overcome problems arising from manual operation. The resulting computer-assisted hyphenated technique for automatic, on-line, routine analysis should be, however, fully optimized to satisfy the demand of environmental analysis in terms of simplicity, reproducibility, accuracy, sample throughput and economic cost.[13] Tseng *et al.*[15] have proposed a fully automated on-line hyphenated system for Hg speciation in environmental samples, both solid (sediments, bio-tissues and suspended matter) and liquid (sea and fresh water). The samples were derivatized by hydride generation or ethylation, pre-concentrated by cryo-trapping and determined by GC–quartz furnace (QF)-AAS. Automation of the hyphenated technique resulted in improved reproducibility of the analysis. Although the system was specifically optimized for the determination of mercury compounds, it is also potentially applicable to other hydride-, or ethylated derivative, forming elements. Detection limits using the method were 0.5, 3 and $0.1 \, ng l^{-1}$ for the mercury species in dry sediments, biological tissues and aqueous samples, respectively. A sample throughput of three or six samples per hour was achieved for ethylation and hydride generation, respectively. Metal ions or any other inorganic compounds from sediments and large organic molecules from biological tissues, may interfere in the detection step. Sodium chloride or other minor components in seawater samples also strongly affected the sensitivity, reproducibility and selectivity of the analysis. The presence of metal ions strongly decreased sensitivity in the determination of both Hg^{2+} and $MeHg^{+}$ hydride generation, but had no effect if ethylation was used as the derivatization technique.[240] EDTA partially eliminated the interfering effect of metals in the determination of $MeHg^{+}$ by hydride generation and avoided the decomposition of $MeHg^{+}$ to HgO, promoted by NaCl.

2.8.4 Other Fast Methods

With the aim of reducing the sample work-up time and hence the time of analysis, other approaches have been followed, generally based on the use of HPLC for species separation, which eliminates the derivatization step for polar compounds, required when using GC. In addition, some modifications of the chromatographic separation conditions have been found to reduce retention times, without loss of peak resolution. Le *et al.*[241] proposed the use of elevated column temperatures in an HPLC system, followed by dual-element detection by ICP-MS, for the determination of thirteen arsenic species and four selenium compounds. The use of a constant column temperature of 70 °C resulted in an improved resolution and faster separation of these species and six arseno-sugar metabolites in human urine was completed in 19 min compared with 37 min at room temperature. Woller *et al.*[242] illustrated a rapid and accurate analysis technique for redox species of arsenic [As(III) and As(V)] and selenium [Se(IV) and Se(VI)], by coupling an HPLC micro-bore column (< 2 mm i.d.) with a micro-concentric nebulizer (MCN) and ICP-MS.[243] Micro-bore columns have the advantages of good resolving power, low solvent consumption and small sample size. These columns operate at low mobile phase flow rate; hence specific flow rate nebulizers such as the MCN may be used when coupling with ICP-AES or ICP-MS.[244,245] With this coupling, it was possible to achieve the speciation of redox species of arsenic and selenium in less than 4 min, using ICP-MS as detector. The use of cation-exchange chromatography with a solid phase, based on continuous bed chromatography (CBC), permits higher flow rates than conventional columns, by a factor of 4, resulting in a drastic reduction in retention times without any loss of resolution. With this type of column AsB, AsCh and DMA can be separated from more toxic arsenic species [As(III), As(V) and MMA] within 4 min.[246] Finally, Taniguchi *et al.*[247] used ion exclusion chromatography to separate As(III), As(V) and MMA from water samples, with a guard column and a six-way motorized valve (Rheodyne, Model 9750), inserted before the separation column. The valve introduced As(III) and As(V) eluted from the guard column into the separation column and then changed to intro-duce MMA directly into the hydride generator (HG)–ICP-MS system and then the valve returned to the original position to complete the separation and detection of As(III) and As(V). Peak broadening of these species as result of stopping the flow in the separation column was negligible. Using this column switching method the analysis time was shortened to 9 min.

2.9   Speciation of Volatile Species

Multi-elemental speciation analysis of volatile metal and metalloid compounds in air (Me_4Pb, Et_4Pb; Me_4Sn, Et_4Sn, HgO, Me_2Hg, Et_2Hg, Me_2Se) can be achieved[248] by simultaneous sampling of the analytes in the field by cryo-focusing on a small glass wool-packed column at − 175 °C. Detection was performed in the laboratory by low temperature GC coupled with ICP-MS. If other volatile carbon-containing species are present in the samples, the addition of oxygen to the carrier gas is necessary for interference reduction. A similar procedure for the determination of volatile antimony, tin and bismuth in landfill and fermentation gases has been proposed by Feldmann *et al.*[249] Gases were directly sampled into Tedlar bags by using a membrane pump and cryogenically pre-concentrated by trapping on a U-shaped trap packed with Chromosorb (10% SP-2100) at − 78 °C (dry ice–acetone slush). This relatively high tempera-ture was chosen to avoid condensation of carbon dioxide and methane, the major components of landfill gas. The species (Me_3Sb, Me_4Sn and Et_2Me_2Sn) were separated and detected with a GC–ICP-MS system.

The speciation of both inorganic [Se(IV) and Se(VI)] and organic (DMSe, DMDSe, DESe and DEDSe) selenium species in sediments has been studied by Gómez-Ariza *et al.*[250] Inorganic selenium species were quantitatively leached from sediments using 2 M sodium hydroxide under sonication for 4 h in a polystyrene centrifuge tube. Separation of selenite and selenate was carried out using anion-exchange phases packed in cartridges. Selenite was determined by GC–MS after derivatization using 4-chloro-*o*-phenyleneamine in 0.1 M HCl at 75 °C. Selenate was determined after conversion into selenite by boiling for 30 min with 5 M HCl. Volatile selenium species were desorbed from sediments using a dynamic headspace desorption method with activated carbon as the sorptive trap. The species were eluted with carbon disulfide. The four organic selenium species can be directly analysed by GC–MS, but the sensitivity of the method was improved for dialkyl diselenides after derivatization with 1-fluoro-2,4-dinitrobenzene, followed by extraction into ethyl acetate and evaporation to dryness under a nitrogen stream.

Recently, Luque de Castro and colleagues[251–253] have applied pervaporation in conjunction with flow injection analysis for the separation of volatile analytes from aqueous and solid samples. This technique places the sample in the donor chamber of a laboratory-designed and -built pervaporation cell, which is then heated, either in a conventional water bath or by microwaves. The volatile analytes or reaction products then evaporate to a gaseous gap above the sample, from where they diffuse through a membrane to a static or dynamic acceptor stream, to reach the detector or a FI manifold for derivatization. Pervaporation has been used for mercury speciation (Me_2Hg, Et_2Hg and $MeHgCl$) coupled with GC and atomic fluorescence detection in soils and sewage samples.[254] The mercury species pervaporate through the membrane and are collected in a stream of argon to be pre-concentrated in a mini-column packed with Tenax. The recoveries of the species are quantitative and procedure can be considered as an alternative to headspace GC.

3   Sample Preservation and Storage

Species stability in the samples has to be carefully considered, to avoid their alteration during sample handling and storage in long-term environmental campaigns. This is also of paramount importance in the preparation of reliable certified reference materials (CRMs) for method validation. A number of sources can be cited as causes of species instability – chemical reactions between species, interactions with the container material, microbial activity, temperature, pH, light action, *etc.* It is therefore very important to consider the influence of these factors on species stability, so that the analytical speciation is representative of the sample studied and, consequently, of the environmental problem assessed. Additionally, optimum conditions for sample preservation during long-term storage and sample delivery may be assessed.[255] The variables and conditions related to chemical species preservation and storage in environmental samples have recently been reviewed in detail for As, Hg, Sb, Se, Pb and Cr. Some further questions related to this issue are now considered.

3.1   Mercury Species

Mercury is preserved from losses for 10 days with 0.5 ml of 20% w/v potassium dichromate dissolved in 1:1 nitric acid. Skawara *et al.*[256] have tested the stability of ionic mercury and methyl- and ethyl-mercury in unprocessed river water for 14 days. Other authors, however, recommended acidification with dilute nitric acid to preserve $1000-10000\,\mathrm{mg\,l^{-1}}$ mercury in polyethylene containers, although stability is higher in glass containers.[257] Gadner and Gunn[258] found that $6\,\mathrm{ml\,l^{-1}}$ of concentrated HNO_3 was adequate for the preservation of this element in seawater.

Wet cod liver and wet plaice muscle tissue have been stored in screw cap jars at -5, -25 and $-70\,^\circ$C and freeze-dried plaice in the dark at ambient temperature. The content of Hg was measured in the samples after 0, 4, 12 and 24 months.[259] Changes in the matrix were studied by measuring the total and extractable lipid content, peroxide number, rancidity and HPLC fat fingerprints. Homogeneity was tested by determination of K and Na concentrations in a number of samples. Significant changes were observed in the Hg concentration in the cod liver. Deterioration of the matrix (evaluated by peroxide numbers that denote lipid oxidation) in wet samples at -5 and $-25\,^\circ$C indicated that sample storage should be at $-70\,^\circ$C or lower.

Due to the presence of mercury in liquid hydrocarbons, the speciation of this element in these matrices requires studies on species stability. There is currently a lack of knowledge on this topic and, as a consequence, there is an uncertainty in the reliability of sampling procedures and in particular, sample storage. The stability of elemental mercury, methylmercury and inorganic mercury species has been evaluated in heptane, toluene and mixed hydrocarbon solutions and significant losses of inorganic and elemental mercury have been shown to occur *via* two pathways: (i) by adsorption on to the container wall and (ii) by reactions to form low-solubility mercury(I) compounds. This latter pathway produces rapid losses of dissolved elemental mercury and mercury(II) chloride species when they are present in the organic solution.[260] Samples should be analysed as soon as possible after collection, but may contain Hg_2^{2+} precipitates, which could lead to erroneous determination of mercury present. Ultrasonic treatment of the sample container, prior to the determination of mercury, is likely to increase the recovery.

3.2   Arsenic Species

Studies on the stability of arsenic compounds in water samples chiefly concern the inorganic forms of this element, arsenite and arsenate. However, the information about the redox stability of inorganic arsenic is controversial. Feldman[261] asserts that As(III) solutions are unstable and undergo spontaneous and complete conversion into As(V) within 4 days at the 1 and $10\,\mu\mathrm{g\,l^{-1}}$ level. Other authors found essentially no change in As(III) after 20–33 days of storage.[262] A pH range of 2–10 was evaluated for this latter experiment, in the presence of Fe^{3+} or H_2S. However, pH-dependent oxidation of As(III) or reduction of As(V) in this mixed sample occurred within a few hours under acidic conditions and within days at pH 7. Although most arsenic oxidation state changes are attributed to the presence of dissolved oxygen, Agget and Kiegman[263] asserted that deoxygenation was unnecessary, but that the effects of temperature and pH were very significant. The stability of As(III) in interstitial sediment waters containing ppm concentrations of iron and manganese was maintained for up to 6 weeks, by immediate acidification to pH 2 (HCl) and storage near 0 °C, without freezing. Acidification was necessary to prevent iron precipitating as the hydrous oxide, which would co-precipitate arsenic, but it can also alter the species distribution. In the absence of refrigeration, samples could be safely stored for up to 2 weeks, provided that they were de-oxygenated and kept in well sealed glass containers. Hall *et al.*[264] have tested the behaviour of As(III) and As(V) solutions in deionized water. Unless the water samples are kept at 4–5 °C, As(V) is unstable and can convert rapidly (complete within 48–72 h at $1 \mu g l^{-1}$); the lower the concentration of As(V), the faster is the reduction. This reduction can also occur when spiking 'real' samples and is matrix and concentration dependent. Generally, in river water, As(V) is partially converted to As(III), but after 2 days, this is followed by gradual oxidation of As(III) into As(V) to reach an equilibrium where As(V) is in excess (2:1). Storage at 5 °C delays this oxidation by about 6 days but the trend is similar. The reduction of As(V) to As(III) can be bacterially induced and this has been found on both field[265] (open-sea water) and laboratory[266] studies demonstrating the reduction of arsenate by marine algae and bacteria. Nitric acid (0.1%) stabilizes the arsenic species for at least 15 days at 22 °C, but its effect is immediately to alter the species distribution.[264] In river water the effect is to increase the concentration of As(III) substantially, and to a lesser extent that of As(V), at the expense of other forms of the element. Acidification to 0.1% with HCl also produces the same result. However, the behaviour of arsenic depends on the matrix and oxidation of As(III) to As(V) was observed in other river waters under the same experiment.

Methylated species of arsenic are stable for several months when stored with 4 ml of HCl per litre of sample,[267] or sterile-filtered through 0.2 μm filters.[268] Another storage procedure employed for tube-well water samples is based on the use of nitric acid ($1.0 ml l^{-1}$) or ascorbic acid ($100 mg l^{-1}$) in polythene bottles.[269] Some authors[270] considered that no acidification of river water samples is necessary if analysis takes place within a relatively short period of time (20–26 h). They discovered that under these conditions, As(III), As(V), MMA and DMA were stable in most samples for only 3–4 days.

Storage at -20 °C in the dark is recommended to stabilise a mixture of As(V), MMA, DMA, AsB and AsCh in deionized water, urine and dry clean-up residue of urine.[271] If untreated urine is preserved at both 4 °C and room temperature, AsCh suffers oxidation to the more oxidized species, AsB. In deionized water, AsB and AsCh are transformed into other species such as DMA. All the species are stable for 67 days in the dry urine residue at 4 °C and ambient temperature and it may be a good matrix as a reference material for arsenic speciation.

Finally, the influence of food treatment procedures on the stability of arsenic compounds in aqueous media has been studied by van Elteren and Slejkovec.[272] Boiling and microwave treatment produced no degradation, whereas γ-irradiation and dry heating resulted in partial decomposition of several arsenic compounds.

3.3　Selenium Species

The long-term stability of organoselenium compounds, such as selenocysteine, selenomethionine and trimethylselenonium ($TMSe^+$), has been studied over a 1 year period for two analyte concentrations, 25 and $150\,mg\,l^{-1}$, of selenium at pH 4.5 in the dark, under different storage conditions (-20, 4, 20, $40\,°C$ in Pyrex, PTFE, or polyethylene containers) in an aqueous matrix, or in the presence of a chromatographic counter-ion (pentyl sulfonate at a concentration of $10^{-4}\,mol\,l^{-1}$).[273] Organoselenium compounds are stable in the dark over a 1 year period in Pyrex containers at both 4 and $20\,°C$. When the Pyrex vials were exposed to natural sunlight at room temperature a steady decrease of the seleno-amino acid concentrations resulted. PTFE containers caused losses of $< 25\%$ at both 4 and $20\,°C$ in the dark. However, polyethylene vials produced, at all temperatures tested, a rapid decrease of the $TMSe^+$ concentration. The stability of the selenium species considered does not show significant differences between 4 and $20\,°C$ in any container material used. Storage of solutions at $40\,°C$ indicated slight differences between the Pyrex and PTFE containers. However, polyethylene gave rise to a drastic decrease of the three species, over time, at this temperature. Solutions frozen at $-20\,°C$ in polyethylene vials did not indicate stability for $TMSe^+$. Finally, the concentrations and matrices of the samples do not significantly affect the stability of the species.

Selenium stability during the storage of body fluids has been considered in detail by Sanz-Alaejos and Díaz-Romero[274] and is strongly dependent on the type of fluid to be analysed – urine, blood, milk, semen and others. Body fluids must be sampled according to standard procedures using plastic containers to prevent adsorption losses. These containers must be thoroughly cleaned with dilute nitric acid in order to eliminate possible selenium adsorption. Anticoagulants, such as heparin or EDTA, must be added when whole blood or plasma has to be analysed. Centrifugation is used to separate plasma or serum and ultracentrifugation is necessary for separating platelets and leucocytes from whole blood. Refrigeration can be used to store blood and urine samples for a few days, but most authors prefer to store the samples of body fluids frozen to $-20\,°C$. Storage after lyophilization and desiccation techniques is also used.

Acknowledgement

The authors express their thanks to 'Dirección General de Enseñanza Superior e Investigación Científica', FEDER Projets, Grant no. 1FD97-0610-C03-02.

4　References

1. *Research Trends in the Field of Environmental Analysis*, EUR Report 16000 EN, p. 53.

2. P.J. Craig, *Organometallic Compounds in the Environment, Principles and Reactions*, Longman, Essex, 1986.
3. O.F.X. Donard and P. Michel, *Analusis*, 1992, **20**, M45.
4. R. Lobinski, *Appl. Spectrosc.*, 1997, **51**, 260A.
5. M. Berlin, *Specific Metals*, in *Handbook on the Toxicology of Metals*, ed. L. Friberg, G.F. Norberg and V.B. Vouk, Vol. II, Elsevier, Amsterdam, 1986, p. 387.
6. FAO-WHO, *Technical Report Series*, 1972, No. 505.
7. T. Kjellström, P. Kennedy, S. Wallis and C. Mantell, *Nat. Swedish Environ. Protec. Board*, 1989, PM 3642.
8. O.F.X. Donard and R. Ritsema, *Hyphenated Techniques Applied to the Speciation of Organometallic Compounds in the Environment*, in *Environmental Analysis (Techniques, Application and Quality Assurance)*, ed. D. Barceló, Elsevier, Amsterdam, 2nd edition, 1993, p. 549.
9. R. Lobinski, I.R. Pereiro, H. Chassaigne, A. Wasik and J. Szpunar, *J. Anal. At. Spectrom.*, 1998, **13**, 859.
10. Ph. Quevauviller, O.F.X. Donard, E.A. Maier and B. Griepink, *Mikrochim. Acta*, 1992, **109**, 169.
11. R.W.P.M. Laane, J. Marqueni, R. Ritsema, K.C.J. Van den Ende, O.F.X. Donard and Ph. Quevauviller, *Rijkswaterstaat Rep.*, 1989, GWAO-89.024.
12. M. Ceulemans and F.C. Adams, *J. Anal. At. Spectrom.*, 1996, **11**, 201.
13. A. de Diego, C. Pecheyran, C.M. Tseng and O.F.X. Donard, in *Flow Analysis with Atomic Spectrometric Detectors*, ed. A. Sanz Medel, Elsevier, Amsterdam, 1999.
14. J. Szpunar, V.O. Schmitt, R. Lobinski and J.L. Monod, *J. Anal. At. Spectrom.*, 1996, **11**, 193.
15. C.M. Tseng, A. de Diego, H. Pinaly, D. Amouroux and O.F.X. Donard, *J. Anal. At. Spectrom.*, 1998, **13**, 755.
16. M. Abalos, J.M. Bayona, R. Compañó, M. Granados, C. Leal and M.D. Prat, *J. Chromatogr. A*, 1997, **788**, 1.
17. Y.K. Chau, P.T.S. Wong and G.A. Bengert, *Anal. Chem.*, 1982, **54**, 246.
18. Y.K. Chau, F. Yang and R.J. Maguire, *Anal. Chim. Acta*, 1996, **320**, 165.
19. M.D. Müller, *Anal. Chem.*, 1987, **59**, 617.
20. R.J. Maguire and H. Huneault, *J. Chromatogr.*, 1981, **209**, 458.
21. M. Abalos, J.M. Bayona and Ph. Quevauviller, *Appl. Organomet. Chem.*, 1998, **12**, 541.
22. T. Tsuda, H. Nakanishi, T. Morita and J. Takebayashi, *J. Assoc. Off. Anal. Chem.*, 1986, **69**, 981.
23. M.D. Müller, *Fresenius' J. Anal. Chem.*, 1986, **317**, 32.
24. Y. Hattori, A. Kobayashi, S. Takemoto, K. Takami, Y. Kuge, A. Sugimae and M. Nakamoto, *J. Chromatog.*, 1984, **315**, 341.
25. J.R. Ashby and P.J. Craig, *Sci. Total Environ.*, 1989, **78**, 219.
26. I. Tolosa, J.M. Bayona, J. Albaigés, L.F. Alencastro and J. Tarradellas, *Fresenius' J. Anal. Chem.*, 1991, **339**, 646.
27. M. Astruc, R. Levigne, V. Desauziers, A. Astruc, R. Pinel and O.F.X. Donard, *Proc. Int. Conf. Heavy Metals in the Hydrocycle*, Lisbon, 1988, p. 447.
28. Ph. Quevauviller and O.X.F. Donard, *Fresenius' J. Anal. Chem.*, 1991, **339**, 6.
29. Ph. Quevauviller and O.X.F. Donard, *Appl. Organomet. Chem.*, 1990, **4**, 353.
30. J.L. Gómez-Ariza, E. Morales, R. Beltran, I. Giráldez and M. Ruiz-Benitez, *Analyst*, 1995, **120**, 1171.
31. H.A. Meinema, T. Burger-Wiersma, G. Versluis-de Haan and E.C. Gevers, *Environ. Sci. Technol.*, 1978, **12**, 288.
32. I. Martin-Landa, F. Pablos and I.L. Marr, *Appl. Organomet. Chem.*, 1991, **5**, 399.

33. J.L. Gómez-Ariza, M. Morales and M. Ruiz-Benitez, *Analyst*, 1992, **117**, 641.
34. J.L. Gómez-Ariza, M. Morales and M. Ruiz-Benitez, *Appl. Organomet. Chem.*, 1992, **6**, 279.
35. J.L. Gómez-Ariza, E. Morales, I. Giraldez and R. Beltran, *Intern. J. Environ. Anal. Chem.*, 1997, **66**, 1.
36. J.L. Gómez-Ariza, E. Morales, I. Giraldez, R. Beltran and J.A. Pozas-Escobar, *Fresenius' J. Anal. Chem.*, 1997, **357**, 1007.
37. G. Bressa, F. Cima, P. Fonti and E. Sisti, *Fresenius' Environ. Bull.*, 1997, **6**, 16.
38. Y.K. Chau and F. Yang, *Appl. Organomet. Chem.*, 1997, **11**, 851.
39. P. Bermejo-Barrera, M. Tubio-Franco, J.M. Aguiar-Paz, R.N. Soto-Ferreiro and A. Bermejo-Barrera, *Microchem. J.*, 1996, **53**, 395.
40. R. François and J.H. Weber, *Mar. Chem.*, 1988, **25**, 279.
41. S. Rapsomanikis and R.M. Harrison, *Appl. Organomet. Chem.*, 1988, **2**, 151.
42. L. Randall, J.S. Han and J.H. Weber, *Environ. Technol. Lett.*, 1986, **7**, 571.
43. S. Shawky, H. Emons and H.W. Dürbeck, *Anal. Commun.*, 1996, **33**, 107.
44. V. Desauziers, F. Leguille, R. Lavigne, M. Astruc and R. Pinel, *Appl. Organomet. Chem.*, 1989, **3**, 469.
45. I.S. Krull, K.W. Panaro, J. Noonam and D. Erickson, *Appl. Organomet. Chem.*, 1989, **3**, 295.
46. F. Pannier, A. Astruc and M. Astruc, *Appl. Organomet. Chem.*, 1994, **8**, 595.
47. K. Sasaki, T. Ishizaka, T. Suzuki and Y. Saito, *J. Assoc. Off. Anal. Chem.*, 1988, **71**, 360.
48. C.L. Matthias, S.J. Bushong, L.W. Hall, J.M. Bellama and F.E. Brickman, *Appl. Organomet. Chem.*, 1988, **2**, 547.
49. Y. Cai, S. Rapsomanikis and M.O. Andreae, *Talanta*, 1994, **41**, 589.
50. Y. Cai, S. Rapsomanikis and M.O. Andreae, *J. Anal. At. Spectrom.*, 1993, **8**, 119.
51. Y. Cai, S. Rapsomanikis and M.O. Andreae, *Anal. Chim. Acta*, 1993, **274**, 243.
52. Y. Cai, S. Rapsomanikis and M.O. Andreae, *Mikrochim. Acta*, 1992, **109**, 67.
53. O.F.X. Donard, B. Lalère, F. Martin and R. Lobinski, *Anal. Chem.*, 1995, **67**, 4250.
54. B. Lalère, J. Szpunar, H. Budzinski, Ph. Garrigues and O.F.X. Donard, *Analyst*, 1995, **120**, 2665.
55. Ph. Quevauviller, S. Chiavarini, C. Cremisini, R. Morabito, M. Bianchi and H. Muntau, *Mikrochim. Acta*, 1995, **120**, 281.
56. D. Milde, Z. Plzak and M. Suchanek, *Collect. Czech., Chem. Commun.*, 1997, **62**, 1403.
57. X. Dauchy, R. Cottier, A. Batel, M. Borsier, A. Astruc and M. Astruc, *Environ. Technol*, 1994, **15**, 569.
58. Ph. Quevauviller, R. Lavigne, R. Pinel, M. Astruc and M. Astruc, *Environm. Pollut.*, 1989, **57**, 149.
59. Ph. Quevauviller, F. Martin, C. Belin and O.F.X. Donard, *Appl. Organomet. Chem.*, 1993, **7**, 149.
60. L. Cortez, Ph. Quevauviller, F. Martin and O.F.X. Donard, *Environ. Pollut.*, 1993, **82**, 57.
61. P.H. Dowson, J.M. Bubb and J.N. Lester, *Chemosphere*, 1994, **28**, 905.
62. P.M. Sarradin, A. Astruc, R. Sabrier and M. Astruc, *Mar. Pollut. Bull.*, 1994, **28**, 621.
63. C. Rivas, L. Ebdon, E.H. Evans and S.J. Hill, *Appl. Organomet. Chem.*, 1996, **10**, 61.
64. A.M. Caricchia, S. Chiavarini, C. Cremisini, R. Morabito and C. Ubaldi, *Int. J. Environ. Anal. Chem.*, 1993, **53**, 37.
65. R. Companó, M. Granados, C. Leal and M.D. Prat, *Anal. Chim. Acta*, 1995, **314**, 175.
66. M.J. Waldock and M.E. Waite, *Appl. Organomet. Chem.*, 1994, **8**, 649.
67. M. Heisterkamp, T. De Smaele, J.P. Candelone, L. Moens, R. Dams and F.C. Adams, *J. Anal. At. Spectrom.*, 1997, **12**, 1077.

68. S.A. Estes, P.C. Uden and R. Barnes, *Anal. Chem.*, 1981, **53**, 1336.
69. D. Chakraborti, W.R.A. De Jonghe, W.E. Van Mol, R.J.A. Van Cleuvenbergen and F.C. Adams, *Anal. Chem.*, 1984, **56**, 2692.
70. G. Westöö, *Acta Chem. Scand.*, 1967, **20**, 1790; G. Westöö, *Acta Chem. Scand.*, 1968, **22**, 2277.
71. K. May, M. Stoeppler and K. Reisinger, *Toxicol. Environ. Chem.*, 1987, **13**, 153.
72. M. Hempel, H. Hintelmann and R.D. Wilken, *Analyst*, 1992, **117**, 669.
73. H. Hintelmann and R.D. Wilken, *Appl. Organomet. Chem.*, 1993, **7**, 173.
74. Ph. Quevauviller, O.F.X. Donard, J.C. Wasserman, F.M. Martin and J. Schneider, *Appl. Organomet. Chem.*, 1992, **6**, 221.
75. Y. Cai, G. Tang, R. Jaffe and R. Jones, *Int. J. Environ. Anal. Chem.*, 1998, **68**, 331.
76. Z. Huang, *Fenxi Ceshi Xuebao*, 1998, **17**, 22.
77. N.S. Bloom, *Can. J. Fish. Aquat. Sci.*, 1989, **46**, 1131.
78. L. Liang, M. Horvat, E. Cernichiari, B. Gelein and S. Balogh, *Talanta*, 1996, **43**, 1883.
79. M. Horvat, N.S. Bloom and L. Liang, *Anal. Chim. Acta*, 1993, **281**, 135.
80. Y.H. Lee, J. Munthe and A. Iverfeldt, *Appl. Organomet. Chem.*, 1994, **8**, 659.
81. M. Horvat, V. Mandic, L. Liang, N.S. Bloom, S. Padberg, Y.H. Lee, H. Hintelmann and J. Benoit, *Appl. Organomet. Chem.*, 1994, **8**, 533.
82. M. Floyd and L.E. Sommer, *Anal. Lett.*, 1975, **8**, 525.
83. M. Horvat, K. May, M. Stoeppler and A.R. Byrne, *Appl. Organomet. Chem.*, 1988, **2**, 515.
84. C.J. Cappon and J.C. Smith, *Anal. Chem.*, 1977, **49**, 365.
85. R. Cela, R.A. Lorenzo, M.C. Mejuto, M.H. Bollain, M.C. Casais, A. Botana, E. Rubi and M.L. Medina, *Mikrochim. Acta*, 1992, **109**, 111.
86. M.H. Arbab-Zavar, PhD Thesis, University of Southampton, 1982.
87. A.W. Fitchett, E.H. Daughtrey and P. Mushak, *Anal. Chim. Acta*, 1975, **79**, 93.
88. A. Yatsui, C. Tsutsumi and S. Toda, *Agric. Biol. Chem.*, 1978, **42**, 2139.
89. A. Shinagawa, K. Shiomi, H. Yamanaka and T. Kikuchi, *Bull. Jpn. Soc. Sci. Fish.*, 1983, **49**, 75.
90. E.A. Crescelius, N.S. Bloom, C.E. Cowan and E.A. Jenne, in *Arsenic Speciation*, EPRI, Batelle Northwest Laboratories, Washington DC, 1986, Vol. 2, p. 18.
91. J. Dedina and D.L. Tsalev, in *Hydride Generation Atomic Absorption Spectrometry*, ed. J.D. Winefordner, Wiley, Chichester, 1995, p. 213.
92. B. Amran, F. Lagarde, M.J.F. Leroy, A. Lamotte, C. Desmeney, M. Ollé, M. Albert, G. Rauret and J.F. Lopez-Sanchez, in *Quality Assurance for Environmental Analysis*, ed. Ph. Quevauviller, E.A. Mayer and B. Griepink, Elsevier, Amsterdam, 1995, Vol. 17, p. 285.
93. W.R. Cullen and K.J. Reimer, *Chem. Rev.*, 1989, **89**, 713.
94. W. Maher and E. Butler, *Appl. Organomet. Chem.*, 1988, **2**, 191.
95. J.L. Gómez-Ariza, D. Sánchez-Rodas and I. Giráldez, *J. Anal. At. Spectrom.*, 1998, **13**, 1375.
96. M. Dodd, S.A. Pergantis, W.R. Cullen, H. Li, G.R. Eigendorf and K.J. Reimer, *J. Anal. At. Spectrom.*, 1996, **121**, 223.
97. R. Muñoz-Olivas, O.F.X. Donard, N. Gilon and M. Potin-Gautier, *J. Anal. At. Spectrom.*, 1996, **11**, 1171.
98. J.L. Martin and M.L. Gerlach, *Anal. Biochem.*, 1969, **29**, 257.
99. M. Potin-Gautier, C. Boucharat, A. Astruc and M. Astruc, *Appl. Organomet. Chem.*, 1993, **7**, 593.
100. H. Emteborg, G. Bordin and A.R. Rodriguez, *Analyst*, 1998, **123**, 245.
101. J. Alberti, R. Rubio and G. Rauret, *Fresenius' J. Anal. Chem.*, 1995, **351**, 420.

102. V.G. Torgov, M.G. Demidova and A.D. Kosolapov, *J. Anal. Chem.*, 1998, **53**, 846.
103. A.J. Aller and C. Robles, *J. Anal. At. Spectrom.*, 1998, **13**, 469.
104. M.M. Gómez, T. Gasparic, M.A. Palacios and C. Camara, *Anal. Chim. Acta*, 1998, **374**, 241.
105. S. Zhang, Y.K. Chau, W.C. Li and A.S.Y. Chau, *Appl. Organomet. Chem.*, 1991, **5**, 431.
106. J. Kuballa, R.D. Wilken, E. Jantzen, K.K. Kwan and Y.K. Chau, *Analyst*, **120**, 667.
107. M. Ceulemans, C. Witte, R. Lobinski and F.C. Adams, *Appl. Organomet. Chem.*, 1994, **8**, 451.
108. M. Nagase and K. Hasebe, *Anal. Sci.*, 1993, **9**, 517.
109. M. Nagase, H. Kondo and K. Hasebe, *Analyst*, 1995, **120**, 1923.
110. J.A. Stäb, U.A. Th. Brinkman and W.P. Cofino, *Appl. Organomet. Chem.*, 1994, **8**, 577.
111. L.N. Mackey and T.A. Beck, *J. Chromatogr.*, 1982, **240**, 455.
112. F. Pannier, A. Astruc and M. Astruc, *Anal. Chim. Acta*, 1996, **327**, 287.
113. D.S. Forsyth and R. Iyengar, *J. Organomet. Chem.*, 1998, **3**, 211.
114. N. Gilon, A. Astruc, M. Astruc and M. Potin-Gautier, *Appl. Organomet. Chem.*, 1995, **9**, 633.
115. D.S. Forsyth and C. Cloroux, *Talanta*, 1991, **38**, 951.
116. Y. Tan and W.D. Marshall, *Analyst*, 1997, **122**, 13.
117. F.R. Abou-Shakra, M.P. Rayman, N.I. Ward, V. Hotton and G. Bastian, *J. Anal. At. Spectrom.*, 1997, **12**, 429.
118. *Analytical Supercritical Fluid Chromatography and Extraction*, ed. M.L. Lee and K.E. Markides, Chromatography Conferences, Provo, UT, 1990.
119. S.B. Hawthorne, *Anal. Chem.*, 1990, **62**, 633A.
120. L.T. Taylor, *Anal. Chem.*, 1995, **67**, 364A.
121. K.E. Laintz, C.M. Wai, C.R. Yonker and R.D. Smith, *Anal. Chem.*, 1992, **64**, 2875.
122. S. Wang, S. Elshani and C.M. Wai, *Anal. Chem.*, 1995, **67**, 919.
123. Y. Lin, R.D. Brauer, K.E. Laintz and C.M. Wai, *Anal. Chem.*, 1993, **65**, 2549.
124. J. Wang and W.D. Marshall, *Anal. Chem.*, 1994, **66**, 1658.
125. J. Wang and W.D. Marshall, *Anal. Chem.*, 1994, **66**, 3900.
126. S.L. Cleland, L.K. Olson, J.A. Caruso and J.M. Carey, *J. Anal. At. Spectrom.*, 1994, **9**, 975.
127. K. Li and S.F.Y. Li, *J. Chromatogr. Sci.*, 1995, **33**, 309.
128. J.W. Oudsema and C.F. Poole, *Fresenius' J. Anal. Chem.*, 1992, **344**, 426.
129. R. Alzaga and J.M. Bayona, *J. Chromatogr. A*, 1993, **655**, 51.
130. J. Dachs, R. Alzaga, J.M. Bayona and P. Quevauviller, *Anal. Chim. Acta*, 1994, **286**, 319.
131. Y. Liu, V. Lopez-Avila, M. Alcaraz and W.F. Beckert, *J. High Resolut. Chromatogr.*, 1993, **16**, 106.
132. Y. Liu, V. Lopez-Avila, M. Alcaraz and W.F. Beckert, *Anal. Chem.*, 1994, **66**, 3788.
133. U. Kumar, N.P. Vela, J.G. Dorsey and J.A. Caruso, *J. Chromatogr. A*, 1993, **655**, 340.
134. J.M. Bayona and Y. Cai, *TrAC*, 1994, **13**, 327.
135. N.P. Vela and J.A. Caruso, *J. Anal. At. Spectrom.*, 1996, **11**, 1129.
136. Y.K. Chau, F. Yang and M. Brown, *Anal. Chim. Acta*, 1995, **304**, 85.
137. Y. Liu, V. Lopez-Avila, M. Alcaraz and W.F. Beckert, *J. Assoc. Off. Anal. Chem.*, 1995, **78**, 1275.
138. Y. Cai, R. Alzaga and J.M. Bayona, *Anal. Chem.*, 1994, **66**, 1161.
139. C.G. Arnold, M. Berg, S.R. Mueller, U. Dommann and R.P. Schwarzenbach, *Anal. Chem.*, 1998, **70**, 3094.

140. C. Maldonado, R. Alzaga, P. Bou and J.M. Bayona, *8th Symposium of Environmental and Biological Samples in Chormatography*, 1997, Almeria (Spain), 1997, pp. 171.
141. D.E. Carrit, *Anal. Chem.*, 1953, **25**, 1972.
142. T.B. Pierce, *Anal. Chim Acta*, 1961, **24**, 1927.
143. T.B. Pierce and P.F. Peck, *Anal. Chim. Acta*, 1962, **26**, 557.
144. T. Braum and A.B. Farag, *Anal. Chim. Acta*, 1974, **69**, 85.
145. A.G. Howard and M.H. Arbab-Zavar, *Talanta*, 1979, **26**, 895.
146. T. Yano, S. Ide, Y. Tobeta, H. Kobayashi and K. Ueno, *Talanta*, 1976, **23**, 457.
147. R. Shah and S. Devi, *Analyst*, 1996, **121**, 807.
148. R. Shah and S. Devi, *Anal. Chim. Acta*, 1997, **341**, 217.
149. J.L. Manzoori, M.H. Sorouraddin and A.M. Haji Shabani, *J. Anal. At. Spectrom.*, 1998, **13**, 305.
150. R. Say, N. Satiroolu, E. Piskin, S. Bektas and O. Genc, *Anal. Lett.*, 1998, **31**, 511.
151. Q. Xu, X. Yin and S. Li, *Fenxi Huaxue*, 1995, **23**, 1305.
152. X. Yin and M. Liu, *Fenxi Huaxue*, 1996, **24**, 1248.
153. Y. Cai, R. Jaffe, A. Alli and R.D. Jones, *Anal. Chim. Acta*, 1996, **334**, 251.
154. M.K. Donais, P.C. Uden, M.M. Schantz and S.A. Wise, *Anal. Chem.*, 1996, **68**, 3859.
155. R. Falter and G. Ilgen, *Fresenius' J. Anal. Chem.*, 1997, **358**, 407.
156. R. Falter and G. Ilgen, *Fresenius' J. Anal. Chem.*, 1997, **358**, 401.
157. X.F. Yin, W. Frech, E. Hoffmann, C. Ludke and J. Skole, *Fresenius' J. Anal. Chem.*, 1998, **361**, 761.
158. X.P. Yan, R. Kerrich and M.J. Hendry, *Anal. Chem.*, 1998, **70**, 4736.
159. J. Chwastowska, E. Sterlinska, W. Zmijerwska and J. Dudek, *Chem. Anal.*, 1996, **41**, 45.
160. P. Bermejo-Barrera, G. Gonzalez-Campos, M. Ferron-Novais and A. Bermejo-Barrera, *Talanta*, 1998, **46**, 1479.
161. V.K. Ososkov, C.C. Chou and B.B. Kebbekus, *Int. J. Environ. Anal. Chem.*, 1998, **69**, 1.
162. K. Kadokami, T. Uehiro, M. Morita and K. Fuwa, *J. Anal. At. Spectrom.*, 1988, **3**, 1887.
163. O. Evans, B.J. Jacobs and A.L. Cohen, *Analyst*, 1991, **116**, 15.
164. S. Chavarini, C. Cremisini, T. Ferri, R. Morabito and C. Ubaldi, *Appl. Organomet. Chem.*, 1992, **6**, 147.
165. J.L. Gómez-Ariza, R. Beltran, E. Morales, I. Giráldez and M. Ruiz-Benitez, *Appl. Organomet. Chem.*, 1994, **8**, 553.
166. A. Lopez, M. Gómez, C. Cámara and M.A. Palacios, *J. Anal. At. Spectrom.*, 1994, **9**, 291.
167. M. Gómez, C. Cámara, M.A. Palacios and A. Lopez-Gonzálvez, *Fresenius' J. Anal. Chem.*, 1997, **357**, 844.
168. S. Yalcin and X.C. Le, *Talanta*, 1998, **47**, 787.
169. C.I. Measures and J.D. Burton, *Anal. Chim. Acta*, 1980, **120**, 177.
170. S. Nielsen, J.J. Sloth and E.H. Hansen, *Analyst*, 1996, **121**, 31.
171. T. Kubota, K. Suzuki and T. Okatani, *Talanta*, 1995, **42**, 949.
172. Y. Cai, M. Cabañas, J.L. Fernandez-Turiel, M. Abalos and J.M. Bayona, *Anal. Chim. Acta*, 1995, **314**, 183.
173. C. Pécheyran, D. Amouroux and O.F.X. Donard, *J. Anal. At. Spectrom.*, 1998, **13**, 615.
174. M.B.de la Calle, M. Ceulemans, C. Witte, R. Lobinski and F.C. Adams, *Mikrochim. Acta*, 1995, **120**, 73.
175. J.L. Gómez-Ariza, J.A. Pozas, I. Giráldez and E. Morales, *Analyst*, 1999, **124**, 75.

176. S.C. Apte, A.G. Howard and A.T. Campbell, *Arsenic and Antimony*, in *Environmental Analysis Using Chromatography Interfaced with Atomic Spectroscopy*, ed. R.M. Harrison and S. Rapsomanikis, Ellis Horwood, Chichester, 1989, p. 259.

177. J.L. Burguera, M. Burguera, C. Rivas and P. Carrero, *Talanta*, 1998, **45**, 531.

178. A.G. Howard, *Selenium Combination Techniques*, in *Environmental Analysis Using Chromatography Interfaced with Atomic Spectroscopy*, ed. R.M. Harrison and S. Rapsomanikis, Ellis Horwood, Chichester, 1989, p. 318.

179. De Q. Zhang, H.W. Sun and L.L. Yang, *Fresenius' J. Anal. Chem.*, 1997, **359**, 492.

180. J. Sanz-Asensio, M. Pérez-Clavijo and M.T. Martínez-Soria, *Anal. Chim. Acta*, 1997, **343**, 39.

181. O.F.X. Donard and Pinel, *Tin and Germanium*, in *Environmental Analysis Using Chromatography Interfaced with Atomic Spectroscopy*, ed. R.M. Harrison and S. Rapsomanikis, Ellis Horwood, Chichester, 1989, p. 196.

182. F.M. Martin, C.M. Tseng, C. Belin, Ph. Quevauviller and O.F.X. Donard, *Anal. Chim. Acta*, 1994, **286**, 343.

183. O.F.X. Donard, S. Rapsomanikis and J.H. Weber, *Anal. Chem.*, 1986, **58**, 772.

184. O.F.X. Donard, L. Randall, S. Rapsomanikis and J.H. Weber, *Int. J. Environ. Anal. Chem.*, 1986, **27**, 55.

185. S. Rapsomanikis, *Analyst*, 1994, **119**, 1429.

186. Ph. Quevauviller, F. Martin, C. Belin and O.F.X. Donard, *Appl. Organomet. Chem.*, 1993, **7**, 149.

187. L. Randall, O.F.X. Donard and J.H. Weber, *Anal. Chim. Acta*, 1986, **184**, 197.

188. A.G. Horward and C. Salou, *Anal. Chim. Acta*, 1996, **333**, 89.

189. X. Yin, E. Hoffmann and C. Ludke, *Fresenius' J. Anal. Chem.*, 1996, **355**, 324.

190. K.J. Lamble and S.J. Hill, *Anal. Chim. Acta*, 1996, **334**, 261.

191. X. Le, W.R. Cullen and K.J. Reimer, *Anal. Chim. Acta*, 1994, **285**, 277.

192. H. Chen, I.D. Brindle and X.C. Le, *Anal. Chem.*, 1992, **64**, 667.

193. A.G. Horward and C. Salou, *J. Anal. At. Spectrom.*, 1998, **13**, 683.

194. S.N. Willie, *Spectrochim. Acta Part B*, 1996, **51**, 1781.

195. S. Nielsen and E.H. Hansen, *Anal. Chim. Acta*, 1997, **343**, 5.

196. M. Heisterkamp, T. De Smaele, J.P. Candelone, L. Moens, R. Dam and F.C. Adams, *J. Anal. At. Spectrom.*, 1997, **12**, 1077.

197. Y. Liu, V. López-Avila, M. Alcaraz and W.F. Beckert, *J. High Resolut. Chromatr.*, 1994, **17**, 527.

198. J.A. Stäb, U.A.Th. Brinkman and W.P. Cofino, *Appl. Organomet. Chem.*, 1994, **8**, 577.

199. I. Fernández-Escobar, M. Gibert, A. Messeguer and J.M. Bayona, *Anal. Chem.*, 1998, **70**, 3703.

200. R.J. Maguire, *Environ. Sci. Technol.*, 1984, **18**, 291.

201. W.M.R. Dirkx, W.E. Van Mol, R.J. Van Cleuvenbergen and F.C. Adams, *Fresenius' J. Anal. Chem.*, 1989, **335**, 769.

202. R. Lobinski and F.C. Adams, *J. Anal. At. Spectrom.*, 1992, **7**, 987.

203. E. Bulska. H. Emteborg, D.C. Baxter, W. Frech, D. Ellingsen and Y.Thomassen, *Analyst*, 1992, **117**, 657.

204. R.D. Wilken, J. Kuballa and E. Jantzen, *Fresenius' J. Anal. Chem.*, 1994, **350**, 77.

205. G. Pritzl, F. Stuer-Lauridsen, L. Carlsen, A.K. Jensen and T.K. Thorsen, *Int. J. Environ. Anal. Chem.*, 1996, **62**, 147.

206. C. Carlier-Pinasseau, G. Lespes and M. Astruc, *Appl. Organomet. Chem.*, 1996, **10**, 505.

207. A.A. Ansari, I.B. Singh and H.J. Tobschall, *Sci. Total Environ.*, 1998, **223**, 157.

208. C. Carlier-Pinasseau, G. Lespes and M. Astruc, *Environ. Technol.*, 1997, **18**, 1179.

209. M.J. Waldock and M.E. Waite, *Appl. Organomet. Chem.*, 1994, **8**, 649.

210. T.A. Sergeeva, M.A. Palacios and P.J. Craig, *Quim. Anal.*, 1997, **16**, 297.

211. T. De Smaele, L. Moens, R. Dams, P. Sandra, J. Van der Eycken and J.Vandyck, *J. Chromatogr. A*, 1998, **793**, 99.

212. K. Bergmann and B. Neidhart, *Fresenius' J. Anal. Chem.*, 1996, **356**, 57.

213. M. Heisterkamp and F.C. Adams, *Fresenius' J. Anal. Chem.*, 1998, **362**, 489.

214. O.X.F. Donard, B. Lalère, F. Martin and R. Lobinski, *Anal. Chem.*, 1995, **67**, 4250.

215. I.R. Pereiro, V.O. Schmitt, J. Szpunar, O.F.X. Donard and R. Lobinski, *Anal. Chem.*, 1996, **68**, 4135.

216. C.M. Tseng, A. de Diego, F.M. Martin and O.X.F. Donard, *J. Anal. At. Spectrom.*, 1997, **12**, 629.

217. C. Gebersmann, M. Heisterkamp, F.C. Adams and J.A.C. Broekaert, *Anal. Chim. Acta*, 1997, **350**, 273.

218. T. Dagnac, A. Padro, R. Rubio and G. Rauret, *Talanta*, 1999, **48**, 763.

219. I. Rodriguez, M. Santamaria, M.H. Bollain, M.C. Mejuto and R. Cela, *J. Chromatogr. A*, 1997, **774**, 379.

220. I. Rodriguez, M. Santamaria, M.H. Bollain, M.C. Mejuto and R. Cela, *Spectroscopy*, 1997, **13**, 51.

221. C.M. Tseng, A. De Diego, F.M. Martin, D. Amouroux and O.F.X. Donard, *J. Anal. At. Spectrom.*, 1997, **12**, 743.

222. I.R. Pereiro, A. Wasik and R. Lobinski, *Anal. Chem.*, 1998, **70**, 4063.

223. V.O. Schmitt, A. De Diego, A. Cosnier, C.M. Tseng, J. Moreau and O.F.X. Donard, *Spectroscopy*, 1997, **13**, 99.

224. V.O. Schmitt, J. Szpunar, O.F.X. Donard and R. Lobinski, *Can. J. Anal. Sci. Spectrosc.*, 1997, **42**, 41.

225. I.R. Pereiro, A. Wasik and R. Lobinski, *Fresenius' J. Anal. Chem.*, 1999, **363**, 460.

226. W.S. Chao and S.J. Jiang, *J. Anal. At. Spectrom.*, 1998, **13**, 1337.

227. Y. Morcillo, Y. Cai and J.M. Bayona, *J. High Resolut. Chromatogr.*, 1995, **18**, 767.

228. Y. Cai and J.M. Bayona, *J. Chromatogr.*, 1995, **696**, 113.

229. S. Tutschku, S. Mothes and R. Wennerich, *Fresenius' J. Anal. Chem.*, 1996, **354**, 587.

230. J. Poerschmann, F.D. Kopinke and J. Pawliszyn, *Environ. Sci. Technol.*, 1997, **31**, 3629.

231. M. Guidotti and M. Vitali, *Ann. Chim.*, 1997, **87**, 497.

232. G. Lespes, V. Desauziers, C. Montigny and M. Potin-Gautier, *J. Chromatogr. A*, 1998, **826**, 67.

233. L. Moens, T. De Smaele, R. Dams, P. Van Den Broeck and P. Sandra, *Anal. Chem.*, 1997, **69**, 1604.

234. M. Guidotti and M. Vitali, *J. High Resolut. Chromatogr.*, 1998, **21**, 665.

235. L. Dunemann, H. Hajimiragha and J. Begerow, *Fresenius' J. Anal. Chem.*, 1999, **363**, 466.

236. C.M. Barshick, S.A. Barshick, Ph.-F. Britt, A.D. Lake, M.A. Vance and E.B. Walsh, *Int. J. Mass. Spectrom.*, 1998, **178**, 31.

237. B. Szostek and J.H. Aldstadt, *J. Chromatogr. A*, 1998, **807**, 253.

238. B. Beckerman, *Anal. Chim. Acta*, 1982, **135**, 77.

239. S. Mothes and R. Wennrich, *Am. Environ. Lab.*, 1997, **9**, 5.

240. A. de Diego, C.M. Tseng, T. Stoichev, D. Amouroux and O.F.X. Donard, *J. Anal. At. Spectrom.*, 1998, **13**, 623.

241. X.C. Le, X.F. Li, V. Lai, M. Ma, S. Yalcin and J. Feldmann, *Spectrochim. Acta, Part B*, 1998, **53**, 899.

242. A. Woller, H. Garraud, J. Boisson, A.M. Dorthe, P. Fodor and O.F.X. Donard, *J. Anal. At. Spectrom.*, 1998, **13**, 141.
243. H. Garraud, A. Woller, P. Fodor and O.F.X. Donard, *Analusis*, 1997, **25**, 25.
244. G. Zorob, M. Tomlinson, J. Wang and J. Caruso, *J. Anal. At. Spectrom.*, 1995, **10**, 853.
245. F. Vanhaecke, M. Van Holderbeke, L. Moens and R. Dams, *J. Anal. At. Spectrom.*, 1996, **11**, 543.
246. R. Sur, J. Begerow and L. Dunemann, *Fresenius' J. Anal. Chem.*, 1999, **363**, 526.
247. T. Taniguchi, H. Tao, M. Tominaga and A. Miyazaki, *J. Anal. At. Spectrom.*, 1999, **14**, 651.
248. C. Pecheyran, C.R. Quetel, F.M. Lecuyer and O.F.X. Donard, *Anal. Chem.*, 1998, **70**, 2639.
249. J. Feldmann, Y. Koch and W.R. Cullen, *Analyst*, 1998, **123**, 815.
250. J.L. Gómez-Ariza, J.A. Pozas, I. Giráldez and E. Morales, *J. Chromatogr. A*, 1998, **823**, 259.
251. I.L. Mattos and M.D. Luque de Castro, *Anal. Chim. Acta*, 1994, **298**, 159.
252. I.L. Mattos, M.D. Luque de Castro and M. Valcárcel, *Talanta*, 1995, **42**, 755.
253. I. Papaefstathiou, M.D. Luque de Castro and M. Valcárcel, *Fresenius' J. Anal. Chem.*, 1996, **354**, 442.
254. D.W. Bryce, A. Izquierdo and M.D. Luque de Castro, *Anal. Chem.*, 1997, **69**, 844.
255. J.L. Gómez-Ariza, E. Morales, D. Sánchez-Rodas and I. Giráldez, *TrAC*, 2000, **19**, 200.
256. W. Skawara, J. Dudek, E. Sterlinska, J. Chwastowska and L. Pszonicki, *Rap. IChTj.*, *Ser. A*, 1998, **1/98**, 1.
257. B. Wampfler and M. Rosslein, *Accredit. Qual. Assur.*, 1998, **3**(11), 468.
258. M. Gadner and A. Gunn, *Anal. Commun.*, 1997, **34**, 245.
259. J. De Boer and F. Smedes, *Mar. Pollut. Bull.*, 1997, **35**, 93.
260. J. Snell, J. Qian, M. Johansson, K. Smit and W. Frech, *Analyst*, 1998, **123**, 905.
261. C. Feldman, *Anal. Chem.*, 1979, **51**, 664.
262. A.A. Al-Sibaai and A.G. Fogg, *Analyst*, 1973, **98**, 732.
263. J. Agget and M.R. Kiegman, *Analyst*, 1987, **112**, 153.
264. G.E. Hall, J.C. Pechat and G. Gauthier, *J. Anal. At. Spectrom.*, 1999, **14**, 205.
265. M.L. Peterson and R. Carpenter, *Mar. Chem.*, 1983, **12**, 295; J.G. Sanders, *Mar. Chem.*, 1985, **17**, 329.
266. F.V. Vidal and V.M.V. Vidal, *Mar. Biol.*, 1980, **60**, 1.
267. E.A. Crescelius, N.S. Bloom, C.E. Cowan and E.A. Jenne, *Speciation of Selenium and Arsenic in Natural Waters and Sediments*, Vol. 2, in *Arsenic Speciation*, EPRI, Battelle Northwest Laboratories, Washington, 1986.
268. M.O. Andreae, *Limnol. Oceanogr.*, 1979, **24**, 440.
269. A. Chatterjee, D. Das, B.K. Mandal, T.R. Chowhury, G. Samanta and D. Chakraborti, *Analyst*, 1995, **120**, 643.
270. R.K. Anderson, M. Thomson and E. Culbard, *Analyst*, 1986, **111**, 1153.
271. M.A. Palacios, M. Gómez, C. Cámara and M.A. López, *Anal. Chim. Acta*, 1997, **340**, 209.
272. J.T. van Elteren and Z. Slejkovec, *J. Chromatogr. A*, 1997, **789**, 339.
273. R. Muñoz Olivas, Ph. Quevauviller and O.F.X. Donard, *Fresenius' J. Anal. Chem.*, 1998, **360**, 512.
274. M. Sanz-Alaejos and C. Díaz-Romero, *Chem. Rev.*, 1995, **95**, 227.

CHAPTER 4

Aspects of the Threshold Limit Concept

BERND NEIDHART

1 Introduction

Threshold limit values (TLVs) are statutory, quantitatively stated, restrictive levels for harmful substances, radiation and noise in the eco-compartments soil, water and air, which should prevent health risks and annoyances and avoid, or at least reduce to an acceptable degree, possible damage to goods and to ecosystems. In this sense, TLVs are instruments of health and ecopolitics which serve the application of the principle of precaution.[1] In relation to the following, the term precaution stands for the limitation of the emissions and immissions of harmful substances with the highest degree of responsibility possible.

TLVs are set. This demands reasons. One of the tasks of science is to provide these reasons. TLVs are scientifically based on toxicology and epidemiology in which scientific methods are applied to establish a causality between the concentration of a harmful substance and its risk for humans or the ecosystem. These methods are part of the so-called risk assessment or risk analysis procedures[2] which can be divided into four phases:

- risk identification
- dose-response analysis
- exposure analysis
- risk characterization

However, before setting a TLV, the entity to be protected must be unequivocally defined. In human toxicology this depends on the population affected, e.g.[3]

- the worker in the work place (occupational health)
- consumers (food)
- the general public (environment)

With this background the possible exposure paths have to be defined:

- oral

- inhalable
- dermal

and the predominant exposure frequencies have to be considered, in particular:

- a single/rare exposure (in the case of an accident)
- regular, partly long-term exposure (*e.g.* the application of pesticides)
- or continuous, long-term exposure (*e.g. via* the ecosystem).

A precondition for setting a TLV is that sufficient toxicological and/or occupational-medical or industrial-hygienic experience in handling the possibly harmful substance exists. In assessing the risks, experience with humans has principally the highest priority, *e.g.* in comparison to experiments with animals.[4]

The matter of harmful substances and their identification and quantification is the domain of analytical chemistry and thus it becomes evident that the TLV concept is embedded in the field of interaction between science, law and politics. On the one hand, the risks to humans and the ecosystem have to be identified and detected, whilst on the other hand these risks have to be assessed and decisions made, in order to prevent danger and limit risk. These decisions should be rational, *i.e.* they should be comprehensible, reasonable and based on facts.

Imperative reasons are *per se* insufficient. In order to ensure that TLVs are observed in practice, directives for measuring procedures are indispensable, and analytical quality assurance systems (AQA) are needed in order to establish comparability of results. If, for example, within the EU the procedures for monitoring the observance of TLVs are not strictly regulated, individual member states could take advantage and develop different, national regulations with the result that judgements would not be comparable or would only be of limited value. Therefore, the setting of TLVs has to be strongly coupled to demanding, and well defined, analytical methods.[5]

The concepts of AQA, which were developed in the wake of the harmonization of the European market and in connection with the globalization of the world's major trading zones, have now been formalized *via* the appropriate directives and norms (EN ISO/IEC 17025). Criteria for analytical quality, designed to ensure comparability of analytical results, have been established. Use of these criteria allows comparability to be achieved *via* the traceability of results to national or international standards, through an unbroken chain of comparisons. Within the framework of an AQA it is essential to be able to identify, unequivocally, the corresponding sample to which a particular, high quality analytical result, pertains; it is proposed to define this criterion as trackability. Validation remains the central task in the development of an analytical method whose analytical capability in specific applications can be estimated with the aid of the measurement uncertainty. Finally, proficiency testing serves to demonstrate the comparability in terms of scatter of the results, *e.g.* in round-robin tests.[6]

From all this it becomes clear that the threshold limit concept has five aspects:[7]

- analytical
- toxicological/medical

- legal, political and administrative
- psychological and sociological
- philosophical

In the following, we will focus on the analytical and the legal aspects of the threshold limit concept, because of the serious differences between science and law that often lead to problems in the courtroom. The three basic differences are:[8]

- science is digital – it focuses on measurement; law is analogue – it depends on precedent
- science is predictive, general and replicable; law is retrospective and particular
- science is objective and universal; law is normative and contingent

Integrating science into legal decision-making involves 'integrating a predictive model limited to variables that can be generalized into a retrospective process that seeks an understanding of the significance of difference'.

2 Analytical Aspects

2.1 Importance of Speciation

International legislation concerned with trace elements in food, in the environment or in occupational health bases most of its regulations on the total element content that is frequently given as maximum limits or guideline levels. In contrast, only a few regulations take into account the (molecular) species of the elements.[9] This, however, is of special importance for the heavy metals and some metalloids, because their toxicological potential is strongly determined by their oxidation state and/or the form of chemical binding. Due to the fact that bioavailability depends upon either special, species-dependent mechanisms of uptake or simply on solubility, which for biological materials increases with the lipophylic character of the respective compounds, speciation must become an indispensable part of the threshold limit concept.

2.2 Sampling

In connection with the TLV concept, sampling has two important aspects. One is concerned with speciation, for which it is essential that the information on the chemical species survive both the sampling procedure and the subsequent analytical method. This is by no means a trivial issue and demands a large effort on the part of the analyst according to the state of his/her art. Nevertheless, this situation is also a great challenge for analytical chemistry as a scientific discipline.

Of comparable importance is the representativeness of sampling. In analytical science, measurements are not usually made on the total amount of the material of interest, but on a much smaller quantity, the sample, which is selected from the target in some manner. As a consequence, metrologists in chemistry have hitherto concentrated on the analytical process in isolation. For the end-user of the data, however, the measurement of interest is the concentration of the analyte in the target. Hence, the uncertainty relevant to the end-user should include the uncertainty contribution introduced by preparing the laboratory sample from the target.[10]

Samples taken from the same target will usually vary in composition, both between one another and with respect to the average composition of the target, partly because of the heterogeneity of the target but also because of shortcomings in the sampling procedure, such as possible contamination, loss of analyte or the use of an inappropriate sampling procedure.

2.2.1   *Time-dependent Inhomogeneous Populations*

In the case of populations which are inhomogeneous in time, *e.g.* gases (fuel gas) and liquids (waste water) time-dependent sampling must be planned and performed.[11] In the ideal case, sampling would be continuous and thus correspond to an optimum frequency of spot checks. As a rule, the results of continuous measurements have low uncertainties. However, such measurements only make sense if the population is spatially homogeneous. This demand is generally fulfilled for populations of gases and single-phase liquids where inhomogeneity in time can be neglected during sampling because it is determined through the measurement.

In many situations, continuous measurements are not available or at least not available with the accuracy required, or the necessary systems are not installed for economic reasons. Consequently, discontinuous measurement procedures are very often used. Spot checks yield collectives of a limited number of more or less independent single measurements, which can only provide limited information. The quality of the information is dependent on the number of measurements. The larger the number of measurements, the lower is the standard deviation of the result. The standard deviation plays an important role when the concentration determined is close to the TLV. In this case an unequivocal decision as to whether the result lies below or above the limit is no longer possible.[12]

Example: The TLV for a harmful substance is set to $0.05\,\mathrm{mg\,m^{-3}}$. The analysis of three spot checks results in a mean concentration of $0.047\,\mathrm{mg\,m^{-3}}$. The standard deviation is $0.005\,\mathrm{mg\,m^{-3}}$ for a confidence level of 95%. The true concentration thus lies with a probability of 95% between 0.052 and $0.042\,\mathrm{mg\,m^{-3}}$. Only if the concentration is determined to lie below 0.045 or above $0.055\,\mathrm{mg\,m^{-3}}$ can adherence to or transgression of the TLV be stated.

2.2.2   *Spatially Inhomogeneous Populations*

In the case of populations that are spatially inhomogeneous, continuous sampling is not possible.[11] The real concentrations can only be determined or estimated from the analyses of spot checks. Samples which are spatially inhomogeneous stem as a rule either from heterogeneous solid targets (soil) or from targets having a complex matrix (sludge); the same problem arises in food analysis. In these cases the overall problem is not only associated with the sampling procedure but also, very often, with the sample preparation and the final analysis. Measurement of a single spot check sample is not acceptable for decision-making.

In practice, however, second or further spot check samples are only taken and analysed if the analyte concentration in the first sample lies above the TLV. In cases where the first measurement is favourable, *i.e.* for a producer, further analyses are not performed although the standard deviation of the result cannot exclude a transgression of the TLV with an acceptable level of probability.

Example: Nuts from areas contaminated with radioactivity by the Chernobyl accident were analysed according to this concept. Only if the radioactivity of a spot check sample turned out to lie above the specified TLV was a second sample taken to test whether the first result was an outlier or not. If the radioactivity in the first spot check sample was below the TLV, however, further analyses were not performed and thus it was not checked whether the low result was an outlier.

Repeated measurements of more spot check samples of the same population in principle reduce the uncertainty caused by systematic errors. Reliability can only be achieved *via* a large number of measurements on the identical sample performed by different laboratories applying different and independent methods. For samples having considerable variability in matrix composition (*e.g.* soil) reliability in the results can only be achieved by the use of a CRM. Unfortunately, for most of the problems to be solved, suitable CRMs are not available.

2.3 Analytical Quality Criteria

As will become clear in the following section, the analytical quality criteria, reproducibility, repeatability[13] and measurement uncertainty, as well as limit of detection and limit of determination,[14] are all indispensable pieces of information for the decision-makers in court. As a consequence, analytical chemists are coming under increasing pressure to demonstrate the quality of their results. This can be achieved by quantifying the above mentioned criteria, among which measurement uncertainty is most useful.[15] Associated with uncertainty of measurement are confidence limits. In quantitative measurements, a scientist can never be 100% confident that the true result lies within the reported range, since there is always a small, but finite, chance that the true value lies outside the reported range. This has to be taken into account when analytical results are taken as a basis for legal judgements.[16]

3 Legal Aspects

Neidhart *et al.* have published an article on 'Analytical and legal aspects of the threshold limit value concept', which has a special focus on the bridge between the disciplines of analytical chemistry and law, and which, from a practical view, contains suggestions for dealing with TLVs in the courtroom.[17] In the following, this concept is presented in a condensed form.

3.1   General View

A statute has to be worded in a general way such that it may be applied to a majority of individual cases. Hence the use of indefinite terms is unavoidable. In practice, indefinite terms such as 'damaging', 'serious' or 'dangerous' need to be defined. The TLV concept thus provides criteria for administrative intervention and is therefore an important instrument in the implementation of environmental policy.

In the legal literature, justification for the TLV concept is founded on claims that its application generates legal certainty, allows more rapid judgements and, furthermore, holds out the promise of greater civil acceptance of official decisions.[18]

3.2   Legal Reasoning for TLVs

Legal decisions are founded on the principle that a particular factual situation is associated with a more general factual situation to which an appropriate legal consequence applies. In this way it is possible to infer, predictively, the legal consequences in an individual case, for which the factual situation is not explicitly stated in the law. Application of the law does, however, presume a concrete interpretation of the general terms used.[19]

A further problem is that it is difficult to estimate the consequences of the exposure of a worker to an occupational hazard, or of a pollution event on the environment. Demands for a ruling as to whether such events should be tolerated by the public are directed in the first instance to the public authority responsible, whose personnel do not usually have the necessary knowledge at their disposal for taking such decisions.[20] At the present time, the public authorities are completely overloaded, for example, with cases concerned with refuse dumps.

Since it is too expensive and time consuming to allow each particular case to be investigated by experts, TLVs are consulted instead.[21] *TLVs should make it possible for public authorities to take decisions without further recourse to expert knowledge.*

The public authorities therefore no longer have to find their own yardsticks for the decision-making process; by comparing the measured concentration value with the TLV they can simply establish the result. Instead of deciding between 'damaging' and 'non-damaging' the distinction is made on the basis of 'TLV exceeded' or 'TLV not exceeded'.[21]

A further consequence of the TLV concept is that provided the measured value is available, the administration's decision is predictable and consistent. *Threshold limit values thus generate legal certainty.*[22]

Moreover, the use of threshold limit values by public authorities serves to give their decision-making processes more transparency. Lists of threshold limit values are accessible to the public so that anybody can make the comparison between a measured concentration value and the corresponding TLV and will arrive at the same result as the authority.[20] *This generates public acceptance of rulings.*

3.3 Legal Reasoning and Analytical Reality

The TLV concept presumes the availability of the measured concentration value, with the manner in which the result is determined being only partly of concern.*
However, threshold limit values can only fulfil the demands placed upon them if their application takes account of the real circumstances of analytical chemistry. It is often impossible to establish unequivocally whether a threshold limit value has been exceeded, by comparison of a measured concentration value with the appropriate TLV. Investigation of the analytical chemistry problems associated with giving an answer to this question shows that the three legal demands made on the TLV concept, as described above, can only be fulfilled within certain limitations.

3.4 Legal Certainty

'At best, the comparison of well resolved numerical values allows the legal consequences to be predicted exactly. It is the predictability of the legal system that generates certainty in law for the public at large'.[18] This statement presumes *that measurements are reproducible and that the measured value is always an exact reflection of the true concentration.* However, the analyst can only establish a range or domain in which, for a particular chosen statistical probability, the true concentration lies. Supposing that in the discontinuous monitoring of the emissions from a technical plant three independent measurements are made. For a statistical confidence level of 90%, it is first possible to demonstrate adherence to or transgression of the TLV if the average value from the three test samples is less than 64% or greater than 35 % of the TLV respectively.

For a given concentration measurement, the TLV concept enables the decision taken by a public authority or the ruling of a court of law to be predicted. If the measured emissions correspond to concentrations between 64% and 135% of the TLV, a situation which often arises as a result of the tendency on the part of the emitter to utilize as much of the allowance up to the TLV as possible, then the judgement becomes dependent on the random measurement result and is no longer reproducible or predictable in the legal sense. Legal uncertainty arises, however, not only from the absence of suitable standards for the assessment but is also a direct consequence of the analytical difficulties involved in the determination of the concentration values on which the assessment is based. It is thus ascertained that use of the TLV concept can only guarantee legal certainty if the

---

* The TA-Luft[22] specifies exact instructions for measurement procedures.

measured concentration value is both reproducible and accurate in the analytical chemistry sense.

The legal uncertainty arising from the absence of TLVs because of the lack of a uniform set of yardsticks for decision-making is simply transferred into the analytical chemistry regime. Non-reproducibility of the measurements means that the judgement no longer depends on the arbitrariness of the particular public authority but rather on the haphazard outcome of the analytical procedure/method involved.

In assessing the results of chemical analysis, consideration has to be given to the circumstances in which the data are generated. Thus, legal certainty within the framework of the TLV concept can only be achieved, if the experimentally determined concentration values are subjected to a complex assessment procedure prior to making comparisons with TLVs.

3.5   More Rapid Judgement

Legal certainty requires that a comprehensive assessment of the analytical results be performed by specialists. This, of course, is not in keeping with the expectation that public authorities lacking qualified personnel should be able to make more rapid judgements with the help of the TLV concept. Before using a TLV, it is necessary to decide whether it is still up-to-date and whether due consideration to the danger to be assessed in the case concerned has been given during its specification.[19] Such decisions can only be made by experts.

Prior to making comparisons with the TLV, the measurement uncertainty associated with each individual analytical result must be considered. It has to be established whether the analytical result to be compared should be corrected by adding or subtracting the measurement uncertainty. Since a part of the uncertainty depends on the standard deviation of the measurement, the magnitude of the uncertainty is dependent on the number of measurements. For this reason the number of measurements to be performed should also be stipulated.

In order to enable the TLV concept to be applied quickly and easily, such crude simplifications have to be made that the individual case can no longer be adequately evaluated. Application of the TLV concept as a substitute for expert advice must presume first a refinement and extension of existing TLV lists and secondly, continuous monitoring of their currency.

The degree of cost and effort needed for this would probably exceed that required in seeking the opinions of individual experts for each special case.

3.6   Public Acceptance

It is presumed that use of the TLV concept will ensure that decisions taken by public authorities on whether a particular contamination be declared permissible will be accepted by the public. Civil acceptance will disappear, however, as soon as it becomes apparent that damage has been caused by a substance, even though its concentration is less than the TLV, regardless of whether the cause of this be an inadequately specified TLV or an inaccurate analytical measurement. Given the known relationships between speciation and toxicity, there must be concern over the lack of TLVs described in terms of chemical species.

From the standpoint of the analytical chemist, no guarantee can be given that the TLV has not actually been exceeded in each and every case, even when the measured concentration lies below the TLV. On the other hand, improvements being made to analytical methods mean that it is becoming easier to detect substances causing damage.

4 Suggestions for Dealing with TLVs

At the present time, it is clearly not possible to do without the TLV concept, first because there is no apparent alternative and secondly because the requirement that TLVs be specified does indeed promote developments which are highly desirable. For example the stipulation of TLVs leads to further developments in waste retention technologies. Moreover, as a result of statutory specifications, industries are compelled to include waste management considerations in their strategic planning.[23]

In specifying a TLV and the corresponding analytical method, it is not only the medical–toxicological and analytical chemistry factors which are important, but also the orderly nature of the decision-making process since, only *via* the latter can a legally binding system be established.[18] This procedure, however, does not allow for the rapid changes needed to keep up with the pace of technical progress. In most cases, TLVs and the specified measurement procedures, are already out-of-date by the time they come into force.[19]

Regardless of the analytical measurement practice there is always a domain in which, as a result of the uncertainty in the measured concentrations, no firm decision can be made. The particular analytical method used only influences the magnitude of this domain. Making additional measurements cannot resolve this problem completely. A few doubtful cases will always remain in which the measurement uncertainty increases all the more, the stronger the temporal or spatial variation of the analyte concentration. Since the producers of emissions may escape the consequences arising from any inaccuracy in the measurement, it is usual to base decisions on the arithmetic mean value. However, this can produce situations in which even the worst offenders remain unpunished. Moreover, the measurement uncertainty is usually taken into account by subtracting the 'range of scatter' from the concentration so determined.[24] This, however, can lead to a situation in which, even though the actual pollutant concentration exceeds the TLV by an amount corresponding to twice the magnitude of the range of scatter, no contravention of the TLV can be proven.

This practice means that it is convenient for the producers of emissions to apply inaccurate methods and to perform the minimum number of measurements, although the reliability of the results increases with each additional measurement.

This kind of practice may be inhibited by laying down precise guidelines for the measurement procedure for each individual case. The application of more modern, more accurate analytical methods is thus hindered, because the so-called 'permissible error' stipulated by the directive in force is coupled to a particular analytical method. It therefore seems questionable whether a legally binding decree is an appropriate instrument for the detailed specification of analytical chemistry procedures.

A number of suggestions have been made which attempt to deal with the problems that arise when scientific means are used to settle legal matters. Nisipeanu[25] has suggested that in addition to the TLV, whose observance should continue to be checked *via* comparison with the arithmetic mean of all the measured values, an absolute maximum value should be introduced above which no single measured value should lie. The introduction of peak or maximum values is an attempt to take into account the fact that an analytical measurement does not produce an accurate value, but rather a more or less broad range of concentration in which the value sought is to be found. Its implementation would mean that criminal punishment of violations of the TLV is facilitated, but at the same time it would not encourage the use of more precise analytical procedures, since the peak value is specified without consideration being given to the range of scatter of the particular analytical method.

An incentive for the use of more precise analytical procedures could be provided if the measurement uncertainty of the method applied in a particular case is taken into account *via* the specification of a threshold limit range (TLR), and if measures could be taken against the producers of emissions, even when the measured concentration exceeds a value given by the TLV minus the measurement uncertainty of the method. Such an approach would have the advantage, compared with present practice, that the use of more precise analytical methods would be encouraged, since a smaller uncertainty in the measurement would mean that a higher concentration value could be tolerated without the risk of legal consequences.

In addition, this approach could make it worthwhile for the producers of emissions to perform more measurements than are absolutely necessary, and to use the most up-to-date analytical methods. In certain cases, it could even be profitable for plant operators to go over to continuous monitoring methods, which provide results that are usually closer to the real concentrations than those of discontinuous methods. This change would make the legal ratification of the use of standard methods and statutory stipulation on their application unnecessary, since the producers of emissions would select the most accurate measurement methods on the grounds of self-interest. As a result, more modern methods that produce more accurate results could be used to demonstrate adherence to the TLV. Moreover, this would encourage the development and application of improved analytical methods and technologies for the prevention of emissions and clean processing techniques which produce less pollution. The analytical methods used in individual cases must, of course, be evaluated by experts.

The use of threshold limit ranges (TLR) would make the decision-making process for the public authority concerned just as simple as in the case of the TLV, but now without the adverse consequences of simplification on the legal certainty.

A further possibility would be to make it the responsibility of the producers of emissions to demonstrate that their emissions do not exceed the TLV. The consequences of such a system are similar to those suggested above. The measurement procedures/analytical methods would no longer have to be specified since it would be in the interests of all plant operators to choose the most accurate procedure/method and to perform enough individual measurements in order to prove, without doubt, adherence to the TLV. Misuse of the TLV would be discouraged since the measurement outlay would increase as the value of the emitted concentration approaches the TLV. The development of improved procedures/methods would be encouraged and these could be applied straight away without the need for a complicated bureaucratic standardization procedure. In order to assess the procedures/methods, it would indeed be necessary to use experts. A validated standard specification for analytical methods could be of assistance in this assessment and would make it understandable to non-experts.

The current prejudice against plant operating with continuous monitoring *vis á vis* discontinuous measurements of emissions could be compensated for by the introduction of a system based on reversal of the burden of proof. Although continuous measurements produce superior results at a cost that is usually higher than in the case of discontinuous measurements, under the present system this has been to the detriment of plant operators performing continuous measurements as a result of the higher accuracy of the measurements.

5 Conclusion

In many cases the consequences of technological progress and effects on the environment cannot be judged with the desirable legal certainty. Difficulties therefore arise in the application of legal norms which demand the proof of a causal correlation. In order to overcome these problems, the legal profession is urgently requested to recognize the structure of scientific, technical and medical statements and incorporate these into the categories of legal evidence. A move in this direction will mean that the findings of scientific theory as well as statistical aspects will have to be taken into account. This could be of use especially for the description of error sources which threaten the legal assessment of non-legal information.[26]

The stipulation of TLVs and the association of legal consequences with their contravention involves the tacit presumption that procedures for monitoring adherence to the TLV are both possible and are applied. However, monitoring is an analytical chemistry problem and hence the limitations of the TLV concept are determined by the realities of analytical chemistry.

Analytical chemists should take the lead in this process and should interpret the still confused situation as a challenge to return responsibility to the analyst. The only prerequisite for success in this task is a willingness based on sound analytical knowledge.

6   References

1. P.M. Wiedemann, in *Chemische Grenzwerte*, eds. P. Janich, P.C. Thieme and N. Psarros, Wiley-VCH, Weinheim, 1999, pp. 67ff.
2. *Umweltgutachten 1996*, eds. Der Rat von Sachverständigen für Umweltfragen, Verlag Metzler-Poeschel, Stuttgart, 1996, ISBN 3-8246-0545-7.
3. H.P. Gelbke, in *Chemische Grenzwerte*, eds. P. Janich, P.C. Thieme and N. Psarros, Wiley-VCH, Weinheim, 1999, pp. 67ff.
4. H.M. Bolt, in *Chemische Grenzwerte*, eds. P. Janich, P.C. Thieme and N. Psarros, Wiley-VCH, Weinheim, 1999, pp. 67ff.
5. M. Schröder, in *Chemische Grenzwerte*, eds. P. Janich, P.C. Thieme and N. Psarros, Wiley-VCH, Weinheim, 1999, pp. 67ff.
6. B. Neidhart, *Fresenius' J. Anal. Chem.*, 1997, **216**, 1'97 (p. M23–M25).
7. P. Janich, in *Chemische Grenzwerte*, eds. P. Janich, P.C. Thieme and N. Psarros, Wiley-VCH, Weinheim, 1999, pp. 67ff.
8. R. Baum, *Chem. Eng. News.*, 1997, 38.
9. T. Berg and E.H. Larsen, *Fresenius' J. Anal. Chem.*, 1999, **363**, 431.
10. M. Thompson, *Accred. Qual. Assur.*, 1998, **3**, 117/118.
11. A. Schmolke, Diploma Thesis, University of Marburg, 1994.
12. B. King, *Accred. Qual. Assur.*, 1999, **4**, 27.
13. J. Fleming, B. Neidhart, Ch. Tausch and W. Wegscheider, *Accred. Qual. Assur.*, 1996, **1**, 41.
14. J. Fleming, H. Albus, B. Neidhart and W. Wegscheider, *Accred. Qual. Assur.*, 1997, **2**, 51.
15. EURACHEM, *Quantifying Uncertainty in Analytical Measurements*, EURACHEM, P.O. Box 46, Teddington, UK, 1995.
16. R. Treble, VAM Bulletin, No 20, LGC, Teddington, UK, 1999.
17. B. Neidhart, W. Mummenhoff, A. Schmolke and P. Beaven, *Accred. Qual. Assur.*, 1998, **3**, 44.
18. C. Schrader, *Natur und Recht*, 1989, 288.
19. C. Gusy, *Administrativer Vollzugsauftrag und justizielle Kontrolldichte im Recht der Technik*, DVBI, 1987, 497.
20. Judgement of the First Senate of the Federal Administrative Court from 17th February 1978 (1 C102.76) BvwGE 55, 250, 257.
21. N. Hoppe and M. Beckmann, *Umweltrecht*, C.H. Beck, Munich, 1989, §3 Rn, 2 (p. 40).
22. The first general administrative regulation concerning the German Federal Air Pollution Control Act (Technical preservation of clean air) from 27th February 1986, *GMBl*, 1986, **37**(7), 95.
23. G. Zenk, *Öko-Audits nach der Verordnung der EU*, Gabler, Wiesbaden, 1995.
24. B. Peters, *Natur und Recht*, 1989, 167.
25. P. Nisipeanu, *Natur und Recht*, 1988, 225.
26. W. Mummenhoff, *Erfahrungssätze im Beweis der Kausalität*, Carl Heymanns Verlag KG, Köln, Berlin, Bonn, München, 1997.

CHAPTER 5

Considerations of the Legislative Aspects of the Data Quality Requirements in Trace Element Analysis

ROGER WOOD AND HELEN CREWS

1 Introduction

In food laboratories a considerable range of analyses, from proximate measurements, *i.e.* water, protein and fat in the percentage concentration range, to trace determinations (normally in the $0.01-1000 \, \text{mg} \, \text{kg}^{-1}$ range), may be required. In addition 'food' should not be regarded as a single matrix but as a complex series of matrices, *e.g.* 100% oil, 100% carbohydrate or 100% protein with all variants of composition in between (*i.e.* the so-called 'food triangle' concept). This paper stresses the legislative and contractual requirements with respect to analytical data quality that laboratories undertaking such work are required to observe. These requirements apply to all analyses including the determination of total trace element concentrations and their chemical species. Since the chemical form of an element can dictate whether or not elevated levels of it may pose a threat to human health (*e.g.* organic *versus* inorganic arsenic; see chapters by Urieta *et al.* and by Larsen and Berg in this book), an ability to provide accurate speciation data is increasingly important to national and international trade. The following is not intended to be a comprehensive guide to the many legislative aspects of trace elements in food but does attempt to highlight the main requirements for providing data of acceptable quality of which practising analysts should be aware.

2 Legislative Requirements

Formal laboratory quality and methodology requirements for the foodstuff sector have been adopted by the European Union and the Codex Alimentarius

Commission. In essence, laboratories will have to:

- use validated methods of analysis (*i.e.* have been validated through collaborative trials)
- become accredited to ISO/IEC Guide 25 [now superseded by ISO standard 17025]
- participate in proficiency testing schemes
- introduce appropriate internal quality control procedures.

2.1 The European Union

For analytical laboratories in the food sector there are legislative requirements regarding analytical data that have been adopted by the European Union. In particular, methods of analysis have been prescribed by legislation for a number of foodstuffs since the UK acceded to the European Community in 1972. However, the Union now recognises that the competency of a laboratory (*i.e.* how well it can use a method) is equally as important as the 'quality' of the method used to obtain results.

2.1.1 The Council Directive on the Official Control of Foodstuffs

The Council Directive on the Official Control of Foodstuffs, which was adopted by the Community in 1989,[1] looks forward to the establishment of laboratory quality standards, by stating that 'In order to ensure that the application of this Directive is uniform throughout the Member States, the Commission shall, within one year of its adoption, make a report to the European Parliament and to the Council on the possibility of establishing Community quality standards for all laboratories involved in inspection and sampling under this Directive' (Article 13).

Following that, the Commission, in September 1990, produced a report which recommended establishing Community quality standards for all laboratories involved in inspections and sampling under the OCF Directive. Proposals on this have now been adopted by the Community in the Directive on Additional Measures Concerning the Food Control of Foodstuffs[2] and the key statements are given below.

2.1.2 Directive on Additional Measures Concerning the Control of Foodstuffs (AMFC)

Article 3 of the AMFC Directive states:

1 Member States shall take all measures necessary to ensure that the laboratories referred to in Article 7 of Directive 89/397/EEC[1] comply with the general criteria for the operation of testing laboratories laid down in European standard EN 45001[3] supplemented by Standard Operating Pro-

cedures and the random audit of their compliance by quality assurance personnel, in accordance with the OECD principles Nos. 2 and 7 of good laboratory practice as set out in Section II of Annex 2 of the Decision of the Council of the OECD of 12 Mar 1981 concerning the mutual acceptance of data in the assessment of chemicals.[4]

2   In assessing the laboratories referred to in Article 7 of Directive 89/397/EEC Member States shall:

(a) apply the criteria laid down in European standard EN 45002;[5] and

(b) require the use of proficiency testing schemes as far as appropriate.

Laboratories meeting the assessment criteria shall be presumed to fulfil the criteria referred to in paragraph 1. Laboratories which do not meet the assessment criteria shall not be considered as laboratories referred to in Article 7 of the said Directive.

3   Member States shall designate bodies responsible for the assessment of laboratories as referred to in Article 7 of Directive 89/397/EEC. These bodies shall comply with the general criteria for laboratory accreditation bodies laid down in European Standard EN 45003.[6]

4   The accreditation and assessment of testing laboratories referred to in this article may relate to individual tests or groups of tests. Any appropriate deviation in the way in which the standards referred to in paragraphs 1, 2 and 3 are applied shall be adopted in accordance with the procedure laid down in Article 8.'

Article 4 of the AMFC Directive states:

'Member States shall ensure that the validation of methods of analysis used within the context of official control of foodstuffs by the laboratories referred to in Article 7 of Directive 89/397/EEC comply whenever possible with the provisions of paragraphs 1 and 2 of the Annex to Council Directive 85/591/ EEC of 23 December 1985 concerning the introduction of Community methods of sampling and analysis for the monitoring of foodstuffs intended for human consumption.'[7]

2.2   Codex Alimentarius Commission

2.2.1   *Guidelines for the Assessment of the Competence of Testing Laboratories Involved in the Import and Export Control of Foods*

The decisions of the Codex Alimentarius Commission (CAC) are becoming increasingly important because of the acceptance of Codex Standards in the World Trade Organisation agreements. They may be regarded as being semi-legal in status. Thus, on a world-wide level, the establishment of the World Trade Organisation (WTO), and the formal acceptance of the Agreements on the Application of Sanitary and Phytosanitary Measures (SPS Agreement) and, Technical Barriers to Trade (TBT Agreement), has dramatically increased the status of Codex as a body. As a result, Codex Standards are now seen as *de facto* international standards and are increasingly being adopted by reference into the food law of both developed and developing countries.

Because of the status of the CAC described above, the work that it has carried out in the area of laboratory quality assurance must be carefully considered. One of the CAC Committees, the Codex Committee on Methods of Analysis and Sampling (CCMAS), has developed criteria for assessing the competence of testing laboratories involved in the official import and export control of foods. These were recommended by the Committee at its Twenty-first Session in March 1997[8] and adopted by the Codex Alimentarius Commission at its Twenty-second Session in June 1997.[9] They mirror the EU recommendations for laboratory quality standards and methods of analysis.

The guidelines provide a framework for the implementation of quality assurance measures to ensure the competence of testing laboratories involved in the import and export control of foods. They are intended to assist countries in their fair trade in foodstuffs and to protect consumers.

2.2.2 CAC Criteria for Laboratories Involved in the Import and Export Control of Foods

The criteria for laboratories involved in the import and export control of foods, now adopted by the CAC are:

- to comply with the general criteria for testing laboratories laid down in ISO/IEC Guide 25: 1990 General requirements for the competence of calibration and testing laboratories[10] [*i.e.* effectively accreditation]
- to participate in appropriate proficiency testing schemes for food analysis which conform to the requirements laid down in The International Harmonised Protocol for the Proficiency Testing of (Chemical) Analytical Laboratories,[11] [already adopted for Codex purposes by the CAC at its 21st Session in July 1995]
- to use, whenever available, methods of analysis which have been validated according to the principles laid down by the CAC, and
- to use internal quality control procedures, such as those described in the Harmonised Guidelines for Internal Quality Control in Analytical Chemistry Laboratories.[12]

In addition, the bodies assessing the laboratories should comply with the general criteria for laboratory accreditation, such as those laid down in the ISO/IEC Guide 58:1993: Calibration and testing laboratory accreditation systems – General requirements for operation and recognition.[13]

Thus, as for the European Union, the requirements are based on accreditation, proficiency testing, the use of validated methods of analysis and, in addition, the formal requirement to use internal quality control procedures that comply with the Harmonised Guidelines. Although the EU and Codex Alimentarius Commission refer to different sets of accreditation standards, the ISO/IEC Guide

25:1990 and EN 45000 Series of Standards are similar in intent. It is only through these measures that international trade will be facilitated, and the requirements to allow mutual recognition to be fulfilled, will be achieved.

3 An Example of National Requirements for Food Chemical Surveillance

3.1 UK Food Standards Agency Surveillance Requirements

The UK Food Standards Agency undertakes food chemical surveillance exercises and it has developed information for potential contractors on the analytical quality assurance requirements that are necessary for these exercises. These requirements are outlined below. They emphasise the need for a laboratory to produce and report data of appropriate quality.

The requirements are divided into three parts.

Part A: quality assurance requirements for surveillance projects provided by potential contractors at the time tender documents are completed and when commissioning a survey. Here information is sought on:

- The formal quality system in the laboratory if third party assessed (*i.e.* if UKAS accredited or GLP compliant)
- The quality system if not accredited
- Proficiency testing
- Internal Quality Control
- Method Validation

Part B: information to be defined by the Food Standards Agency customer once the contract has been awarded and to be agreed with the contractor, *e.g.* the sample storage conditions to be used, the methods to be used and a copy of Standard Operating Procedures (SOPs) where accredited, the internal quality control (IQC) procedures to be used, the measurement limits, (*i.e.* limit of detection [LOD], limit of determination/quantification [LOQ] and reporting limits *etc.*), the measurement uncertainty *etc.*

Part C: information to be provided by the contractor on an on-going basis once a contract is awarded and to be agreed with the customer to ensure that the contractor remains in 'analytical control'.

3.2 Consequence of the Legislative Requirements

As a result of the above, very strict quality standards are imposed on analytical laboratories, particularly in the food sector. The customers of such laboratories expect them to meet these standards. These standards apply to *all* types of analyses.

There are a number of other 'quality initiatives' currently taking place on an international basis. These include:

- IUPAC/ISO/AOAC Internal Quality Control Guidelines

- Harmonisation of Reporting of Test Results for Recovery Factors
- Consideration and Development of In-House Method Validation Protocols
- CCMAS Discussion Paper and Criteria for Evaluating Acceptable Methods of Analysis for Codex Purposes

These are discussed further below.

4   IUPAC/ISO/AOAC Internal Quality Control (IQC) Guidelines

International Guidelines on IQC[12] have been developed and internally adopted (see Codex Requirements for laboratories stipulated above). The requirements follow good practice, and are given below.

4.1   Quantitative Chemical Analysis

Short (e.g. n < 20) frequent runs of similar materials. The concentration range of the analyte in the run is relatively small, so a common value of standard deviation can be assumed. Insert a control material at least once per run. Plot either the individual values obtained, or the mean value, on an appropriate control chart. Analyse in duplicate at least half of the test materials, selected at random. Insert at least one blank determination.

Longer (e.g. n > 20) frequent runs of similar materials. Again a common level of standard deviation is assumed. Insert the control material at an approximate frequency of one per ten test materials. If the run size is likely to vary from run to run it is easier to standardise on a fixed number of insertions per run and plot the mean value on a control chart of means. Otherwise plot individual values. Analyse in duplicate a minimum of five test materials selected at random. Insert one blank determination per ten test materials.

Frequent runs containing similar materials but with a wide range of analyte concentration. It cannot be assumed that a single value of standard deviation is applicable. Insert control materials in total numbers approximately as recommended above. However, there should be at least two levels of analyte represented, one close to the median level of typical test materials and the other approximately at the upper or lower decile as appropriate. Enter values for the two control materials on separate control charts. Duplicate a minimum of five test materials, and insert one procedural blank per ten test materials.

Ad hoc analysis. The concept of statistical control is not applicable. It is assumed, however, that the materials in the run are of a single type, *i.e.* sufficiently similar for general conclusions on errors to be made. Carry out duplicate analysis on all of the test materials. Carry out spiking or recovery tests or use a formulated control material, with an appropriate number of insertions (see above), and with different concentrations of analyte if appropriate. Carry out blank determinations. As no control limits are available, compare the bias and precision with fitness for purpose limits or other established criteria.

5 Harmonised Guidelines for the Use of Recovery Information in Analytical Measurement

The IUPAC Interdivisional Working Group on Quality Assurance has developed Harmonised Guidelines on the use of Recovery Information.[14] Their publication has highlighted the whole issue of the use of recovery information.

The following recommendations are made regarding the use of recovery information in the Guidelines.

1   In general results should be corrected for recovery, unless there are overriding reasons for not doing so. Such reasons would include the situation where a limit (statutory or contractual) has been established using uncorrected data, or where recoveries are close to unity. It is, however, of over-riding importance (a) that all data, when reported, should be clearly identified as to whether or not a recovery correction has been applied and (b) that, if a recovery correction has been applied, the amount of the correction and the method by which it was derived should be included with the report. This will promote direct comparability of data sets. Thus, in all situations, correction functions should be established based on appropriate statistical considerations, documented, archived and available to the client.

2   Recovery values should always be established as part of method validation, whether or not recoveries are reported or results are corrected, so that measured values can be converted into corrected values and *vice versa*.

3   When the use of a recovery factor is justified, the method of calculation should be given in the method.

4   IQC control charts for recovery should be established during method validation and be used in all routine analysis. Runs giving recovery values outside the control range should be considered for re-analysis in the context of acceptable variation, or the results reported as semi-quantitative.

6 In-House Method Validation

It is now accepted that a laboratory must take appropriate quality assurance measures to ensure that it is capable of, and does provide, data of the required quality. Such measures include:

- using validated methods of analysis
- using internal quality control procedures
- participating in proficiency testing schemes, and
- becoming accredited to an International Standard, normally the ISO/IEC Guide 25.

Aspects of the above have been previously addressed by the IUPAC Interdivisional Working Party on Harmonisation of Quality Assurance Schemes for

Analytical Laboratories, specifically by preparing Protocols/Guidelines on method performance (collaborative) studies,[15] proficiency testing[11] and internal quality control.[12]

Thus it may be seen that method validation is one, albeit essential, component of the measures that a laboratory should implement to allow it to produce reliable analytical data. There is a continuing need for reliable analytical methods for use in determining compliance with national regulations as well as international requirements in all areas of analysis. The reliability of a method is determined by some form of a validation procedure.

'Full' validation for an analytical method usually comprises an examination of the characteristics of the method in an inter-laboratory method performance study (also known as a collaborative study or collaborative trial). In some sectors, most notably food, the requirement for control analysts to use methods which have been fully validated is prescribed by legislation.[2,16] Internationally accepted protocols have been established for the 'full' validation of a method of analysis by a collaborative trial, most notably the International Harmonised Protocol[15] and the ISO Procedure.[17] These protocols/standards require a minimum number of laboratories and test materials to be included in the collaborative trial fully to validate the analytical method. However, before a method is subjected to validation by a collaborative trial (*i.e.* to become fully validated) the method must undergo some validation within a single laboratory, usually in the laboratory which develops/modifies the analytical method. This validation may be regarded as 'in-house method validation'.

As stated above, the ideal validated method is one that has progressed fully through a collaborative study in accordance with international harmonised protocols for the design, conduct and interpretation of method performance studies. It is not practical or necessary to require that all analytical methods used in the analytical laboratory be assessed at the ideal level. Limiting factors for completing ideal multi-laboratory validation studies include high costs, lack of sufficient expert laboratories available and willing to participate in such studies, and overall time constraints. These considerations are critical, and indeed overbearing in a number of sectors. Thus 'in-house' method validation should be undertaken to:

1 ensure the validity of the method prior to the costly exercise of a formal collaborative trial,
2 ensure that validated methods are being used correctly when used by analysts prior to undertaking a particular study, and
3 provide evidence of the reliability of analytical methods if supporting trial data is not available, or where the conduct of a formal collaborative trial is not feasible for either practical or economic reasons.

6.1   Existing In-House Method Validation Protocols and Measurement Uncertainty Standards/Guides

A number of protocols/guidelines have been prepared on in-house method validation, most notably the following:

- A Protocol on the Validation of Chemical Analytical Methods developed by the Nordic Committee on Food Analysis,[18]
- A generic laboratory guide developed by EURACHEM produced by the UK Laboratory of the Government Chemist with the support of the UK Department of Trade and Industry Valid Analytical Measurement Initiative,[19]
- An Interlaboratory Analytical Method Validation Short Course developed by the AOAC International,[20]
- A Guide to the Validation of Methods developed by the Dutch Inspectorate for Health Protection,[21]
- A Guide to Analytical Quality Assurance in Public Analyst Laboratories prepared by the UK Association of Public Analysts,[22]
- A Guide to the Fitness for Purpose of Analytical Methods, prepared by a EURACHEM Working Group,[23]
- ISO *Guide to the Expression of Uncertainty in Measurement,*[24]
- The EURACHEM Guide to Quantifying Uncertainty in Analytical Measurement,[25]
- The NMKL Guide to the *Estimation and Expression of Measurement Uncertainty in Chemical Analysis,*[26] and
- Joint FAO/IAEA Expert Consultation, December 1997, on the Validation of Analytical Methods for Food Controls, the Report of which is available.[27]

These issues were also discussed in an International Workshop, held November 1999, under the auspices of IUPAC/IAEA/FAO and AOAC. The Workshop had the following objectives:

1 to enable participants to give information on their approaches to, and problems with, method validation within their particular sector of interest,
2 to develop general harmonised guidelines for the in-house validation of methods of analysis, these to emphasise the difficulties inherent with the traditional inter-laboratory method validation study, and thus to point the way towards the acceptance of in-house validated methods by regulatory authorities in particular, provided these are employed under defined conditions
3 to develop a practical approach to validation of methods of analysis of residues of pesticides and veterinary drugs and trace organic contaminants in foods.

7 Criteria Approach for Evaluating Acceptable Methods of Analysis for Codex Purposes

The necessity of stipulating specific methods of analysis in legislation is being questioned. Such methods are required to meet certain defined criteria. Typically values for the following should be determined: specificity, accuracy, precision, repeatability intra-laboratory (within laboratory), reproducibility inter-laboratory (within laboratory and between laboratories), limit of detection, sensitivity, practicability and applicability under normal laboratory conditions and other criteria which may be selected as required.

In addition, the following need to be considered:

1   the method selected should be chosen on the basis of practicability and preference should be given to methods which have applicability for routine use,
2   all proposed methods of analysis must have direct pertinence to the Codex Standard to which they are directed,
3   methods of analysis which are applicable uniformly to various groups of commodities should be given preference over methods which apply only to individual commodities,
4   official methods of analysis elaborated by international organisations occupying themselves with a food or group of foods should be preferred.

7.1   Method Criteria

There is now a move towards stipulating method criteria. This has been discussed extensively within both the EU and the CCMAS, where the procedure has now been accepted in principle. At recent sessions of the CCMAS papers have been discussed in which the arguments were given for amending the present Codex procedure, whereby the specified numeric values in Codex standards are determined using prescribed methods of sampling and analysis. The methods of analysis and sampling are elaborated and agreed through defined Codex procedures. It was stated that there were a number of criticisms to be made of this Codex procedure, in particular that the analyst is denied freedom of choice and thus may be required to use an inappropriate method in some situations; that the procedure inhibits the use of automation; and that it is administratively difficult to change a method found to be unsatisfactory or inferior to another currently available.

The Committee has accepted in principle an alternative approach whereby a defined set of criteria, to which methods should comply without specifically endorsing specific methods, should be adopted[1]. The Committee agreed the following points.

1   This 'criteria' approach gives greater flexibility than the present procedure adopted by Codex and, in the case of non-defining methods, eliminates the need to consider and endorse a series of Type III methods. The Committee recognised that the endorsement of many Type III methods for any specific determination does, in practice, rarely occur; that this reduces the effectiveness of the present Codex system for the endorsement of methods was appreciated.
2   In some areas of food analysis there are many methods of analysis that are available, which meet Codex requirements as regards method characteristics, but which are not considered by CCMAS and the Commission because of

time constraints on the Committee.

3 The adoption of a more generalised approach would ensure that such methods are brought into the Codex system and does not disadvantage developments being undertaken elsewhere in the analytical community.

4 It may be necessary to continue to prescribe a single Type II reference method for the dispute situation but the criteria approach could certainly be applied to the present Type III methods.

8 Conclusions

For all laboratories undertaking chemical analysis there are defined quality criteria that should be followed. Many of these are laid down through internationally accepted guidelines and protocols. These criteria apply not only to the more routine chemical analyses but also to specialist analysis such as trace element speciation. Many of these latter analyses are still used only in research and development studies. However, an informed awareness at an early stage of the requirements and criteria for methods, which may need to be used in future legislative applications, is not a bad thing.

In the future, legislation will favour a method performance based approach, and this will be applicable to the speciation area. It will ensure that method characteristics are well defined and evaluated.

9 References

1. European Union, *Council Directive 89/397/EEC on the Official Control of Foodstuffs*, O.J. L186 of 30.6.1989.
2. European Union *Council Directive 93/99/EEC on the Subject of Additional Measures Concerning the Official Control of Foodstuffs*, O.J. L290 of 24.11.1993.
3. European Committee for Standardization, *General Criteria for the Operation of Testing Laboratories – European Standard EN* 45001, Brussels, CEN/CENELEC, 1989.
4. Organisation for Economic Co-operation and Development, *Decision of the Council of the OECD of 12 Mar 1981 concerning the mutual acceptance of data in the assessment of chemicals*, Paris, OECD, 1981.
5. European Committee for Standardization, *General Criteria for the Assessment of Testing Laboratories – European Standard EN45002*, Brussels, CEN/CENELEC, 1989.
6. European Committee for Standardization, *General Criteria for Laboratory Accreditation Bodies – European Standard EN45003*, Brussels, CEN/CENELEC, 1989.
7. European Union, *Council Directive 85/591/EEC Concerning the Introduction of Community Methods of Sampling and Analysis for the Monitoring of Foodstuffs Intended for Human Consumption*, O.J. L372 of 31.12.1985.
8. Codex Alimentarius Commission, *Report of the 21st Session of the Codex Committee on Methods of Analysis and Sampling – ALINORM 97/23A*, Rome, FAO, 1997.
9. Codex Alimentarius Commission, *Report of the 22nd Session of the Codex Alimentarius Commission – ALINORM 97/37*, Rome, FAO, 1997.
10. International Organization for Standardization, *General Requirements for the Com-*

petence of Calibration and Testing Laboratories – ISO/IEC Guide 25, Geneva, ISO, 1990 [now superseded by ISO standard 17025].

11. M. Thompson and R. Wood, *Pure Appl. Chem.*, 1993, **65**, 2123 (also published in *J. AOAC Int.*, 1993, **76**, 926).

12. M. Thompson and R. Wood, *Pure Appl. Chem.*, 1995, **67**, 49.

13. International Organization for Standardization, *Calibration and Testing Laboratory Accreditation Systems – General Requirements for Operation and Recognition – ISO/IEC Guide* 58, Geneva, ISO, 1993.

14. M. Thompson, S. L. R. Ellison, A. Fajgelj, P. Willetts and R. Wood, *Pure Appl. Chem.*, 1999, **71**, 337.

15. W. Horwitz, *Pure Appl. Chem.*, 1988, **60**, 855; revised W. Horwitz, *Pure Appl. Chem.*, 1995, **67**, 331.

16. *Procedural Manual of the Codex Alimentarius Commission, 10th Edition'*, FAO, Rome, 1997.

17. *Precision of Test Methods*, Geneva, 1994, ISO 5725; previous editions were issued in 1981 and 1986.

18. *Validation of Chemical Analytical Methods*, NMKL Secretariat, Finland, 1996, NMKL Procedure No. 4.

19. *Method Validation – A Laboratory Guide*, EURACHEM Secretariat, Laboratory of the Government Chemist, Teddington, UK, 1996.

20. *An Interlaboratory Analytical Method Validation Short Course developed by the AOAC International*, AOAC International, Gaithersburg, Maryland, USA, 1996.

21. *Validation of Methods*, Inspectorate for Health Protection, Rijswijk, The Netherlands, Report 95-001.

22. *A Protocol for Analytical Quality Assurance in Public Analysts Laboratories*, Association of Public Analysts, 342 Coleford Road, Sheffield S9 5PH, UK, 1986.

23. *The Fitness for Purpose of Analytical Methods*, EURACHEM Secretariat, Internet Version, 1998.

24. *Guide to the Expression of Uncertainty in Measurement*, ISO, Geneva, 1993.

25. *Quantifying Uncertainty in Analytical Measurement*, EURACHEM Secretariat, Laboratory of the Government Chemist, Teddington, UK, 1995, EURACHEM Guide (under revision).

26. *Estimation and Expression of Measurement Uncertainty in Chemical Analysis*, NMKL Secretariat, Finland, 1997, NMKL Procedure No. 5.

27. *Validation of Analytical Methods for Food Control*, Report of a Joint FAO/IAEA Expert Consultation, December 1997, FAO Food and Nutrition Paper No. 68, FAO, Rome, 1998.

Appendix: Example of Application of Criteria

Introduction

A number of methods of analysis have been recommended as Codex general trace element analysis methods. These have been summarised and the methods evaluated for

* Precision (within and between laboratories)
* Recovery
* Specificity
* Applicability
* Detection limits

Methods Considered

The following methods were considered in the valuation. Numbers refer to subsequent tables.

Method 1:

1. R.J. Gafan, S.G. Gapar, C.A. Subjoc and M. Sanders 'Determination of lead and cadmium in foods by anodic stripping voltammetry: I. Development of method', *J. Assoc. Off. Anal. Chem.* 1982, **65**, 970–977.

2. S.G. Capar, R.J. Gajan, E. Madzsar, R.H. Albert, M. Sanders and J. Zyren 'Determination of lead and cadmium in foods by anodic stripping voltammetry: II. Collaborative study', *J. Assoc. Off. Anal. Chem.* 1982, **65**, 978–986.

3. Association of Official Analytical Chemists. 'Cadmium and lead in food. Anodic stripping voltammetric method.' Official Final Action, 1988. *AOAC Official Methods of Analysis*, 1, (982.23), 1990, pp. 239–241.

Summary: Weigh 5–10 g (5 g dry weight) in quartz beaker, add 5 ml ashing aid (10% potassium sulfate, 2.5% nitric acid), cover, and dry in oven at 110 °C. (Pyrex was included as a material for the ashing beakers, but it should not be used.) Ash 4 hours minimum at 500 °C. If carbon present, add 2 ml HNO_3 and reash 0.5 hour. Dissolve in approximately 10 ml 10% nitric acid, dilute to 50 ml, and mix. Let any precipitate settle. Analyse for Pb and Cd by standard addition DPASV or LSASV. For DPASV add a 5 ml aliquot to electrolysis cell containing 5 ml electrolyte (1.7M HAc, 1.25M NaAc, and 0.01M tartaric acid, pH 4.7). (After revelation of a tin interference, the tartaric acid concentration was increased to 0.1M.) For LSASV, add 2 ml aliquot to 3 ml of above electrolyte. In the suspected presence of thallium, make 5 ml aliquot basic with 3 ml NaOH (NaOH concentration omitted).

Method 2:

1. P.W. Hendrikse, F.J. Slikkerveer, J. Zaalberg and A. Hautfenne, 'Determination of copper, iron and nickel in oils and fats by direct graphite furnace atomic absorption spectrometry', *Pure Appl. Chem.*, 1988, **60**, 893–900.

2. 'Copper, iron and nickel in edible oils and fats. Direct graphite furnace atomic absorption spectrometric method.' AOAC, Official First Action 1990 Method,

AOAC Official Methods of Analysis, 15th Edition: 1st Supplement (1990), AOAC, Arlington, VA, 990.05, pp. 7–8.

Method 3:
P.W. Hendrikse, F.J. Slikkerveer, A. Folkersma, A. Dieffenbacher and IUPAC Commission on Oils, Fats and Derivatives. 'Determination of lead in oils and fats by direct graphite furnace atomic absorption spectrometry. Results of a collaborative study and the standardised method', *Pure Appl. Chem.*, 1991, **63**, 1191–1198.
Summary: Details not available to author. Assumption made that method is same as that described above for Ni, Cu and Fe.

Method 4:
1. 'Tin in canned foods. Atomic absorption spectrophotometric method.' AOAC, Official Final Action 1988, *AOAC Official Methods of Analysis*, AOAC, Arlington, VA, 985.16, pp. 270–271.
2. R.W. Dabeka, A.D. McKenzie and R.H. Albert, 'Atomic absorption spectrophotometric determination of tin in canned foods using nitric acid–hydrochloric acid digestion and nitrous oxide acetylene flame', *J. Assoc. Off. Anal. Chem*, 1985, **68**, 1985, 209–213.
Summary: Sample (5–40 g depending on fat and water content) is digested with nitric acid, then hydrochloric acid. Solution is diluted to 100 ml in flask containing KCl to remove flame AAS interference. Solution is filtered and analysed by flame AAS.

Method 5:
'Metals. Determination by atomic absorption spectrophotometry in foodstuffs.' Nordic Committee on Food Analysis. No. 139, 1991.
Summary: Dry ash 10–20 g sample in platinum or quartz crucible. Add 5 ml 6M HCl and evaporate to dryness. Dissolve ash in exact volume (10–30 ml) 0.1M nitric acid. Determine Pb, Cr, Ni and Cd by GFAAS using the method of standard additions. Determine Fe, Cu and Zn by flame AAS with conventional calibration.

Method 6:
'Determination of total aflatoxin levels in peanut butter by enzyme-linked immunosorbent assay: collaborative study', *JAOAC Int.* 1992, **75**, 693–697.
Summary: Blend 10 g peanut butter with 50 ml (1 + 1) acetonitrile/water, filter and dilute 1:25 in aflatoxin assay diluent concentrate/water (20 + 80). Add 50μ l to wells on plate reader using Biokits total aflatoxin ELISA kit.

Method 7:
'Solvent-efficient thin-layer chromatographic method for the determination of aflatoxins B1, B2, G1 and G2 in corn and peanut products: collaborative study', *J. Assoc. Off. Anal. Chem.*, 1994, **77**, 637–646.
Summary: Shake 50 g 300 mesh corn or peanuts with 200 ml methanol/water (85 + 15), filter, extract with NaCl and hexane, back extract with CHC31, evapor-

ate to dryness. Dissolve in dichloromethane and pass through silica gel column. Elute with CHC31/acetone (9 + 1). Evaporate to dryness, dissolve in CHC31, and analyse by TLC fluorodensitometry.

Method 8:
'Enzyme-linked immunosorbent assay of total aflatoxins B1, B2 and G1 in corn: follow-up collaborative study', *J. Assoc. Off. Anal. Chem.*, 1994, **77**, 655–658.
Summary: Blend 10 g corn with 50 ml (1 + 1) acetonitrile/water, filter and dilute 1:25 in aflatoxin assay diluent concentrate/water (20 + 80). Add 50 μl to wells on plate reader using Biokits total aflatoxin ELISA kit.

Method 9:
'Multifunctional column coupled with liquid chromatography for determination of aflatoxins B1, B2, G1 and G2 in corn, almonds, brazil nuts, peanuts and pistachio nuts: collaborative study', *J. Assoc. Off. Anal. Chem.*, 1994, **77**, 1512–1521.
Summary: AOAC Official Method 991–31.

Method 10:
'Analysis of food for lead, cadmium, copper, zinc, arsenic and selenium using closed system sample digestion: collaborative study', *JAOAC*, 1990, **63**, 485–495.
Summary: 0.3 g food digested with 5 ml HNO_3 2 h in teflon pressure bomb at 150 °C. Dilute to 10 ml. Dry ash aliquot for Pb, Cu and Cd and measure by ASV. Dry ash for As and Se and measure by hydride, AAS. Digest portion with $HClO_4$, evaporate to dry and measure by flame AAS for Zn.

Method 11:
'An analytical method for the determination of copper and nickel in fats and oils by flameless atomic absorption spectroscopy. Results of a ring test', *Rivista Sostanze Grasse*, 1986, **63**, 401–402.
Summary: Dilute 2 ml with 8 ml MIBK. Manually inject into graphic furnace AAS, comparing with standards in MIBK.

Method 12:
'Evaluation of a method for the determination of total cadmium, lead and nickel and foodstuffs using measurement by flame atomic absorption spectrophotometry', *Analyst*, 1978, **103**, 580–594.
Summary: Digest 5–20 g with nitric and sulfuric acids, dilute to 100 ml, extract with APDC and DDTC into 10 ml MIBK, filter MIBK phase. Measure by flame AAS.

Method 13:
'Electrothermal atomic absorption spectroscopic determination for chromium in plant tissue: interlaboratory study', *JAOAC*, 1985, **66**, 850–852.
Summary: Digest 1 g with nitric, perchloric and sulfuric acids, reduce with sulfite, coprecipitate with Fe. Extract Fe with HCl and MIBK. Dissolve silica with HF. Determine by Cr by GFAAS.

Statistical Characteristics

Table 1 *The statistical characteristics of the above methods are given in Table 1.*

Method	Element	Assessed level, μg g⁻¹	Repeatability, mean and range % CV	Reproducibility mean and range % CV	Recovery %	Specificity	Applicability	Detection limit, μg g⁻¹
1	Pb	0.08–0.5	14.8 (7–37)	21(13–38)	90 +	Sn, Tl interfere	All foods	0.025
1	Cd	0.03–0.26	13 (7–26)	18.5 (9–35)	90 +	Good	All foods	0.015
2	Cu	0.03–0.15	9.1 (5.4–15.1)	18 (15–21.4)	90 +	Excellent	Fats and oils	0.005
2	Fe	0.13–0.96	10.8 (6.2–21)	22.3 (19–27)	90 +	Excellent	Fats and oils	0.005
2	Ni	0.13–0.9	11 (6.4–16.8)	18.8(17–22)	85 +	Excellent	Oils	0.005
3	Pb	0.02–0.09	6.4 (4.5–10.1)	18 (10.2–27.7)	95 +	Excellent	Fats and oils	0.005
4	Sn	50–250	5.6 (2.2–12)	7.3 (3.3–15)	95 +	Excellent	All foods	10
5	Pb	0.045–0.25	33 (26–40)	33 (26–40)	80 +	Excellent	All foods	0.006
5	Cr	0.014–0.078	44 (43–46)	46 (43–52)	80 +	Excellent	All foods	0.009
5	Ni	0.03–0.34	44 (25–55)	49 (34–58)	90 +	Excellent	All foods	0.026
5	Cd	0.19–0.53	15 (14–17)	19 (18–21)	80 +	Excellent	All foods	0.004
5	Fe	3.8–212	10.1 (8.2–12)	11.5 (11–12)	90 + +	Excellent	All foods	0.08
5	Cu	7.1–45	12 (3.5–20)	12 (11–22)	90 + +	Excellent	All foods	0.1
5	Zn	6.6–37	4.7 (4.3–5.1)	4.8 (4.3–5.3)	80 +	Excellent	All foods	0.06
6	Total aflatoxin	0.009–0.090	18 (9–30)	30 (24–37)	84–90	Good	Peanut butter	0.08
7	Aflatoxins B_1, B_2, G_1, G_2	0.003–050	39 (21–57)	45 (29–63)	75 +	Excellent	Corn and peanuts	0.006
8	Total aflatoxin B_1, B_2, G_1, G_2	0.000–0.030	NA screening test	NA	NA	Good	Corn	NA
9	Aflatoxins B_1, B_2, G_1, G_2	0.000–0.030	17 (6–23)	30 (12–69)	70 +	Good	Corn, almonds, brazil nuts, peanuts, pistachio nuts	0.006–0015
10	Pb	0.03–2.8	36 (10–98)	51 (17–106)*	90 +	Good	Chicken, apple	1
10	Cd	0.014–1.0	61 (9–127)	61 (16–214)	90 +	Good	Chicken, apple	0.05
10	Zn	0.06	26 (6–214)	58 (7–218)	90 +	Excellent	Chicken, apple	1

10	As	0.017–1.9	26 (9–55)	61 (16–147)	80+	Excellent	Chicken, apple	0.02
10	Se.	0.019–1.6	50 (13–129)	69 (13–155)	80+	Excellent	Chicken, apple	0.02
11	Cu, Ni	0.07–0.2	5 (3.2–8.2)	20 (14–24)	80+	Excellent	Oil	0.005
12	Pb	0.02–2.2	16 (4–43)	28 (8–75)	90+	Excellent	General	0.07
12	Cd	0.003–1	13 (2–39)	24 (9–69)	90+	Excellent	General	0.007
12	Ni	0.03–1.1	15 (8.5–2.8)	18 (11–28)	90+	Excellent	General	0.06
13	Cr	0.04–3.2	26 (6–71)	33 (6–71)	90+	Excellent	General	0.01

Assessment of the Statistical Characteristics

Table 2 *The statistical characteristics of the methods were assessed, and in particular the precision values. These are outlined Table 2 using the Horwitz predicted values.*

Method	Element	Assessed level, $\mu g\ g^{-1}$	Predicted Horwitz RSD_R	$HORRAT_r$	$HORRAT_R$	Conclusion
1	Pb	0.08–0.5	23.4 – 17.8	2.4 – 0.6	1.6 – 0.7	Satisfactory
1	Cd	0.03–0.26	27.1 – 19.6	1.4 – 0.5	1.3 – 0.5	Satisfactory
2	Cu	0.03–0.15	27.1 – 21.3	0.8 – 0.4	0.8 – 0.7	Satisfactory
2	Fe	0.13–0.96	21.8 – 16.1	1.4 – 0.6	1.2 – 1.2	Satisfactory
2	Ni	0.13–0.9	21.8 – 16.3	1.2 – 0.6	1.0 – 1.0	Satisfactory
3	Pb	0.02–0.09	28.8 – 23.0	0.5 – 0.3	1.0 – 0.4	Satisfactory
4	Sn	50–250	8.9 – 7.0	2.0 – 0.5	1.7 – 0.5	Satisfactory
5	Pb	0.045–0.25	25.5 – 19.7	2.4 – 2.0	1.6 – 1.3	Satisfactory
5	Cr	0.014–0.078	30.4 – 23.5	2.3 – 2.7	1.7 – 1.8	Satisfactory
5	Ni	0.03–0.34	27.1 – 18.8	3.0 – 2.0	2.1 – 1.8	Satisfactory
5	Cd	0.19–0.53	20.5 – 17.6	1.2 – 1.2	1.0 – 1.0	Satisfactory
5	Fe	3.8–212	13.1 – 7.1	1.4 – 1.7	0.9 – 1.5	Satisfactory
5	Cu	7.1–45	11.9 – 9.0	2.5 – 0.6	1.8 – 1.2	Satisfactory
5	Zn	6.6–37	12.0 – 9.3	0.6 – 0.7	0.4 – 0.5	Satisfactory
6	Total aflatoxin	0.009–0.090	32.5 – 23.0	1.4 – 0.6	1.1 – 1.0	Satisfactory
7	Aflatoxins B_1, B_2, G_1, G_2	0.003–0.050	38.4 – 25.1	2.2 – 1.3	1.6 – 1.2	Satisfactory
8	Total aflatoxin B_1, B_2, G_1,	0.000–0.030	NA	NA	NA	Screening method
9	Aflatoxins B_1, B_2, G_1, G_2	0.000–0.030	<64 – 27.1	<0.7 – 0.3	<1.3 – 0.4	Satisfactory
10	Pb	0.03–2.8	27.1 – 13.7	5.4 – 1.1	3.9 – 1.2	Unacceptable
10	Cd	0.014–1.0	30.4 – 16.0	6.3 – 0.8	7.0 – 1.0	Unacceptable
10	Zn	0.06	24.4	13.1 – 0.4	8.9 – 0.3	Unacceptable

10	As	0.017–1.9	29.5 – 14.5	2.8 – 0.9	5.0 – 1.1	Unacceptable
10	Se	0.019–1.6	29.1 – 14.9	6.7 – 1.3	5.3 – 0.9	Unacceptable
11	Cu, Ni	0.07–0.2	23.9 – 20.4	0.5 – 0.2	1.0 – 0.7	Satisfactory
						Satisfactory
12	Pb	0.02–2.2	28.8 – 14.2	2.2 – 0.4	2.6 – 0.6	Satisfactory
12	Cd	0.003–1	38.4 – 16.0	1.5 – 0.2	1.8 – 0.6	Satisfactory
12	Ni	0.03–1.1	27.1 – 15.8	0.5 – 0.3	1.0 – 0.7	Satisfactory
13	Cr	0.04–3.2	26.0 – 13.4	4.1 – 0.7	2.7 – 0.4	Satisfactory

Methods/analyte combinations which meet HORRAT recommendations, *i.e.* are less than 2, are assumed to be satisfactory.

Generalised Criteria

From the above it is possible to state that for lead any method of analysis may be used provided it meets the criteria shown in Table 3.

Table 3

Parameter	Value/Comment
Applicability	All foods
Detection limit	No more than one twentieth of the value of specification
Determination limit	No more than one tenth of the value of specification
Precision	HORRAT$_r$ and HORRAT$_R$ values of less than 2 in the validation collaborative trial
Recovery	80–105%
Specificity	No cross interferences permitted

Environment

CHAPTER 6

Metal Speciation for Improved Environmental Management

O.F.X. DONARD

1 The Occurrence and Importance of Metal Species in the Environment

The fate and impact of trace elements are directly linked to their chemical form, which may be as free ions, small organometallic moieties, or larger biomolecules incorporated into biological systems. This was well established in the 1970s and 1980s when, by employing electrochemical techniques, it was possible to differentiate between free ions and complexed metals, attached to either organic or inorganic ligands. One drawback of this approach is that it is only capable of dealing with a minority of metallic elements, but it is able to determine what is directly 'bioavailable' to the organism. We now know that the total percentage of 'free ions' that are really free in solution in the environment is very low (< 2%), due to the numerous opportunities for complexation with naturally occurring ligands.

Organometals are present in the environment either as a result of direct methylation (*in situ* as for example with Hg) or as a result of direct anthropogenic inputs. Here, in solution, they represent a very minor proportion of the total metallic burden. Natural methylation reactions occur under specific conditions and the yield of these reactions is, in general, very low, both in natural waters and in sediments. However, the new form of the element can be very much more toxic, as is the case for mercury (methylmercury being more toxic than metallic mercury), or less harmful as in the case of arsenic, for which toxicity diminishes as the degree of methylation increases. Nonetheless, these changes in the form in which the element is present result in profound changes to its physico-chemical properties, and can dramatically affect the toxicity and translocation between the different compartments of the environment.

For mercury, it is now clearly established that concentrations in water of a few $ng\,l^{-1}$ of methylmercury, representing less than 1% of the total Hg content, translates into $mg\,kg^{-1}$ of methylmercury in the top predators of the food chain.

In these instances, the methylmercury burden represents between 90% and 100% of the total mercury content. Other elements, such as As or Se, when incorporated into biological systems, become part of the backbone of larger molecular structures (*e.g.* seleno-amino acids or metalloproteins). In these cases, well-characterised molecular structures are in evidence, but in natural waters all of these species can react with ill-defined humic or fulvic substances, becoming incorporated into obscure macromolecules.

The ever-changing formulation of an element under natural conditions may be illustrated by considering the case of selenium (Figure 1). In seawater, selenium is taken up by biological cells as inorganic ions [Se^{IV} or Se^{VI}], depending upon the conditions. These selenium species are biologically incorporated and transformed into seleno-amino acids. They, in turn, can either be directly released back into the environment *via* direct excretion or, during cell lysis, decay to metallic Se^0 or to seleno-amino acids. Photolytic reactions with these seleno-amino acids will produce dimethyl selenide or dimethyl diselenide, which will evaporate from aqueous media.[1]

The same applies to other metals or metalloids. The number of arsenic species in the environment is continuously growing, due to direct economic interest. Indeed, the relatively high levels of 'water soluble' arsenic, detected in the Australian lobster, needed to be identified, in order to facilitate the export of this highly valuable commodity. The remarkable efforts of Edmonds and Francesconi, using traditional structural chemical methods, revealed that most of the arsenic present was in the form of a non-toxic arsenic compound 'arsenobetaine'.[2] Since then, due to the ever increasing consumption of seafood world-wide, tremendous efforts have been made by numerous research teams

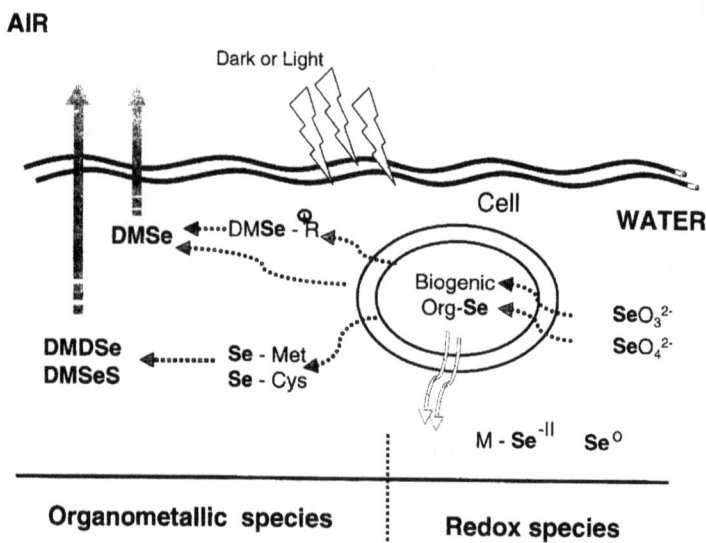

Figure 1 *Bio-geochemical transformations of selenium in marine waters* (adapted from ref. 1)

who have characterised more than 30 arsenic-containing compounds. Similarly, the cases of Hg and organotin compounds in the environment have been well documented, and here also new species are continually being discovered and characterised, allowing environmental scientists to improve their understanding of the fate and pathways of these species in the various compartments of ecosystems. Indeed, the key environmental parameters used to model the fate of metals in the environment, such as particulates, water partition (K_p), bioaccumulation factor (K_{ow}) or liquid to gas partition (Henry's Law constants), are highly affected by the formulation of the metal-containing compound. Tin can be methylated in water, but the degree of methylation favours the adsorption of monomethylated species, whilst the trimethylated form remains in solution and is highly toxic.[3] Similar facts have been demonstrated for mercury species.[4]

As already indicated, the toxicity of metal species is also directly related to their chemical formulation. The recognised active, toxic form of copper in water is Cu^{2+}, and it is used as such in anti-fouling paint formulations. Other highly alkylated organometallic compounds exhibit high toxicity. These aspects are very well understood and this knowledge is used, for example, in organotin based, anti-fouling paints. As stated above, the trisubstituted moieties are often far more toxic than the monoalkylated forms. These differences in toxicity are of crucial importance. Organotin compounds are very important catalysts and are at the crossroads of major organometallic syntheses. The less toxic mono- or di-substituted organotin species are found in many industrial products (*e.g.* stabilisers for PVC), whilst the series of trisubstituted compounds are highly toxic and widely utilised in biocide formulations. Since they are directly introduced into the environment, they represent a serious hazard. They are sometimes too efficient and persistent for the targeted organisms, and may directly affect non-targeted species, resulting in a serious threat to the whole food chain. They are today suspected of being endrocrine disrupters, and are of great environmental concern, due to their widespread use throughout the world.

2 Analytical Responses and Legislative Implications

Since the case for considering speciation in the construction and implementation of legislation is now well proven, the question has to be asked as to why it is not so considered. A comparison between the different attitudes held by the organic and inorganic analytical chemists may throw some light on the reasoning. In the case of organic determinations, instrumentation and subsequent advances were developed directly to address the molecular structure of compounds, produced primarily through anthropogenic activities (*e.g.* identification of the formulation of a pesticide). The methodologies employed usually rely on hyphenated techniques, to achieve separation followed by detection. Characterisation of the organic moieties is obtained by determination of the fragments of the analytes by mass spectrometry. Structural identification has always been a more important issue than sensitivity in the organic analytical domain. It could be argued that inorganic analysis did not develop in a logical and systematic manner, due to the early discoveries of the spectroscopic emission lines of elements in flames, made during the 19th century. These early and spectacular findings encouraged analytical scientists to use the spectroscopic emission lines of elements. Since they were obtained *via* combustion in a flame, all structural information was already lost. This analytical field developed rapidly, using absorption, emission or fluorescence techniques. A rapid development of instrumentation, such as the atomic absorption spectrometer, using either flame or graphite furnace atomisation, took place. The subsequent introduction of plasma-based emission spectroscopy, which allowed simultaneous determinations, and more recently plasma-mass spectrometry, to detect and quantify metallic ions in various matrices has advanced inorganic analysis and made possible ever lower detection limits. These techniques also, however, destroy the initial chemical formulation of the analytes. The evolutions of the two analytical domains are compared in Figure 2.

Information Level

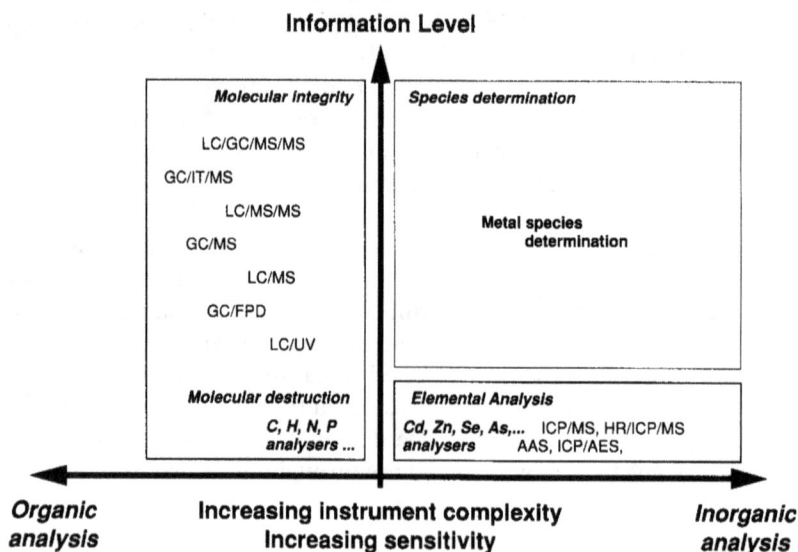

Figure 2   *Position of metal species determination amongst the different analytical trends*

Metals species determination is now a pressing requirement and the future will bring the two domains ever closer. Work is now in progress in some research laboratories to use mild, glow-discharge techniques to recover structural information after chromatographic separation and prior to mass spectrometric detection.[5,6]

It is still true that most environmental regulations are currently based on 'total metal' determination in the various matrices. It must be remembered, however, that in all of the most severe poisoning cases worldwide it was the chemical form of the element which was finally identified as the main issue. In all cases, major macroscopic biological incidents occurred, which caused regulations to be changed. One of the first major 'environmental accidents', the Minamata poisoning case in the 1960s, by methylmercury, indicated the need for correct species identification. Since then, a large range of analytical sample preparation and detection procedures have been developed, for the determination of mercury and its compounds.

Another famous example arose in the 1980s in France, when a severe decrease in the production of oysters was observed. Following lengthy and difficult investigations, employing mainly eco-toxicological tests, organotin compounds, released from anti-fouling paints, were finally identified as the culprits. Analytical confirmation came only after biological markers had alerted scientists to there being a problem. Since that time, tremendous efforts by the analytical research fraternity have produced methods, which have now been further simplified, and these can now be implemented in routine laboratories. This has allowed us, some 20 years later, to assess the extent of the world wide contamination.

The question of the chronic effect of highly toxic compounds at low doses within the food chain is ever present. There are now serious concerns regarding inorganic arsenic species in drinking water. 'Blackfoot disease', which affects large populations in India and Bangladesh, is a result of such an exposure, and calls for urgent action to be taken. When these biological responses become evident on the macroscopic scale, it suggests that the problem has already existed for some years. Indeed, in most cases, at least ten years will have elapsed during which the environmental problem has become well established. A detailed examination of the population dynamic of the bottom dwelling bivalve *Scorbicularia plana* proved that the biological impact of organotin compounds, caused by their introduction in the anti-fouling paints, had immediately impacted and reduced the populations of this bivalve in the Bay of Arcachon.[7,8]

An ambiguous situation regarding the legal positions of different legislative bodies now exists. This strange situation is illustrated in Figure 3.

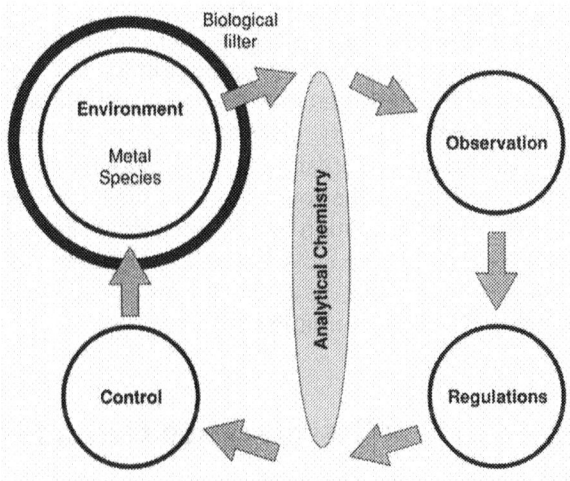

Figure 3   *The essential role of analytical chemistry in promoting the appropriate information for improved environmental management* (adapted from ref. 14)

Biology is definitely the primary source of information and the ultimate regulation alarm. However, even though we now have the necessary analytical solutions to correctly generate the appropriate data, most environmental regulations are still based on 'total metal' content. The situation is not clear in the legislative texts of the EU. Indeed, legislative texts can be rather vague, mentioning the need for speciated determination, by referring to one or more metals '*and their compounds*'. In other instances, the degree of analytical precision is too high to be directly implemented by routine laboratories.

The *Council decision 75/437/EEC* on Marine Pollution from Land-based Sources aims at the prevention of pollution by three categories of contaminants. The mention of the chemical formulation of the elements appears in the Annex of the documents. This general phrasing can be found in many legislative acts, such as the *Council decision 77/585/EEC* on the protection of the Mediterranean Sea together with additional decisions 81/420/EEC, 83/101/EEC and 84/132/EEC, where additional contaminants are listed ('the dumping of which requires special care – namely arsenic, lead, copper, zinc, beryllium, nickel, vanadium, selenium, antimony *and their compounds*'). Similar statements also occur in the annexe of various other Council decisions (77/586/EEC on prevention of contamination of the Rhine river or 80/68/EEC on groundwater pollution).

Conversely, in some instances, the requirement for full species determination can be much too specific, and be inconsistent with the present analytical state of the art available. *Council decision 76/464/EEC* recommends the monitoring of a list of 132 substances in industrial effluents. In this list, organic compounds are clearly listed, but the recommendations regarding arsenic require the determination of arsenic '*and its mineral compounds*'. On the other hand, requirements for cadmium and mercury make a vague mention of '*its compounds*'. While such statements are of great interest and importance to industry, the environmental scientist would like to see much greater clarity.

Ironically, all organotin species are listed, with their full formulation and their counter-ions. It would be unrealistic, given the current level of expertise in most analytical laboratories, to suppose that these compounds could be determined or that determination is even necessary, given that the analyte of interest is the organotin moiety. In most cases, organotin compounds are determined as the equivalent of the cationic form. The counter-ion is usually removed during the analytical protocol, by derivatisation which prepares the analytes for gas chromatographic separation. These facts illustrate a serious lack of understanding, and an urgent need to improve the communication of information, between the scientific community and the legislating bodies.

3   Existing Analytical Responses and Current Needs

If it is accepted that the regulations concerning environmental issues are poorly specified, the status and understanding of the role of metal species determination in environmental matters is currently changing. At present, the overall situation is biased, since most environmental regulations are based on the 'total metal definition' resulting in an improper definition of the problem, and the subsequent enforcement of ill defined legislation, due in part, to the lack of suitable instrumentation (Figure 3).

In the industrial hygiene and food domains, the importance of speciation is both understood and applied. In these areas, speciation is associated with positive benefits. In the field of industrial hygiene, the direct advantages of improving the safety of workers are obvious, and speciation measurement is progressing within the limitations of current analytical knowledge. In the food industry, metal speciation translates into beneficial supplementation, and relates to the direct bioavailability of metal species to the organisms. In the cases of iron or selenium, for example, they are both directly related to the health of the individuals receiving them. Supplementation of selected metal species is of paramount importance for whole populations at a national level, when the natural diet is deficient. In contrast, when speciation concepts apply to environmental issues they have always been associated with negative aspects. Indeed, some metal species may introduce severe disorders in the biological food chain. Therefore, despite the fact that metal species determination has long been understood, since at least the date of the Minamata poisoning, few efforts have been made for its implementation in every day life. Firstly, analytical techniques were more complex than traditional atomic spectrometry methods used to perform 'total determinations'. Secondly, the requirement for metal species determination in environmental issues has always been as a result of the enforcement of regulations. Major industries have not been receptive to these new constraints, which they could not easily implement in their environmental management programmes.

The whole situation has now changed, due largely to the considerable effort of the analytical community. There is now a great array of analytical solutions allowing simple and effective metal determination. This results from a tremendous effort by laboratories within the EU in producing reference materials for metal species in environmental matrices. Continuous critical discussions regarding the results, during inter-calibration studies, have allowed critical evaluation of each of the analytical steps. The overall result has been that the whole analytical community of the EU has gained an in-depth understanding of the analytical steps required for speciation analysis. At present, most analytical methods developed for Cr, Sn, Hg or As species determination can be implemented in routine laboratories. The overall analytical steps have been simplified, and the existence of a large range of certified reference material allows the production of high quality, assured results. It can be stated with certainly that the determination of organotin, organomercury and arsenic species is no more expensive than the cost of routine organic pesticide determinations. In most cases, the analytical protocols are very similar, and could use the same instrumentation.

A second important fact, that will certainly allow the promotion of the use of species determination in environmental regulation, is that this evolution has been closely followed by the major instrument manufacturers. Some of them have long been aware of this new 'analytical discipline'. There is an ever increasing number of publications in the literature in the field of metal species determination. At present, several companies clearly advertise metal species determination solutions. All others are evaluating these issues and follow them closely as an up and coming new market.

The final and most important point is related to the change in attitude towards metal species determination by major industrial companies. In most cases, they have understood that metal species determination helps them to address correctly an issue and find an appropriate solution to it, in the most cost-effective manner. '*Metal species determination*' is now understood as a form of improved '*risk assessment*'.

Indeed, as mentioned earlier, when an environmental agency reduces an effluent level in an environmental regulation, this usually results in an enormous additional cost for those companies required to comply with it. In many cases, the lowering of such a level reflects the poor understanding which the agency has of the case under consideration. Usually, the reduction reflects a precautionary measure. It could be said that levels for total metal content decrease by a factor of 10 every ten years as a consequence of the progress and efficiency of modern instrumentation. Recently, the acceptable level of As in drinking water ('as total') was lowered from 50 to $10 \, \text{mg} \, \text{l}^{-1}$. The US Environmental Protection Agency (EPA) is expected to be the first governing body to enforce it. This trend will certainly follow in Europe. A similar situation may be observed for the total concentration of lead in drinking water, for which the maximum permissible content has decreased from $10 \, \mu\text{g} \, \text{l}^{-1}$ in the new EC Framework Drinking Water Directive. This now poses some serious analytical challenges for routine analytical instrumentation, and it will also translate into tremendous additional costs for the water companies, who will have to modify and adapt their drinking water production and decontamination processes to match these new limits.

Many of these environmental regulations could, and should, be based on the appropriate metal species identification, resulting in a far better definition of the problem, and hence better solutions. These aspects are now better understood in industry, and hence many companies are introducing metal species determination to increase the efficiency and the cost effectiveness of their industrial processes.

The combination of the three main issues, discussed above, allows one to anticipate that the introduction of 'metal species' determination in environmental issues will become more common in the future, but initially, probably, under an 'improved risk assessment' perspective. If existing regulations still require improved definition and understanding from governing bodies, the negative aspect of the metal species determination concept is now slowly translating into higher cost effectiveness in environmental issues, and is slowly being integrated by industries concerned with environmental management. The same also applies to the main regulating bodies, who have also made considerable efforts during the last decade with regards to these issues, and in the position of correctly formulating their new legislative guidelines. Several standardisation procedures are in progress in various organisations on 'metal species determination'. This is the case for organotin compounds, which are currently under discussion at ISO.

The current understanding of environmental issues allows us to redefine the requirements for instrumentation, particularly with respect to detection limits. Initially, most of the analytical efforts were targeted towards method development with elegant analytical solutions. This led to a race towards ever-lower detection limits which usually resulted in high-cost hardware solutions. At present, sensitivity requirements are being re-evaluated. Besides the performance requirements of analytical instruments, the quality control aspects (internal and external validation of methods) are also of increasing concern. It is clear that high sensitivity will be required in the area of water monitoring. This is also the case for organotin and organomercury determinations in environmental waters; due to their high toxicity at low levels and high bioaccumulation factors, they should certainly be monitored in environmental waters at low $ng\,l^{-1}$ (ppt). There is now a wide range of analytical protocols and instrumentation that will allow the continuous monitoring of these species. In biological samples, both biological and environmental progress has still to be made, and sub $\mu g\,kg^{-1}$ detection limits will be required. In many industrial issues related to food supplementation, such sensitivity will not be required. The same applies to environmental solids such as sediments and soils. In most cases, the regulations that will apply to them will address the direct levels of contaminants, and in most cases the samples generating problems will contain high levels of these species. The task of both the analytical protocols and instrumentation used will not be related to low detection limits, but will instead have to deal with highly polluted, complex matrices, and will require high selectivity. The need for these routine detection limits in various environmental matrices is summarised in Figure 4.

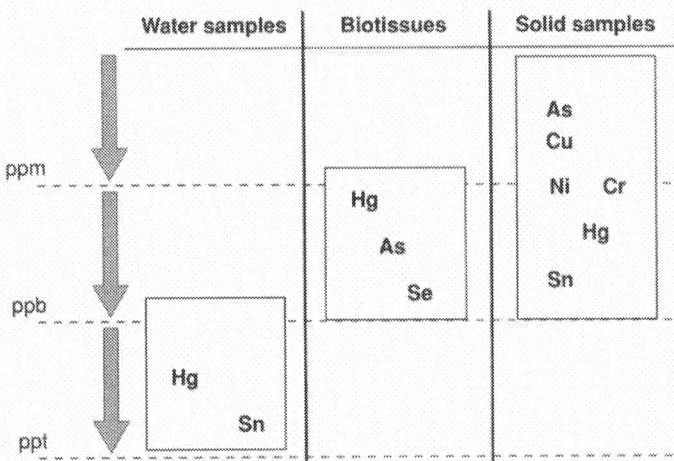

Figure 4   *Analytical detection limits requirements for environmental analysis*

4   Current Status and Future Needs

Most analytical developments will allow routine laboratories to directly implement these requirements. Instrumental solutions for specific determinations are being developed by a number of manufacturers. Also, as mentioned earlier, industry awareness of 'metal species related issues' has increased and been translated into 'improved environmental management' or 'risk assessment' strategies. However, there are several important steps still to be taken into consideration, to enable appropriate environmental management decisions to be implemented based on sound environmental understanding.

The first point concerns the fact that environmental metal species determination relies on a continuous analytical procedure, going from

step 1: sample collection;
step 2: sample storage;
step 3: sample preparation and
step 4: determination.

These steps are shown in Figure 5.

Until now, most efforts have been targeted towards the sample preparation and determination stages. However, the first two steps should really be closely considered, if meaningful environmental data are to be obtained. Organotin determinations, for example, require careful attention to the sampling time, sample handling and preservation if appropriate data is to be obtained.[8] In addition, at the determination stage, there is a need to obtain new certified reference materials, covering the largest array of matrices possible. These ma-

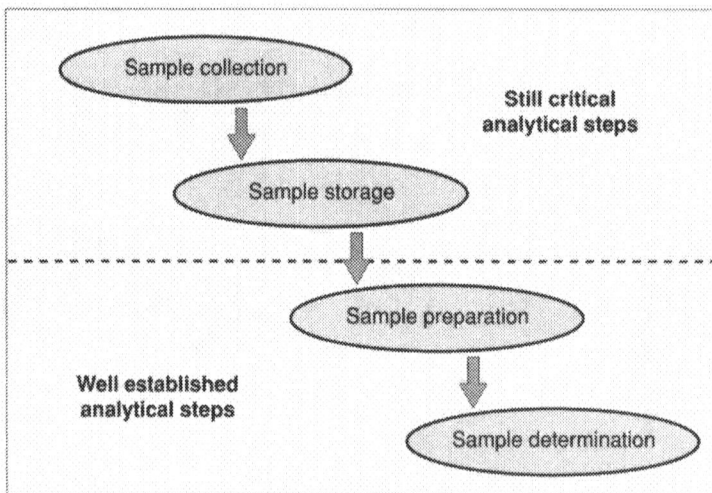

Figure 5   *Current status of analytical expertise in the domain of metal species determination*

terials are the cornerstone of sound data generation. There is also a continuous need to organise laboratory inter-comparison exercises (*e.g.* proficiency testing schemes) to enable laboratories to perform an external validation of their methods, and share expertise with others within the analytical community. Furthermore, the use of certified reference materials in metal species determination (and/or the verification of methods through inter-comparison exercises) should be promoted, in preference to the 'recommended (standardised) methodologies', such as those proposed by the US Environmental Protection Agency.

There are numerous reasons for this recommendation. Metal species determinations are not as well established as many 'total metal' or 'organic compound' determinations, and recommending a specific method and instrumentation will certainly block important progress regarding improvements in the understanding of metal species determination. If the early methods developed by Westöö[9] in the 1960s for methylmercury determination are examined, the detector used (an electron capture detector, ECD) was only sensitive to methylmercury. This analytical solution was quickly adopted, and for a long time methylmercury was the major, if not the only species reported. When the analytical techniques using hyphenated systems coupled to various atomic spectrometers as detectors were employed, other methylated Hg species (*e.g.* dimethylmercury) were reported[10] and this contributed to the improvement in the understanding of the global cycling of mercury.

When dealing with speciation, it should be remembered that the certified reference materials for 'metal species determinations', are the best available for a certain period and represent materials offering the greatest consensus values based on the 'state-of-the-art' at the time of certification.[11] Indeed, in a number of cases, improvements in sample extraction procedures have established that initial values reported for some organometallic species could be higher than previously reported. This was the case for the certification level of monobutyltin in the Canadian CRM PACS 1. Improved extraction, using microwave-assisted procedures, obtained a higher yield of this polar compound.[12] The same considerations apply to mercury species determination in sediments. There is an ongoing controversy suggesting that an 'artificial methylation' of mercury may occur during the sample preparation procedure.[13] This point is still under discussion and has been largely refuted.[11] Research on this point is still ongoing. These two cases indicate the need for the production of CRMs for metal species in a wide variety of matrices, and call for freedom in the development of analytical solutions to achieve the CRMs certified values. Promoting 'standardised analytical protocols' will certainly contribute to a blocking of the system, and slow down the introduction of sound 'metal speciation based' environmental management.

Finally, in addition to achieving better analytical recommendations or improvements in analytical solutions, greater effort should be made towards the better informing of our legislative bodies.[14] As mentioned earlier, the speciation concept has been partly understood and some environmental regulations have been issued in such a way as to ignore the present analytical solutions available. Despite the voluntary efforts made to improve environmental management, this will most probably translate into solutions that are extremely difficult for industry to implement, and result in adverse effects giving totally opposite results to those expected.

In most cases, legislative bodies are keen to get the correct information. It is now the duty of environmental scientists to explain and promote their knowledge of 'metals species determination' to these institutions, so that the environment receives the care that it so urgently needs and deserves.

5 References

1. D. Amouroux, C. Pecheyran and O.F.X. Donard, *Appl. Organomet. Chem.*, 2000, **14**, 236.
2. J.H. Edmons and K.A. Francesconi, *Mar. Poll. Bull.*, 1993, **26**, 665.
3. O.F.X. Donard and J.H. Weber, *Environ. Sci. Technol.*, 1985, **19**, 1104.
4. A. De Diego, N. Dimov, C.M. Tseng, D. Amouroux and O.F.X. Donard, *Appl. Organomet. Chem.*, 2001, **15**, 490.
5. J.H. Waggoner, L.S. Milstein, M. Belkin, K.L. Sutton, J.A. Caruso and H.B. Fannin, *J. Anal. At. Spectrom.*, 2000, **15**, 13.
6. V. Majidi, M. Moser, C. Lewis, W. Hang and F.L. King, *J. Anal. At. Spectrom.*, 2000, **15**, 19.
7. J.M. Ruiz, G. Bachelet, P. Caumette and O.F.X. Donard, *Environ. Poll.*, 1996, **90**, 195.
8. Ph. Quevauviller and O.F.X. Donard, in *Element Speciation in Bioinorganic Chemistry*, S. Caroli, ed., John Willey & Sons, Inc. New York, p. 331.
9. G. Westöö, *Acta Chem. Scand.*, 1996, **20**, 2131.
10. Ph. Quevauviller, O.F.X. Donard, F.M. Martin, J. Schneider and J.C. Wasserman, *Appl. Organomet. Chem.*, 1992, **6**, 221.
11. Ph. Quevauviller, F. Adams, J. Caruso, M. Coquery, R. Cornelis, O.F.X. Donard, L. Ebdon, M. Horvat, R. Lobinski, H. Muntau, R. Morabito and M. Valcárcel. *Anal. Chem.*, http://pubs.acs.news Dec. 1999.
12. J. Szpunar, V. Schmitt, O.F.X. Donard and R. Lobinski, *Trends Anal. Chem.*, 1996, **15**, 181.
13. R. Falter and G. Ilgen, *Fresenius' J. Anal. Chem.*, 1997, **358**, 407.
14. O.F.X. Donard and J. Caruso, *J. Spectrochim. Acta Part B*, 1998, **53**, 157.

CHAPTER 7

Mercury – Do We Know Enough?

MILENA HORVAT

1 Introduction

Mercury is widely considered to be amongst the highest priority environmental pollutants of continuing concern on the global scale. Although there is a continuous problem of occupational exposure to inorganic mercury, predominantly elemental mercury, possible effects on broader sections of the population due to widespread dispersal of mercury in the environment, have become the major concern in recent years. Mercury is one of the most highly bioconcentrated trace metals in the human food chain, and many national and international agencies and organizations have targeted mercury for possible emission control. Mercury toxicity depends on its chemical form, among which alkylmercury compounds are the most toxic. The biogeochemistry of mercury has received considerable attention because of the toxicity of methylmercury compounds (MeHg), the accumulation of mercury in biota, and its biomagnification in aquatic food chains. Concerns about mercury are based on its effects both on ecosystems and human health. The principal pathway for human exposure is the consumption of contaminated fish. Numerous recent studies have concluded that the majority, if not all, of the mercury that is bioaccumulated through the food chain is as MeHg. Therefore, knowledge of the concentration, transport, transformation and dynamics of MeHg in aquatic ecosystems is needed to predict potential impact on humans, as well as on aquatic life.

One of the most comprehensive reviews of current understanding of mercury is summarised in the USA Environmental Protection Agency's (EPA) Mercury Report to Congress[1] that is available on the internet at http://www.epa.gov/ttn/uatw/112nmerc/mercury.html.

2 Chronology of Mercury Investigation

Mercury toxicity was reported in ancient times; the present environmental and health interest in mercury issues, however, started largely in the 1950s with the

Minamata disaster caused by MeHg poisoning, which occurred by ingestion of large quantities of fish and/or shellfish polluted by industrial effluent containing MeHg in Minamata, Japan. The second epidemic occurred in Niigata, Japan for the same reason. It should be stressed that in both cases mercury was discharged from chemical factories in the form of MeHg (up to 70% mercury was in a methylated form), which was formed from inorganic mercury during the technological process for production of acetaldehyde.[2] Both epidemics accounted for several hundred deaths and a large number of victims. Contamination of terrestrial systems by mercury compounds used in agriculture as fungicides and seed disinfectants represented a large scale problem in terrestrial ecosystems that caused severe declines in bird populations.[3] A large epidemic was also reported in Iraq, the cause of which was the ingestion of seed treated with alkylmercury fungicide. Over 450 people died and many more people were permanently affected. Since these epidemics and the fact that MeHg can be formed from inorganic mercury in the environment, methylmercury pollution has been suspected in a large number of mercury polluted areas, but fortunately no other large epidemics have been reported. During the 1960s, accumulation of mercury was noted in wild life and fish in Sweden and it was becoming apparent that the mercury problem was far more extensive than an isolated acute local situation, such as Minamata and Niigata. Similar observations were noted in the Great Lakes of the United States and Canada. Since the 1970s both countries have banned the sale of fish containing mercury levels above $0.5\,mg\,kg^{-1}$ fresh weight. In the 1970s a number of countries introduced regulations and measures by which mercury emissions into the environment were significantly reduced. Extensive investigations were again evoked in the late 1990s based on the discovery that, in both Scandinavia and North America, long-range transport of Hg^0 and the potential role of acidification have become major facts in the future exposure of humans to MeHg. As a result of these hazards, MeHg exceeded regulatory limits in fish ($0.5–1.0\,mg\,kg^{-1}$, fresh weight) in a large number of remote lakes. It should be emphasized that the role of the development of sensitive and specific methods for mercury speciation and analysis played an important role in the better understanding of the mercury biogeochemical cycle (Figure 1).[4]

A large number of laboratories have commenced intensive investigations during the last decade. This has resulted in a clear need to provide a forum for the presentation of results, discussions and the exchange of innovative ideas for future research and to communicate with policy makers, legislators, experts from industry and public representatives in order to promote the direct use of scientific and technical data in the field of environmental protection and control. As a response to these needs, the international conference series on *Mercury as a Global Pollutant* was initiated in 1990. Growing interest in mercury issues is evidenced in the increasing number of participants world-wide: 200 in Gävle, Sweden in 1990; 300 in Monterey, California, USA in 1992; 400 in Whistler, British Columbia, Canada in 1994; over 400 in Hamburg, Germany in 1996; and over 450 in Rio de Janeiro in 1999. The global nature of mercury issues is also shown by increasing international participation, from a few nations at Gävle, mostly from the Northern hemisphere, to more than 40 states in Rio de Janeiro,

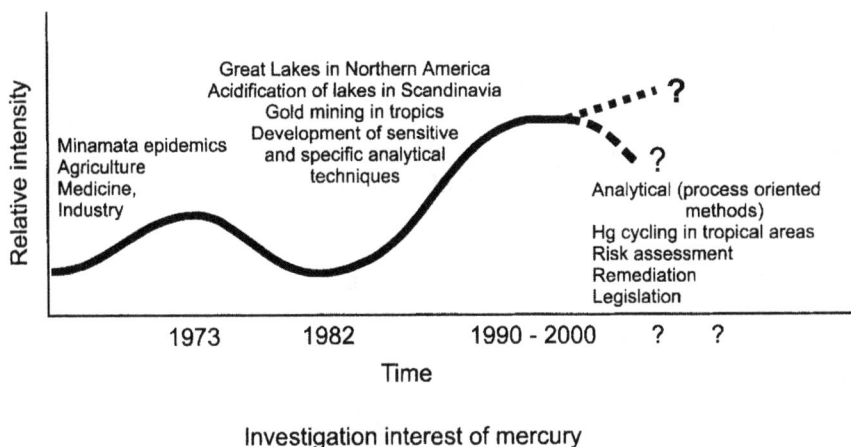

Investigation interest of mercury

Figure 1 *Temporal relative intensity of mercury research in the environment and human health*

where almost half of the participants were from developing countries in the tropics. One of the reasons is related to a revival of gold exploration in developing countries in Latin America and Africa, in China, Vietnam, Cambodia, Thailand, Indonesia, Philippines and the former Soviet Union, and in which mercury is used to extract the gold. This has caused an increase in mercury research in order to set environmental quality criteria in these areas. It is estimated that about 10 million people are currently engaged in gold extraction activities. Latin America, particularly the Amazon region, is considered to be the most seriously impacted by mercury releases from gold production, with over half a million people directly involved in the activity, and at least three times that number indirectly dependent upon it.[5] These observations can be further supported by the number of published papers in peer reviewed international journals.[6]

The interest in current and future mercury investigations will be focused on those areas where the most uncertainty is currently noted. Those are related to (1) environmental cycling of mercury in various environmental conditions, (2) toxicological – long-term exposure to low concentrations of mercury compounds, (3) risk assessment, (4) remediation and (5) legislation.

3 The Role of Analytical Chemistry

During recent years, new analytical techniques have become available that have contributed significantly to the understanding of mercury chemistry in natural systems. In particular, these include ultrasensitive and specific analytical equipment and contamination-free methodologies. These improvements have eventually permitted the determination of total and major species of mercury to be made in air, water, sediments and biota.

The techniques developed for the determination of MeHg in biological and sediment materials involve, in most cases, a succession of analytical steps (extraction, derivatization, separation, detection) which may all be prone to systematic errors. Over ten years ago, these methods were poorly validated due to a lack of evaluation programmes and of certified reference materials.[7,8] The situation has improved recently, thanks to the development of more sensitive and specific analytical techniques, the organization of inter-laboratory studies and the availability of certified reference materials (CRMs), produced by a number of well recognized producers. Unfortunately, there are only a few biological samples and two sediment samples certified for MeHg compounds; these materials do not cover present needs.[8] Controversially, most of the currently available CRMs are of marine origin, whilst most work is conducted in terrestrial ecosystems and fresh-water environments. In addition, CRMs for human exposure assessment, such as blood, urine and hair at several levels of concentration are far from satisfactory. In the absence of CRMs, many other actions should be undertaken to achieve, improve and/or maintain quality of data, including participation in inter-laboratory studies, proficiency testing and production of laboratory reference materials.

The most obvious sources of error are related to incomplete extraction, low and irreproducible recoveries, insufficient peak resolution in chromatography and transformation of mercury species that may lead to artifacts. In the case of readily soluble samples such as fish and mussels, speciation analysis has achieved most success. However, with solids, techniques to remove or solubilize MeHg are difficult to validate by spiking or tracer approaches, as it is difficult to prove that complete extraction/separation has been achieved.[9] A classical example of this difficulty is the speciation of MeHg in sediments and soils. The only feasible approach adopted in the certification of MeHg reference materials was to use different analytical approaches, *i.e.* various extraction/separation schemes and detection methods. Good agreement of the results obtained not only provides assurance that the data are meaningfully comparable but is, of course, also strong evidence of the correctness of data and represents an excellent reference point for all users dealing with the determination of MeHg in sediments.[8,10] However, there is a suspicion that MeHg may be formed during an isolation procedure from inorganic mercury present in the sample, especially when separations are performed at elevated temperatures.[11] These conclusions were based on limited spiking experiments with Hg(II). So far, the conclusion is that there is a lack of evidence of such a source of analytical error, due to the question of whether Hg^{2+} in acidic solution added to the sample, is equivalent to endogenous, inorganic mercury. From the biogeochemistry of mercury, it is well known that free ionic Hg^{2+} is a precursor for mercury transformation mechanisms, including methylation. Normally, the quantity of free ionic Hg^{2+} in the environment is low; probably most of the endogenous inorganic mercury in the sample is bound as humates, to sulfur *etc.*[12] Moreover, by spiking the sample with high quantities of Hg^{2+}, the sample is seriously altered, resulting in the transformation of the mercury species of interest. Therefore, such an analytical approach is in contradiction to the basic principles of speciation analysis, which requires preservation of the mercury species originally present (endogenous) in the sample. So far, no experiments have been conducted with enriched (or radioactive) isotopes of Hg^{2+} incorporated into the test material by a natural process, which could then be used to test for artifact MeHg formation, representing a much more valid experimental design for testing the accuracy of analytical procedures.

Future work in the field of mercury speciation will have to answer the question of possible artifact formation during analytical methods. In particular, research efforts are needed to understand more fully the extraction chemistry in the area of speciation analysis. However, taking into the account the complexity of the mercury cycle presented in Figure 3 (p. 133), much more needs to be done in the validation and optimization of analytical methods for mercury flux measurements at the water/air, sediment/water and soil/air interfaces. The number and nature of variables, such as experimental design, spatial heterogeneity and temporally changing environmental conditions, can significantly influence field flux measurement data.[13] In support of these measurements there is a strong need to develop techniques for *in situ* or on-line measurements, including specific bio-sensor techniques.

4   Mercury as a Global Pollutant

Mercury, one of the most toxic of the heavy metals, is released into the environment from a variety of natural sources (*e.g.* the ocean surface and other water surfaces, soils, minerals, vegetation located on land, forest fires and volcanoes) and through anthropogenic emissions (metal extraction processes, agricultural uses, paints, waste disposal, incinerators, smelting of ores and other industrial minerals, power plants burning fossil fuels for electricity generation). Recent studies[14,15] show that human activity contributes about 50–75% (*e.g.* 3500–4500 tons) of the total yearly input from all sources ($7000 \, \text{ton} \, y^{-1}$) (Figure 2). About half of the anthropogenic emissions, (*ca.* $2000 \, \text{ton} \, y^{-1}$) appear to enter the global mercury cycle, while the other half is deposited locally. As a consequence, human activities have tripled the concentrations of mercury in the atmosphere and in the surface layers of the oceans. It is estimated that 60% (*ca.* 5000 tons) of the total mercury is deposited in the atmosphere on terrestrial environments, and the remainder into the ocean. This refers to the oxidation and adsorption of mercury in abundant terrestrial aerosols. The ocean receives about 90% of its mercury through wet and dry deposition as Hg(II), and the remaining ($200 \, \text{ton} \, y^{-1}$) from river inflows. Particulate scavenging and removal to the deep ocean is equal to the riverine mercury flux ($200 \, \text{ton} \, y^{-1}$). Due to biological reduction of deposited Hg(II) in the mixed layer of the ocean and its volatility, most Hg^0 deposited ($2000 \, \text{ton} \, y^{-1}$) is re-emitted to the atmosphere. This active process, and the minimal removal of mercury to the deep ocean, makes terrestrial systems the dominant sink. The model presented in Figure 2 is based on rather limited data, and uncertainties may account for a factor of two or more.[16]

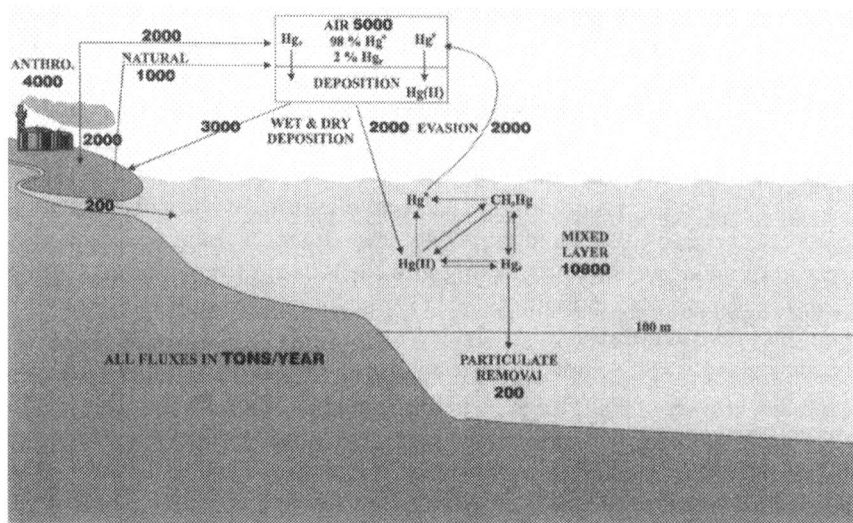

Figure 2   *The current global Hg cycle*
(adapted from Fitzgerald and Mason[14])

5   The Mercury Cycle

Mercury is a heavy metal (density of $13.5 \, g \, ml^{-1}$ at $200 \, °C$) and is the only metal to exist as a liquid at room temperature. In the inorganic form mercury exists in three oxidation states: Hg^0 is referred to as elemental mercury, Hg_2^{2+} is best known as mercurous mercury and is very little used today and Hg^{2+}, divalent or mercuric mercury, forms a wide range of salts and organic compounds. Organic mercury compounds are those in which mercury is covalently linked to at least one carbon atom. A number of those compounds can be found in nature, but the most important ones are monomethylmercury compounds (CH_3Hg^+) and dimethylmercury (CH_3HgCH_3). There are a number of other commercially used mercury compounds and environmental problems with these are usually localized. In general, however, production and usage of organomercurials is declining.

In the environment, mercury can also exist in a number of different physical and chemical forms exhibiting a wide range of properties. Biogeochemical conversion between these different forms provides the basis for the complex distribution pattern of mercury in local and global cycles, and for biological enrichment and effects. The most important chemical forms are elemental mercury (Hg^0), divalent inorganic mercury [Hg(II)], monomethylmercury (MeHg), and dimethylmercury (DMHg). There is a general biogeochemical cycle (Figure 3) by which these different forms may interchange in the atmospheric, aquatic and terrestrial environments. Microbial and abiological transformations play a key role in this cycle by altering the chemical form of mercury compounds.

Environmental problems are mostly related to mercury compounds formed naturally, of which monomethylmercury compounds are the most toxic. In most cases, the complete identity of these compounds is unknown, except for the

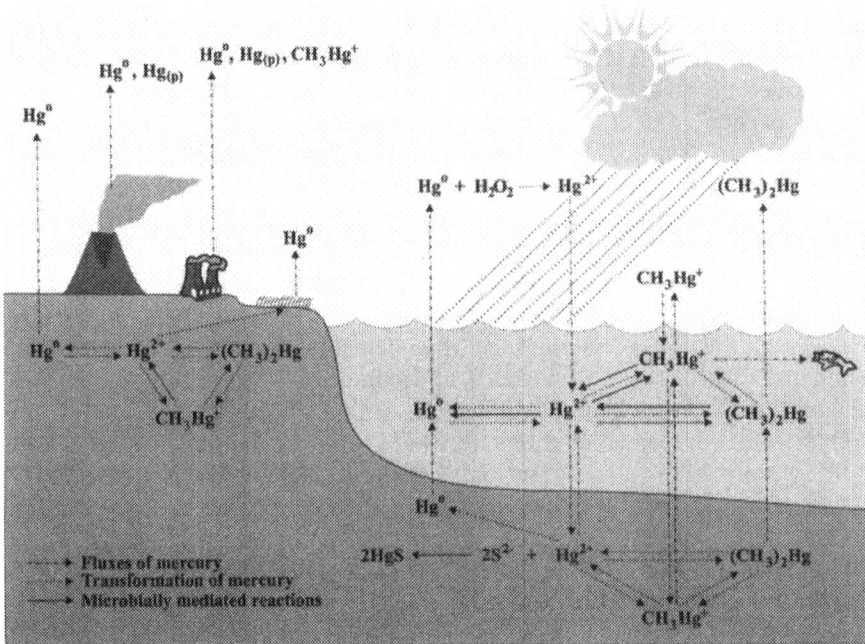

Figure 3 *Mercury cycle in the biosphere*
(adapted from Barkey[34])

monomethylmercury cation CH_3Hg^+, which may be associated with a simple anion, or a large, charged molecule (*e.g.* proteins, organic ligands *etc.*).

Most studies of mercury cycling have so far been made in freshwater systems,[17,18] even though it is well known that the primary exposure of humans to MeHg is through the consumption of marine fish and fish products.[19] In principle, mercury cycles in freshwater and marine aquatic systems are basically similar, but with some distinct differences. The most important feature in both systems is the *in situ* bacterial conversion of inorganic mercury species into the more toxic MeHg, which concentrates in fish muscle.[20-24]

The key feature that influences mercury distribution in aquatic environments is the high stability of association with sulfur and carbon, the stability of the volatile elemental form and its strong affinity for particulates. Consequently, most inorganic and organic mercury appears to be bound to particles, colloids and high molecular weight organic matter, where it is probably co-ordinated with sulfur ligands on particles (distribution coefficients are in the 10^5-10^6 ml g^{-1} range). In turbid rivers most mercury is, therefore, transported by suspended matter. Only a small part of the mercury in fresh, estuarine and sea water is likely to be present in a dissolved form.[25-27]

After entering the atmosphere, mercury exchanges and cycles, to be deposited in the ecosystem almost exclusively as Hg(II). When mercury enters a surface (soil or water) Hg(II) can be methylated to MeHg and perhaps later de-methylated. This is the first step in the aquatic and terrestrial bioaccumulation processes of MeHg. The mechanisms of synthesis/decomposition of MeHg are not very well understood. The main factors that affect the levels of MeHg in fish are the dietary trophic level of the species, the age of the fish, microbial activity and the mercury concentration in the upper layer of the local sediment, dissolved organic carbon content, salinity, pH and redox potential.[20-23] Recent work suggests that sulfato-reducing bacteria are the most important methylating agents, along with environmental conditions, existing in transition regions between oxygenated and anoxic conditions.[28] Methylation–demethylation reactions are assumed to be widespread in the environment, and each ecosystem attains its own steady-state equilibrium, with respect to the individual species of mercury. However, due to bioaccumulation of MeHg, methylation is more prevalent in the aquatic environment than demethylation. In areas highly contaminated with mercury, a biochemical decontamination process seems to be responsible for the minimal net production and accumulation of MeHg.

Once MeHg is formed, it enters the food chain by rapid diffusion and tight binding to proteins in aquatic biota, and attains its highest concentrations in the tissues of fish at the top of the aquatic food chain, due to biomagnification through the trophic levels. For example, most predatory fish species show MeHg values above $1 \, mg \, kg^{-1}$, while concentrations in water are commonly less than $1 \, ng \, l^{-1}$ (*e.g.* an amplification factor of about 10^6).[21,29,30] Although MeHg is the dominant form of mercury in higher organisms, it represents only a very small amount of the total mercury in aquatic ecosystems and in the atmosphere.

Mercury bioaccumulation and trophic transfer have been extensively studied in fresh and marine water systems, but less so in the terrestrial environment.[31] The humic layer of soil is the primary site of mercury accumulation in forest soil, while the uptake from soils by terrestrial vegetation is usually poor, due to the protective role of roots preventing active uptake. Fungi may actively accumulate mercury from soil, but very few studies have so far been conducted suggesting the food safety problems. Vascular plants, mosses and lichens may absorb mercury directly from the air, while foliage represents a dynamic exchange surface, with stomatal uptake of elemental mercury, dry deposition of particulate matter containing mercury, or adsorption of gaseous Hg(II). Although biomagnification in terrestrial food chains is less pronounced, mercury concentrations are higher in terrestrial carnivore species than in herbivores.[32,33]

6   Mercury Toxicity

All forms of mercury are toxic and this toxicity strongly depends on the chemical form. The properties which make mercury a potent environmental toxicant are: (1) the strong affinity of Hg(II) and organomercurials for thiol groups, (2) a tendency to form covalent bonds with organic molecules, (3) the high stability of the Hg–C bond that results from a low affinity for oxygen and (4) a strong tendency to maximize bonding to two ligands in linear stereochemistry.[34]

The kidney is a target tissue for retention of mercury in populations exposed to inorganic mercury compounds or mercury vapour whilst the brain is a target organ for exposure to organic mercury and mercury vapour, which both pass blood–brain and placental barriers. MeHg is more toxic due to the irreversibility of its effects on the central nervous system. Organomercury compounds are more readily absorbed in the gastrointestinal tract than inorganic ones, and are subsequently excreted preferentially in faeces rather than urine, which is the dominant route for excretion of inorganic mercury.

In Table 1 a summary of the toxic effects of mercury in humans is given.

Most understanding of mercury toxicity is based principally on the results of experimental animal studies. All forms of mercury cross the placental barrier; however, elemental mercury and alkyl-mercury compounds show a much higher foetal uptake than inorganic mercury compounds. MeHg levels in foetal red blood cells are 30% higher than in maternal red cells. Maternal milk may also represent a significant neonatal exposure to MeHg. Most evidence of clinical signs, symptoms and neuropathology of MeHg toxicity in humans has been obtained from studies of the epidemics in Iraq and Japan[35,36] and of populations eating mercury-contaminated fish, as well as cases of occupational exposure.

The methylated forms of mercury are of most concern for human health. MeHg is a potent toxin that causes impairment of the central nervous system and developmental toxicity, in humans. MeHg readily crosses the walls of the gastrointestinal tract, due to its fast transport through biological membranes, thus accumulating in the envelopes of nerve cells causing neurological damage. MeHg bound to proteins in tissue is relatively stable and only slowly degraded and excreted from the body. As a result MeHg is accumulated by organisms throughout their lifetimes, and the concentrations of MeHg are further magnified through the trophic interactions of the food web. Fish can therefore contain levels that are magnified by a factor of 10^7 above the mercury concentration in water.

It is generally accepted and well known that the major risks to human health arise from the neurotoxic effects of mercury. The assessment of genotoxicity and carcenogenicity of mercury compounds is difficult to interpret, due to the complexity and variability observed in numerous studies.[37] Although different chemical forms of mercury tended to produce qualitatively comparable genetic effects, MeHg derivatives and other ionizable organomercury compounds were more active in short-term tests than either non-ionizable mercury compounds (DMHg), or inorganic salts (mercuric chloride). The results of cytogenetic monitoring in peripheral blood lymphocytes of humans exposed to elemental mercury or its compounds from accidental, occupational and alimentary sources were either negative or uncertain. Both genotoxic and non-genotoxic mechanisms may contribute to the renal carcinogenicity of mercury, demonstrated in male rodents treated with MeHg.[38]

Recent studies suggest[39] that other effects may also be present when accompanied by other risk factors such as a poor diet comprised of saturated animal fats and low antioxidant intake (*e.g.* poor in selenium). It is known that unsaturated lipids are subject to peroxidative degradation, a process that plays a key role in the development of the arteriosclerotic process. Mercury can reduce the anti-oxidative capacity of selenium superoxide dismutase, catalase activity and glutathione peroxidase activity, and can promote free radical stress and lipid peroxidation. A correlation between increased accumulation of MeHg and the

Table 1 *Summary of toxic properties of different forms of mercury in humans*

Form	Exposure	Effects	Biological indication	Disposition and mechanism of action
Hg⁰	*Occupational:* chlor-alkali industry, production of thermometers, thermostats and fluorescent bulbs, mercury mining, dentistry *Non-occupational:* Dental amalgam fillings	*Severe exposure:* Tremor, gingivitis, erethism, loss of memory, emotional and psychological disturbance, damage to kidneys *Lower exposure:* Cognitive deficits, mild proteinuria, insomnia, loss of appetite, immunological disturbances. Damage is reversible	*Hg in urine:* chronic exposure, indication of Hg levels in kidney *Hg in blood:* indicator of short term exposure *Exhaled air:* short term exposure No good indicators for brain	Inhaled Hg^0 is absorbed in the lung and enters the blood stream; high lipid solubility allows it to cross the blood/brain and placenta barriers easily. Dissolved Hg^0 is oxidized in red cells, brain, liver, lung, and other tissues where it may inhibit the activity of some enzymes that contain SH groups, denaturate proteins, damage cell membranes. At high concentrations it causes cell death and destruction of tissue. Mechanism of damage to the CNS is still not well understood $T_{1/2}$: 1–2 months
Hg(I)	Removed from medical use. Rare in use	Acrodynia, pink disease. Damage is reversible	—	Inhibition of enzymes; its action is poorly understood
Hg(II)	Antiseptic, leather industry, production of batteries, fungicides, use in bleaching soaps and creams	Chronic toxicity: neurological disorders similar to the effects of Hg^0. Repeated exposure to low doses affects the immune systems. Acute exposure may cause irreversible damage of kidney and indirectly cardiovascular collapse	Urine	Similar to Hg^0, except direct passage through the above mentioned biological barriers is more difficult $T_{1/2}$: 1–2 months

| MeHg | Fungicide, slimicide, food – mainly fish and other marine products | Immediate damage of neuronal cells and delayed symptoms of sensory disturbance, constriction of visual field, deafness, motor aberrations, mental disorders, cramps, paralysis | Blood and hair | MeHg is distributed in all tissues including brain. Prenatal damage occurs in all parts of the brain while in adults the damage is local. Inhibition of protein synthesis. It affects cell division and abnormal neuronal migration. It causes the destruction of microtubules in neuronal and astrocystic cells. MeHg damage to the CNS is still unexplained $T_{1/2}$: 70 days. Damage is irreversible |

incidence of mortality from ischaemic heart disease was found in the inhabitants of eastern Finland, whose nutrition is primarily comprised of fish, meat, and saturated animal fat and a low selenium intake.[39] The preliminary, unpublished data of an international study conducted by IARC,[40] which also included 1589 workers from mercury mines, show, among other things, a relation between long-term work in European mercury mines with exposure to elemental mercury (Hg^0) vapour, and increased mortality due to ischaemic heart disease. This result is specific only for workers in the mercury mine of Idrija, Slovenia, and not for workers from the mercury mines and mills of Spain and Italy. It is assumed that the differences in mortality due to ischaemic cardiac disease in mines and mills are due to differences in nutrition in particular countries. It could also be a consequence of higher exposure to elemental mercury in the Idrija mercury mine (where it occurs naturally as native mercury as well as cinnabar), leading to depletion of catalase and lowered anti-oxidative capacity.

7   What is Safe?

The joint FAO/WHO Expert Committee on Food Additives[35,41] provisionally recommended that total mercury intake should not exceed $5 \, \mu g \, kg^{-1}$ of body weight per week, with no more than $3.3 \, \mu g \, kg^{-1}$ per week as MeHg. Of 26 nations reviewed,[42] none showed mean dietary intakes approaching these values.

In the absence of occupational exposure, human intake of mercury is dominated by two sources: the diet and amalgam dental fillings. MeHg in fish and fish products is the major source of mercury and represents up to 85% of mercury in total intake from the diet.[43] Typically less than 1% of total inorganic mercury intake is derived from drinking-water. Amalgam fillings contain 50% of their weight as metallic mercury. Chewing causes a mechanical release of Hg(II) and volatilization of Hg^0, part of which is inhaled and absorbed into the lungs. This is the predominant source of inorganic mercury in the non-exposed population.

Consumption of contaminated fish is the primary route of exposure to MeHg for humans and wildlife. Public health regulations in the USA prohibit consumption of fish with tissue mercury concentrations of $> 1 \, mg \, kg^{-1}$ fresh weight, while some other countries regulate at the $0.5 \, mg \, kg^{-1}$ level.

In 1997, the US Environmental Protection Agency (EPA) set a new guideline for methylmercury in the diet of 0.1 microgram of mercury per kilogram of body weight per day ($0.1 \, \mu g \, kg^{-1} \, day^{-1}$). This is 4.7 times as strict as the World Health Organization's (WHO's) standard of $0.47 \, \mu g \, kg^{-1} \, day^{-1}$. The average concentration of mercury in fresh and marine fish is about $0.2 \, mg \, kg^{-1}$. In practical terms this means that an average person weighing 60 kg can only consume about 30 g of fish per day.[44] This recommendation is based on epidemiological studies conducted in Iraq after the MeHg epidemic. The value for a reference dose is currently under review, before being officially adopted. Two large epidemiological studies in the Seychelles[45] and the Faroe Islands[46] that were designed to evaluate childhood development and neurotoxicity in relation to foetal exposures to MeHg will provide more information for further refinement of reference dose values.

8 Conclusions

The state-of-the-art for mercury is continuously and rapidly evolving. The EPA report[1] and comments by its reviewers clearly show that, "while we do not know nearly as much as we would like to about environmental mercury, we know a lot" (Paul Mushak, Mercury Study Peer Review Workshop Chair for the EPA Mercury Report to Congress). Risk from mercury exposure cannot as yet be quantified, and future research should be focused in areas where the most uncertainties are currently encountered. Our current scientific understanding of the fate and transport of mercury is not sufficient to predict how much MeHg in fish consumed by humans in certain areas is related to anthropogenic emissions of mercury as compared to natural sources. Better information on the consumption patterns of fish and other foods containing mercury in the populations concerned is also needed to better estimate MeHg exposure. Special attention should be oriented towards the most sensitive sectors of the population, such as women of childbearing age, pregnant women and children. More knowledge is also needed to understand better the effects of long-term, low-level exposure to Hg^0 and other forms of mercury. In the field of environmental cycling of mercury and parameterization of various environmental factors, mercury fluxes and transformation mechanisms are needed for better prediction of mercury behaviour in different environments, and for the development of more reliable modelling. The quality of analytical results is still an issue that deserves great attention. More reference materials are needed to cover various environmental matrices and concentration ranges. Of those measurements for which reference materials are not available, particularly in 'dynamic' measurements such as fluxes and transformations at phase boundaries, comparability of measurement data on a global scale is urgently needed. Only with an integrated approach, where basic science and socioeconomic factors are combined, will it then be possible to formulate sound control measures.

9 References

1. EPA Mercury Study Report to the Congress, EPA 452/R-97-0003, EPA, USA, December 1997.
2. T. Takeuchi and K. Eto, *The Pathology of Minamata Disease, a Tragic Story of Water Pollution*, Kyushu University Press. Inc., Okinawa, Yamaguchi, 1999.
3. N. Fimreite, in *The Biogeochemistry of Mercury in the Environment*, ed. J.O. Nriagu, Elsevier, Amsterdam, 1979.
4. O. Lindquist, K. Johansson, M. Aastrup *et al.*, *Water, Air Soil Pollut.*, 1991, **55**, 1.
5. C. Ramel, in *Global and Regional Mercury Cycles: Sources, Fluxes and Mass Balances*, eds. W. Baeyens, R. Ebinghaus and O. Vasiliev, Kluwer Academic Publishers, Netherlands, 1996, pp. 505–514.

6. M. Sperling, *Historical Perspectives of Trace Element Speciation: a Manufacturer's Point of View*, Presentation at the Final Meeting, Speciation 21, 26–27 February, 2000, Pau, France.
7. M. Horvat, in *Global and Regional Mercury Cycles: Sources, Fluxes and Mass Balances*, eds. W. Baeyens, R. Ebinghaus and O. Vasiliev, Kluwer Academic Publishers, Netherlands, 1996, pp. 1–31.
8. M. Horvat, *Chemosphere*, 1999, **39**, 1167.
9. A.R. Byrne, *Analyst*, 1992, **117**, 251.
10. Ph. Quevauviller and M. Horvat, *Anal. Chem.*, 1999, **71**, 155A, Letter to the Editor; *Chemosphere*, Special issue, 1999, **39**(7).
11. G.M. Varsal, T.K. Buachidze, N.S. Velyukhanova and D.N. Chkhetia, in *Global and Regional Mercury Cycles: Sources, Fluxes and Mass Balances*, eds. W. Baeyens, R. Ebinghaus and O. Vasiliev, Kluwer Academic Publishers, Netherlands, 1996, pp. 403–414.
12. M.S. Gustin and S.E. Lindberg, Assessing the contribution of natural sources to the global mercury cycle: The importance of intercomaparing dynamic flux measurement, *Fresenius' J. Anal. Chem.*, 2000, **366**, 417.
13. W.F. Fitzgerald and R.P. Mason, in *Global and Regional Mercury Cycles: Sources, Fluxes and Mass Balances*, eds. W. Baeyens, R. Ebinghaus and O. Vasiliev, Kluwer Academic Publishers, Netherlands, 1996, pp. 85–108.
14. J.M. Pacyna, in *Global and Regional Mercury Cycles: Sources, Fluxes and Mass Balances*, eds. W. Baeyens, R. Ebinghaus and O. Vasiliev, Kluwer Academic Publishers, Netherlands, 1996, pp. 161–178.
15. *Global and Regional Mercury Cycles: Sources, Fluxes and Mass Balances*, eds. W. Baeyens, R. Ebinghaus and O. Vasiliev, Kluwer Academic Publishers, Netherlands, 1996; *Mercury Pollution – Integration and Synthesis*, eds. J.C. Watras and J.W. Huckabee, Lewis Publishers, Boca Raton, FL, 1994.
16. *Mercury as a Global Pollutant*, eds. D. Porcela, J.W. Huckabee and B. Wheatley, Kluwer Academic Publishers, Dordrecht, The Netherlands, 1995, reprinted from *Water, Air, Soil Pollut.*, **80**, 1.
17. W.F. Fitzgerald and T.W. Clarkson, *Environ. Health Perspect.*, 1991, **96**, 159.
18. C.C. Gilmour and E.A. Henry, *Environ. Pollut.*, 1991, **71**, 131.
19. K.R. Rolfhus and W.F. Fitzgerald, *Water, Air, Soil Pollut.*, 1995, **80**, 291.
20. R.P. Mason and W.F. Fitzgerald, *Nature (London)*, 1990, **347**, 457.
21. R.P. Mason, W.F. Fitzgerald, J. Hurley, A.K. Hanson, Jr., P.L. Donaghay and J. Sieburth, *Limnol. Oceanogr.*, 1993, **38**, 1227.
22. T. Barkey, R.R. Turner, E. Saouter and J. Horn, *Biodegradation*, 1992, **3**, 147.
23. D. Cossa, M. Coquery, C. Gobeil and J.-M. Martin, in *Global and Regional Mercury Cycles: Sources, Fluxes and Mass Balances*, eds. W. Baeyens, R. Ebinghaus and O. Vasiliev, Kluwer Academic Publishers, Netherlands, 1996, pp. 229–247.
24. M. Coquery, D.Cossa and J.-M. Martin, *Water, Air, Soil Pollut.*, 1995, **80**, 653.
25. R.F.C. Mantoura, A. Dickon and J.P. Riley, *Estuarine Coastal Mar. Sci.*, 1978, **6**, 387.
26. R.S. Oremland, C.W. Culbertson and M.R. Winfrey, *Appl. Environ. Microbiol.*, 1991, **57**, 130.
27. R.P. Mason and W.F. Fitzgerald, *Water, Air, Soil Pollut.*, 1991, **56**, 745.
28. R.C. Lathrop, P.W.R. Rasmussen and D. Knauer, *Water, Air, Soil Pollut.*, 1991, **56**, 295.
29. A. Boudou and F. Ribeyre, in *Metal Ions in Biological Systems*, eds. A. Sigel and H. Sigel, Vol. 44: *Mercury and Its Effects on Environment and Biology*, Marcel Dekker, New York, Basel, Hong Kong, 1997, pp. 289–319.

30. A. Gnamuš and M. Horvat, in *Mercury Contaminated Sites*, eds. R. Ebinghouse, R.R. Turner, L.D. de Lacerda, O. Vasiliev and W. Solomons, Springer-Verlag, Berlin, Heildelberg, 1999, pp. 281–320.

31. M. Lodenius, in *Mercury Pollution: Integration and Synthesis*, eds. C.J. Watras and J.W. Huckabee, Lewis Publishers, Boca Raton, FL, 1994, pp. 343–354.

32. IPCS/WHO, *Environmental Health Criteria 101: Methylmercury*, World Health Organization, Geneva, 1990.

33. IPCS/WHO, *Environmental Health Criteria 118, Inorganic Mercury*, World Health Organization, Geneva, Switzerland, 1991.

34. S. De Flora, C. Bennicelli and M. Bagnasco, *Mut. Res.*, 1994, **317**, 57.

35. P. Boffeta, E. Merler and H. Vainio, *Scand. J. Work Environ. Health*, 1993, **19**, 1.

36. J.T. Salonen, K. Seppanen, K. Nyyssonen, H. Korpela, J. Kuahanen, M. Kantola, J. Tuomilehto and H. Esterbauer, *Int. J. Epidemiol.*, 1995, **91**, 645.

37. P. Boffetta, M. Garcia-Gomez, V. Pompe-Kirn, D. Zaridze, T. Bellander *et al.*, *Cancer Causes Control*, 1998, **9**, 591.

38. *Evaluation of Certain Food Additives and Contaminants.* Twenty-second report of the Joint FAO/WHO Expert Committeee on Food Additives, World Health Organization, Geneva, 1978 (WHO Technical Report Series, No. 631).

39. R.M. Parr *et al.*, *Human Dietary Intakes of Trace Elements: a Global Literature Survey for the Period 1970–1991*, IAEA, Vienna, 1992 (NAHRES-12).

40. *Trace Elements in Human Nutrition and Health*, WHO, in collaboration with FAO and IAEA, Geneva, 1996, ISBN 92 4 156173 4.

41. R.K. Mahafey, G.E. Rice and R. Schoeny, *Mercury Study Report to Congress Volume IV: Characterisation of Human Health and Wildlife Risk from Mercury Exposure in the United States*, EPA-452/R-97-009, Washington, DC, December 1997.

42. J. Myers, W. Davidson, C. Cox, C.F. Shamlaye, M.A. Tanner, D.O. Marsh, E. Cernishiari, L.W. Lapman, M. Berlin and T.W. Clarkson, *Neurotoxicology*, 1995, **16**, 711.

43. P. Grandjean, P. Weihe, R.F. White, F. Deves, S. Araki, K. Yokoyama, K. Murata, N. Sorensen, R. Dahl and P.J. Jorgensen, *Neurotoxicol. Teratol.*, 1997, **20**, 1.

CHAPTER 8

Organotin Compounds in the Environment: Still a Critical Issue

O.F.X. DONARD, G. LESPES, D. AMOUROUX AND
R. MORABITO

1 General Background

'Sustainable development' is often referred to as one aim of society today. There are consequently attempts to achieve controlled industrial development, with increased protection of the environment. These objectives are well established for a large variety of chemical products used and released into the environment. During the 1960s there was great concern about the fate and impact of persistent organic pollutants (POPs). Although the issues concerning such materials have been well identified, it does not follow that environmental solutions have been applied worldwide. In spite of the considerable efforts that have been made during the last 30 years with regard to these issues, it is clear that our understanding is lacking and our reaction time in dealing with these problems is far too long.

Some 25–40 years ago, the introduction of organotin compounds into a large variety of industrial products contributed to the development of a number of important sectors within the chemical industry. The high toxicity of the trisubstituted species to a variety of organisms was quickly established, and the materials were immediately used in a large variety of biocidal applications. The most successful of these was tributyltin (TBT, see Figure 1) in self-polishing, anti-fouling paint formulations, and these materials were employed on major sea-going vessels. Leisure boats also adopted this most efficient paint, and have, unwittingly, contributed to the contamination of coastlines throughout the world.

Historically, copper oxides were used in early anti-fouling paints but, due to their high price and poor efficiency, organoarsenic, -lead or -mercury compounds replaced them. These organometallic species were quickly withdrawn, and then forbidden, due to human poisoning problems. The first patent concern-

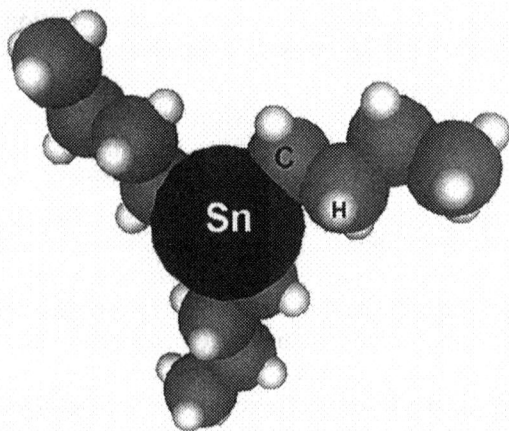

Figure 1 *The tributyltin molecule*

ing organotins in boat anti-fouling paints was British and dated 1947. By the end of the 1960s, the use of tributyltin (TBT) was extensive and by the start of the 1980s 140 000 tons of TBT-based anti-fouling paints were consumed each year in boating and the merchant marine in the United States. During the same period, it was estimated that such paints covered 70% of the global fleet.[1] This success was due to their high efficiency and long lifetime (4–7 years).[2] Between the 1980s and 1990s, their sale and use was severely restricted in Europe, the United States and New Zealand, because of their toxicity. Today, organotin-based anti-fouling paints are still used in many countries including France (for boats over 25 m long), Canada and others. The various applications and the increasing production and release of organotin compounds into the environment, over the past three decades, have led the international community to consider these species as 'global pollutants'.[3,4] Most attention has been paid to issues relating to the direct introduction of these biocides, mainly triorganotins, and more specifically tributyltin (TBT), into the aquatic environment, from the leaching of anti-fouling paints.[5,6] These matters were brought to public attention in the 1980s, after the severe decline of French oyster farming activities in the Bay of Arcachon.[7] Although the ecotoxicological mechanism of organotins was not clearly understood, and the analytical methodologies were, at this time, was rather limited, the French authorities rapidly regulated the use of TBT on boats. This rapid decision, and application of the precautionary principle, permitted the start of recovery of the oyster fisheries in the Bay of Arcachon.[7]

These findings triggered worldwide awareness, and similar precautions have been adopted throughout the rest of the world.[8] The recording and understanding of the impact and toxicity of TBT on marine bivalves was followed by the discovery of their effects as an endocrine disruptor on marine snails in 1990. Indeed, the most dramatic noticeable toxic effect of TBT is imposex, and although this was first recorded on European coastal lines, it has now been reported throughout the world. This phenomenon appears when only $2 \, ng \, l^{-1}$ of

TBT (as Sn) are present in the water. This very low concentration is sufficient to induce changes in the sexual characteristics of marine snails (dog-whelks), leading to sterility and a decline in the populations.

As in many environmental crises, biological responses provide the earliest evidence of a major disorder. Tremendous efforts were made by the international community on various issues fully to address this problem. Progress was made by paint manufacturers, resulting in considerably reduced releases of these compounds into the environment. Simultaneously, significant advances were made in the analytical field. This aspect was of paramount importance, since it took at least a decade to establish reliable sample preparation and determination procedures. This slow development hampered our understanding of the impact, fate and stability of these compounds in the environment.

Some 20 years on, there are still many unanswered questions. The learning process concerning organotins can be considered as a good example of the outcome and management of a major environmental crisis. Questions and uncertainties have been raised throughout the years, yet, in spite of the improvement in knowledge of the various degradation pathways and the evidence of decreasing concentrations in the water column in some places around the world, TBT contamination is still considered to be one of the most significant ecotoxicological problems of the last two decades.[9-11]

Improvements have been made regarding analytical methods, providing reliable information, and allowing a better understanding of the different fates and pathways of these compounds. However, the marine environmental impact was so acute that it has distracted attention from other sources of organotins, their fate, and impacts on fresh water systems. Results published in the last five years demonstrate that these systems, including terrestrial environments (such as soils and amended soils) are now also of concern.

2   Organotins in Industrial and Domestic Products

Organotin sources are very variable, and have a wide range of entry modes into the environment. In general, organotin is mainly used in its forms $R_nSnX_{(4-n)}$ (R = alkyl or aryl group, X = anionic group). Global organotin production is estimated to exceed 50 000 tons per year. Approximately 25% of this total are trisubstituted biocides, tributyl-, triphenyl- and tricyclohexyl-tins (TBT, TPhT and TcHT respectively)[12] of which 4000 tons are TBT.

The properties of organotin compounds differ widely between mono- and tri-substituted moieties. They are directly linked to the number (*n*) and nature of the organic group (R) covalently bound to the tin atom. In general, the anionic group or counter-ion does not present any particular influence on the organotin properties, other than its own properties.[13] However, it may affect the octanol–water partition coefficient and its adsorption potential, but there is very little information available on these issues.

The industrial use of organotins is very different, and is directly related to the chemical formulation of the compounds. Organotins are mainly used as stabilisers, catalysts, biocides or pesticides. Their chemical formulation and their direct industrial applications are summarised in Table 1.

The mono- and di-substituted species exhibit low toxicities and are frequently combined for the synthesis of PVC and other plastics, including packaging material used for food. Many domestic products contain mono- and di- butyl- or octyl-tins: kitchen and toilet papers, sanitary towels, textile products such as socks.[14] The disubstituted species (*e.g.* DBT) are also used as therapeutic products, as stabilisers in lubricating oils and as waterproofing agents.[15]

The trisubstituted compounds display a very high toxicity towards organisms. These highly efficient biocidal applications have been extensively used in anti-fouling paint formulations and fungicides. TBT has therefore been used in anti-fouling paint formulations for marine waters as well as in freshwater boating activities.[16] Different paint formulations have been developed and may also contain TPhT chloride, fluoride or ethanoate as molluscicides.[17]

These species are also widely used in agriculture. TPhT is employed as a fungicide in the culture of potatoes, celery, sugar beet, coffee and rice.[12] It is directly included in the composition of Duter (Ph_3SnOH) or Brestan (Ph_3SnAc).[18] The Fenbutatin oxide (bis[tris(2-methyl-2-phenylpropyl)tin) is

Table 1 *Properties and main uses of the organotin compounds in industrial products*

Organotins	Properties	Main uses
Monosubstituted		
Methyl-, butyl-, octyl-tins (MMT, MBT, MOT)	increase temperature resistance	PVC
Butylin (MBT)	catalyst	industrial processes
Disubstituted		
Methyltin (DMT)	increase light resistance, formation of SnO_2 film	PVC, glasses
Butyltin (DBT)	catalyst, vermifuge	polyurethane, chicken breeding
Butyl-, octyl-tins (DBT, DOT)	stabilisers	rigid PCV and plastics
Trisubstituted		
Butyltin (TBT)	biocide, fungicide, insecticide	antifouling paints, industries of wood, textile, paper and leather, industrial cooling systems, domestic detergents
Cyclohexyltin (TcHT)	insecticide	agriculture
Phenyltin (TPhT)	fungicide, pesticide	agriculture, antifouling paints
Tetrasubstituted		
Phenyltin (TePhT)	anti-corrosive agent	condensers, electrical systems

used as a preventive acaricide in the culture of various vegetables and fruits, such as tomatoes, cucumbers and bananas.[19] In France, Fentin (containing TPhT) is well recognised as an efficient fungicide. It is sprayed onto vegetables and fruits, such as those previously cited, but also onto beans, gherkins, melon, grape and strawberries. In the United States, at least twenty tributylated tin compounds are listed as active constituents in pesticides.[20] Tricyclohexyltin is also used as an agricultural pesticide, known commercially as Plictran.

The tetrasubstituted compounds are not generally used for their own properties, but as intermediate reagents in the synthesis of other organotin species.[15] They are, therefore, very often present as impurities in trisubstituted compounds [*e.g.* tetrabutyltin (TeBT) in TBT], and may reach up to 10% in concentration of the final product. Similarly, trioctyltin (TOT) is not directly used, but is often contained in the commercially available dioctyltin (DOT) as an impurity, and hence can be found in various other products.[14]

3   Occurrence of Organotins in the Environment

At present, we can consider that organotin compounds are ubiquitous in all compartments of the environment, and originate from a variety of different and distinct sources. Currently, most research efforts have been targeted towards maritime and coastal environments. In practice, introduction pathways come from various origins, and the most toxic, trisubstituted moieties are widely present. It has been estimated that, at the beginning of the 1980s, 10% of the biocides used were directly discharged into the environment, and ended up in the aquatic ecosystem.[17] However, the first primary concern arose from issues related to anti-fouling paints in coastal environments.[16]

3.1   Environmental Behaviour of Organotin Compounds

3.1.1   Marine Environments

The environmental behaviour and persistence of organotin compounds varies significantly according to a wide array of environmental conditions. The fate of organotin compounds released into the marine aquatic environment has been extensively studied, since it was, and still is, the primary environmental compartment impacted by the negative effects of these biocides, mainly through anti-fouling paint releases. These major issues have been detailed and reviewed in a monograph edited by Champ and Seligman.[21]

The behaviour of organotin compounds entering coastal waters is governed by both hydrodynamic and biogeochemical conditions, which regulate their dispersion and transformations. Trisubstituted organotin species are, in general, hydrophobic, and their fate is closely associated with the dynamics of particulates in the aquatic environment.[22,23] This implies that TBT would be unlikely to remain truly dissolved, but would either partition on the suspended load or accumulate in the surface microlayer at the air–water interface.[24–26]

These two features have major implications regarding fate and persistence. The highly hydrophobic behaviour of TBT will rapidly remove it from the truly dissolved phase and will allow rapid settlement. This trapping in the sediments leads to long-term storage of the contaminant. This permanent sorption behaviour of TBT on suspended particles can be a reversible process, depending on the conditions.[27]

The hydrophobic behaviour has impacted on the quality of environmental data produced over time. The pre-treatment of water samples, particularly removal or not of the suspended solids by direct filtration, resulted in a large scatter of the data and little inter-comparability between different sampling surveys, performed in different countries by various research organisations. The variability of results and monitoring strategies associated with organotin surveys in coastal environments has been discussed by Quevauviller and Donard.[28]

There are some wide discrepancies reported on major environmental parameters and issues, particularly on the partitioning of these biocides in water. Partition coefficients between dissolved TBT and that adsorbed onto the particulates (K_p values), which are vital parameters for further modelling and predicted scenarios, display large discrepancies between the studies performed. Large arrays of experimental data have been obtained, and range from 2000 up to 2 orders of magnitude.[28] This extreme variability is related to the different conditions observed in the different experimental settings (concentration of suspended load, ionic strength of the reaction media, *etc.*). These discrepancies are due to the different environmental conditions under which the experiments were performed. They can also be attributed to poor analytical performance, and a lack of understanding of the whole analytical chain, starting from poor sampling methods, storage and preparation, prior to analysis. However, these issues are now better understood and this has contributed to the tremendous progress made during the last decade through vigorous analytical efforts. One such was the experience gained through the production of certified reference materials (CRMs) in a wide array of matrices in Canada, USA, Japan and the European Union.

Different mechanisms influence organotin persistence in the environment. Rapid degradation pathways, with successive de-alkylations or de-arylations (rupture of the Sn–C bond) result in less toxic species. These processes may take place through photochemical cleavage (UV irradiation, chemical cleavage), but also by biochemical phenomena (biological degradation).[29] The hydrophobicity of trisubstituted species may favour their accumulation in the surface microlayer, and shorten their life expectancy *via* rapid photolytic degradation.[30] This process, however, appears to be very limited in the environment, and sorption onto particles and settling appears to be the major route and explains the high levels of organotins recorded in sediments.

Biological degradation may also accelerate the degradation of TBT in the environment to its less toxic breakdown products, DBT and MBT. Phytoplankton are the most common effective agents,[31] whilst bacteria can also actively degrade it to inorganic tin.[32] Macro-algae and phanerogams can also be relevant for TBT elimination in some shallow water areas.[33,34] This biodegradability, which was one of the reasons behind the promotion, development and use of this highly active fungicide, may, however, be minimised in open water systems, since biological action is limited by light input and nutrient availability. It therefore only occurs in a small section of the water column, limited, at most, to the photic zone.

There has been extensive debate regarding the half-life of organotins once entrapped in sediments, reported in the literature. Although there is still some controversy on these issues, there is a concurrence of opinion that organotin compounds are more persistent in sediments that was ever anticipated. Early sets of data, produced in the early 1980s, were certainly over-optimistic, due to lack of analytical experience and/or overall poor quality control of the data. Most results agreed that in aerated water, and in the presence of light, TBT can only last from a few hours to a few days. Their persistence in different compartments of the deep water column of the Mediterranean Sea,[35] and later accumulation in the sediment, has produced a large array of half-life estimates, ranging from 1 year to more than a decade.[21,36]

Several reasons may account for this poor quality of data. As mentioned earlier, data from the early 1980s were of questionable quality. It is difficult to obtain reliable historical data due to the wide array of sampling and analytical processes employed. This will affect the continuous recording of organotin levels in sediments. The most interesting places to record the continuous input of organotins into sediments would be harbours. They are, however, frequently dredged, and do not, therefore, preserve a continuous record of organotin settlement.[37] Bioturbation may also significantly alter the organotin levels in the sediments. These approaches were studied in the Bay of Arcachon with little success.[37] Following archival data, the declining population of the bivalves of Arcachon Bay indicated that the impact of TBT on this ecosystem started in the seventies simultaneously with the introduction of TBT-based paints.[38]

Recently, novel findings, based on new analytical developments, have allowed new insights on the remobilisation of organotins *via* other routes, rather than direct resuspension of contaminated sediments.[39] These results report the ubiquitous occurrence of volatile tin compounds in coastal areas. They were obtained by using highly sensitive and selective analytical methods (multi-isotopic determination using gas chromatography coupled to inductively coupled plasma-mass spectrometry). Several volatile tin species have been identified in all of the coastal and marine environments studied. The volatile organic tin compounds, determined in both sediment and water, could result from both natural methylation and hydridisation processes of inorganic tin ($R_nR'_{4-n}Sn$; R = Me, R' = H, n = 0–4), as well as anthropogenic butyltin derivatives released from ship anti-fouling paints, which have accumulated in sediments ($R_nR'_{4-n}Sn$; R = Bu, R' = H or Me, n = 0–3). The most ubiquitous species were found to be the methylated forms of butyltin derivatives (Bu_nSnMe_{4-n}; n = 0–3). These results suggest that continuous diffusion and volatilisation of organotin compounds would occur in estuarine and coastal environments, leading to a slow remobilisation of organotin species entrapped in the sediments. Diffusion rates were calculated from sediments to water, and from water to air, providing a pathway for organotin release.

In the case of sediment to water transfer, the calculated flux densities ranged from 50 to 470 nmol m^{-2} y^{-1}, when considering low turbulent mixing, as in the case of the Arcachon harbour. In the Scheldt estuary, the flux densities obtained were higher and range from 80 to 790 nmol m^{-2} y^{-1}, due to higher turbulent mixing. For sediment to water exchanges, it was assumed that all volatile tin species were produced by chemical or biological mechanisms in bulk sediments and were passively released into pore waters.

Similarly, water to air fluxes were estimated to range from 20 to 510 nmol m^{-2} y^{-1} for estuarine environments such as the Gironde and the Scheldt estuaries respectively, and 90 nmol m^{-2} y^{-1} in the Arcachon harbour.

These new findings have direct global implications regarding the fate of organotin compounds trapped in sediments. In turbulent systems, such as estuaries, diffusion processes of volatile organotin species to the water column, and thence to the atmosphere, may be a significant pathway for the removal of organotin compounds from sediments. It may also account for significant inputs of organotin compounds to the air, possibly explaining the occurrence of organotins in rainwater. In addition, sediments have long been considered to be a sink for these highly toxic constituents. In the case of minimal external inputs resulting from increasingly efficient regulation, sediments should now be considered a significant source of organotin compounds to overlying waters and a potential source to the atmosphere.[39]

Finally, there is an important corollary to these findings. Despite the fact that there is continuous formation and methylation of butyltin compounds occurring in the environment, albeit at a very low rates, the toxicity and bioaccumulation properties of these moieties on organisms should now be evaluated, since they should be more lipophilic than hydrolysed trisubstituted species. The impacts and fates of these new species on organisms should be evaluated.

3.1.2 Freshwater and Terrestrial Environments

Very little is known of the fate and mechanisms regulating the persistence or degradation of organotins in freshwater or in terrestrial ecosystems. It is assumed that the trends observed in marine environments may be directly translated to freshwater. In general, few data are available on the fate of TBT in such systems. Freshwater systems contain high levels of humic substances to which organotin species will bind, thus affecting the amount in solution and altering overall recoveries and analytical performance. Humic substances in freshwater systems have been found to partially dissolve and accumulate organotin species. These processes are directly related to the physico-chemical parameters of the system, and are strongly pH dependent.[40]

The half-life of TBT has been estimated to vary between 4 days and 4 weeks in marine and estuarine waters. In freshwater environments (waters and sediments), it has been estimated to be 4–5 months. In anoxic media, this time is much longer

– it can sometimes be over 12 months and even reach several years in sediments.[41,42]

Decreases in the TBT levels in rivers have been recorded by several authors. They have been related to continuous dilution, together with river flow, but have also been related to degradation processes.[43] Degradation kinetics in freshwater systems are also highly variable, since they depend on environmental conditions such as temperature or the nature of micro-organisms present.[29,43] If these data consider only surface waters, they would predict that TBT would be stable in ground waters. Photolytic degradation may take place, but will be limited to the surface waters or the top layers of soils.

Very little data are available on other organotin species (phenyl- and cyclohexyl-tins), which are frequently reported in these aquatic environments, but are much less stable than TBT. Their input occurs primarily *via* direct spraying of fungicides during agricultural practice. However, the quality of the data available is, in general, low, since they are more difficult to detect than butyltins. They require the highest levels of analytical performance with excellent gas chromatographic conditions, due to their very high boiling points. These conditions do not prevail in most routine analytical laboratories.

The degradation of TPhT leads to the formation of mono- and di-phenyltins, for which the half-life times are reported to be very short in freshwaters, and range from 2 to 3 days.[29] DPhT decomposition can be expected to be the fastest of the three species since they are seldom found in river water.[44,45] In sediments, oxic biodegradation is considered to be the main process involved. The estimated half-life varies between 60 and 140 days, according to the conditions. In sediments, DPhT is the main product of degradation. The kinetics of the degradation of TPhT seem to be rapid and the reaction is complete after a few days.

Few studies have been performed on tricyclohexyl- (TcHT) and octyl-tins. In soils, the TcHT half-life estimates range from 50 days up to 3 years.[17,29] It can be photodegraded in surface waters, inducing the formation of mono- (McHT) and di-cyclohexyltins (DcHT). This enormous span illustrates important gaps in our knowledge in this area, and is again probably due to poor analytical performance.

Finally, direct volatilisation after mechanical spraying of organotins onto fields contributes to their wide dispersion in the atmosphere and surrounding aquatic environments. This contributes to the relatively high levels of organotin compounds, detected in Dutch fresh water-ways.[46] Indeed, organotin pesticide traces were found 20 km from sprayed, cultivated fields.[46] This phenomenon should be considered as significant.[29]

3.2 Organotin Compounds in Marine Environments and Related Questions

Sediments act as the ultimate sink of organotin inputs. Therefore, there has been a very large array of studies which have looked at the concentration of the various organotin species in this compartment of the ecosystem. Concentrations are extremely diverse and range from not detected, through a few $\mu g\,kg^{-1}$ (as Sn) to extremely high concentrations, some hundreds of $mg\,kg^{-1}$ (as Sn). These extremely high values should be handled with caution, since they are always reported in harbours, which could be located close to dry docking areas and therefore include TBT-based paint chips. In a recent report for the French IFREMER, the range of TBT concentrations in different sediments from French harbours ranged from 10 to $21\,300\,\mu g\,kg^{-1}$ (expressed as the TBT ion). They compared locations in different countries, and found they were similar to concentrations found in other harbours in Thailand,[47] Canada (Vancouver)[48] and Japan (Osaka).[49] In a recent study of organotin concentrations in surface sediments of Greek sites[50] very high levels of TBT were detected in all of the coastal sites studied, with average concentrations of TBT higher than $400\,\mu g\,kg^{-1}$ as Sn. In most cases TBT was also the most abundant species, suggesting either little degradation or continuous inputs. The most surprising data were obtained on an open-sea sediment, considered to be a reference site, where the concentration recorded was estimated to be $300\,\mu g\,kg^{-1}$ (as Sn). This study was initiated since Greece has the largest shipping fleet in the world, and its waterways support intense shipping traffic.

It is beyond the scope of this chapter to review and discuss the variability of TBT levels found in different locations through throughout the world, since the concentrations will depend on the location of the sampling site (bay, leisure harbour or commercial harbour). In most published studies, similar average concentrations and ranges of concentrations are reported. These data, however, indicate the same trends where the following decreasing levels can be found with respect to the location. TBT concentrations in sediments are always lower in open bays compared with those of commercial and fishing harbours, where the use of TBT-based paint is still legal. However, these trends should be viewed with caution since in most cases, little to no comparative sedimentological assessments are performed by the authors (see below):

open bays $\ll$ leisure harbours $<$ commercial and fishing harbours

The persistence and half-life of TBT in sediments has been discussed earlier in this chapter. However, what is important are the temporal trends that one would expect in surface sediments, following the enforcement of legislative actions taken throughout the world, and leading to the ban of TBT-based paints for boats smaller than 25 metres. In a very recent paper by Diez et al.[51] the authors tried to compare the levels of organotins in sediments of various bays, marinas and fishing harbours in the Western Mediterranean. They concluded that the banning of TBT-based anti-fouling paints has been effectively recorded in marinas, where a decrease of the concentrations has been observed, but not in commercial or in fishing harbours. Many other studies have attempted to illustrate the same trends. However, most of these studies did not take into account many important basic geochemical parameters, such as grain size or organic carbon content, and in virtually all cases there is no normalisation of the values to a conservative biogeochemical parameter of sediments, such as the fraction $< 2\,\mu m$ and aluminium or scandium content. These aspects are critical when comparative assessments are to be made. TBT was found to have a bimodal behaviour in sediments.[52] One trend is the association of TBT with the fine part of the sediment, a similar correlation to that of most trace metal elements. Another association is with the coarse fraction of sediments ($> 300\,\mu m$); this was found to be mainly with vegetal fragments, which presented extremely high levels of organotin concentrations. This fraction, however, corresponds to less than 1% of the total amount of the sediment.[53] Nonetheless, comparative studies should include normalisation factors, in order to have a correct assessment of the trends studied, as is conventional for most bio-geo-chemical studies.

Furthermore, if there are few satisfactory data on sediment recording of TBT in harbours and other marine environments, most of the observations record a maximum accumulation of TBT concentrations below the 5–10 cm of the core. Surface concentrations are always lower than those in the buried sediments.[52,54] These profiles have also been obtained in various samplings of the harbours of the Arcachon Bay. After normalisation, these trends are preserved. These results raise several questions that would be applicable to any other sedimentological studies concerning TBT comparisons.

This apparent decrease of concentration in the top sediments could first be attributed to the fact that regulations are effectively enforced and that the amount of TBT released in the environment is now limited in some locations. The role of bioturbation should also be taken into account. Finally, these variations should be considered in the light of new results obtained recently by Amouroux *et al.*[39] These indicated that gaseous, fully substituted TBT species (methylbutyultins) were continuously diffusing from the sediments. These findings could also account for the depletion of the TBT concentrations of the top surface sediments. They do, however, suggest that, if direct inputs of TBT are now limited *via* the phasing out of TBT and from the restricted or abandoned use of TBT paints, sediments which were first considered as a sink are now shown to be a source to the overlying-waters (Figure 2).

If, in some cases, evidence suggests that TBT and organotin concentrations can be considered to have declined, serious attention should be paid to the establishment of comparative sedimentological criteria.

The same attitude applies to water level monitoring and comparisons. A number of studies performed in waters in the early 1990s would tend to establish a significant decrease in the TBT levels in the waters of marinas. This trend was observed in both the UK[55-59] and France[60,61] but not in Dutch coastal waters.[62] In most cases, if a decline was recorded, the levels were still above the water quality criteria that the different countries have set, and which range between approximately 2 and $10\,ng\,l^{-1}$ of TBT as Sn. A more recent report (1998) from the Dutch authorities (published by the Dutch National Institute for Coastal and Marine Management/RIKZ)[63] reported the average calculated concentrations of TBT, using a model which integrates emission data, physical

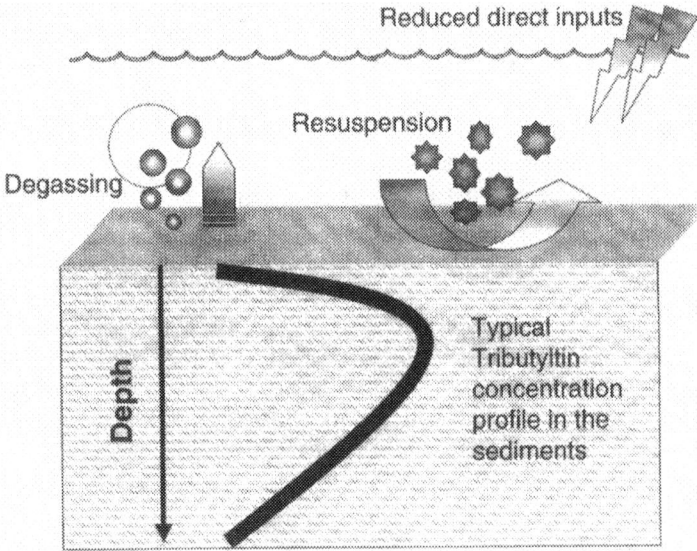

Figure 2 *Sediments as a source of organotin compounds to overlying waters. Typical profile of tributyltin in sediments and mechanisms associated with the depletion of organotin compounds in surface sediments*

and chemical processes and water movements, showed that the target figure set by the Dutch administration for quality standards (maximum of $1\,ng\,l^{-1}$ as TBT) is exceeded all along the coast line and up to $10\,km$ out to sea.

It should be mentioned that serious efforts have been made worldwide to define water quality criteria. By the end of the 1980s, various standards had been promoted by different countries. Most efforts resulted in a similar range of TBT levels for water quality criteria. These levels vary from $10\,ng\,l^{-1}$ down to $1\,ng\,l^{-1}$ in coastal waters. However, there is a lack of harmonisation in the objectives set. These discrepancies of targeted TBT levels are, for example, well illustrated in an EPA report[64] where each US state has different water quality criteria. The US EPA has recommended the following set of values for fresh waters: $149\,ng\,l^{-1}$ for acute and $26\,ng\,l^{-1}$ for chronic contamination (short- and long-term exposures). These values are slightly higher for acute conditions in seawater ($266\,ng\,l^{-1}$) but much lower for chronic exposures in marine waters ($10\,ng\,l^{-1}$). These recommendations are general guidelines, and many states in the USA have adopted their own water quality criteria in case of chronic saltwater issues. Lowest values are set in the state of Virginia ($1\,ng\,l^{-1}$) with the highest been being in Hawaii and Texas (10 and $43\,ng\,l^{-1}$, respectively). In Europe, water quality guidelines aim to achieve the low values of $2–1\,ng\,l^{-1}$ as Sn in saltwaters.[37] These objectives were defined after the evidence of biological impacts of TBT in water at concentration levels as low as $2\,ng\,l^{-1}$. Such levels directly affected oyster larvae reproduction, induced shell malformation in oysters and were found to induce imposex in a gastropod *Nucella lapillus*.[65] These effects have also recently been observed in the common whelk, *Buccinum undatum*.[66]

It should be stressed that, at the time when these water quality values were set, they followed the precautionary principle. At that time, they were usually beyond the reach of routine analytical methods, and equivalent to, if not below, most detection limits of specialist laboratories. Furthermore, there has been much debate on the sampling strategies required to obtain comparable data. Whilst there has been an international effort to achieve harmonisation between countries, more must be done in order to obtain truly comparable data.

During the last decade, results have suggested that, in some harbours or coastal areas, levels of TBT in water have decreased. At the same time, however, there is now growing evidence that organotin contamination is not restricted to the coastal environment and should be of concern for global ocean waters. Here again, biological organisms have acted as indicators and set the agenda for analytical chemistry. Some of the studies can be considered as controversial, with respect to the conclusions promoted by the authors. They do, however, all point out that open ocean systems and organisms living there can be affected by the continuous release of organotin compounds from heavy shipping traffic. Some of the early questions were raised by Hallers-Tjabbes *et al.* in 1994[67] when they correlated the occurrence of imposex developments in the whelk populations on the bottom of the North Sea to the shipping traffic intensities above. Later improvements in analytical chemistry and instrumentation (sensitivity and selectivity of the detector, mainly ICP-MS) confirmed these hypotheses and demonstrated that detectable levels of organotin compounds could be recorded in open seawaters.[68,69] TBT residues have also been detected at the top of the open ocean marine food chain, in large marine mammals and in seabirds.[70,71] This unexpected, apparent persistence of organotin compounds has now been confirmed by excellent analytical developments, which allow reliable results to be obtained at very low, sub-femtogram levels.[72] A recent survey performed in the surface seawaters of East China demonstrated that both TBT and TPhT and all their degradation products were detected with a highly selective and sensitive GC–ICP-MS instrument fitted with a shielded torch, at concentrations ranging between the $pg\,l^{-1}$ (as Sn) range for the butyltins and almost one order of magnitude lower for the phenyltins.[72] These results are also consistent with those of Michel and Averty[35] who reported that TBT and its degradation products could be detected along a 2500 m deep water column profile in the north-western Mediteranean Sea. They recorded that the contamination of the deep waters reached a maximum of $0.04\,ng\,l^{-1}$ as TBT ions, with always significant values at the 2500 m depth. Under these conditions, the authors estimated the half-life of TBT in these environments to be several years. Recent data, also obtained by sensitive GC–ICP-MS, demonstrated that butyltins and phenyltins were found at sub $ng\,l^{-1}$ level in surface seawaters of the Seto inland seas (Japan).[73] They also detected the presence of trioctyltins, probably derived from land-based plastic stabilisers. These recent findings appear then to be quite consistent with the bioaccumulation patterns of organotin found in top predators of open oceans, and suggest that global contamination of oceans by organotin compounds is not restricted to harbours and coastal environments, but under some circumstances could concern global oceanic systems. These questions should be urgently addressed in the future.

3.3 Organotins in Freshwater and Terrestrial Ecosystems

If the magnitude and the extent of organotin contamination is now well recorded and understood in coastal and marine waters and ecosystems, the ubiquitous occurrence of organotin in freshwaters, households, municipal wastes and terrestrial environments is much less well known. Indeed, there are fewer studies concerning tin speciation in freshwaters or even soils, compared with the efforts developed in maritime areas. All of these studies, however, have clearly established the systematic presence of organotins in inland waters.

The first series of important data was produced in the Canadian Great Lake systems in the early 1980s. Of the waters sampled at 250 different sites, 10% appeared to be contaminated by TBT, with levels reaching over $200\,\mathrm{ng}\,l^{-1}$ (as Sn).[29] DBT was also detected in underground waters.[74] The early 1990s did not see the recording of any phenyl- or cyclohexyl-tin in Canadian waters, which may be explained by the fact that, at this time, their use and applications were minimal and/or the sample preparation and detection steps were not sufficiently advanced. Recently, however, TPhT and its degradation products (mono- and di-phenyltins) were found in freshwater sediments at fairly high concentrations, ranging from 4 to $22\,\mu\mathrm{g}\,\mathrm{kg}^{-1}$ (as Sn).[75]

In Europe, a series of studies performed in the 1990s indicated that freshwater environments are also affected by the presence of trisubstituted species. In the different environments studied, the TBT and TPhT concentrations recorded ranged from not detected for TBT and TPhT, respectively, up to 6000 (TBT) and 1200 (TPhT) expressed in $\mathrm{ng}\,l^{-1}$ (as Sn) in river and lake waters and between 5000 (TBT) and 900 (TPhT) $\mu\mathrm{g}\,\mathrm{kg}^{-1}$ (as Sn) in sediments. These data illustrate that freshwater systems can also be subjected to considerable organotin inputs. The detected species are predominantly the three butyltin species, with the less frequent occurrence of phenyltins.[12,29,44,76-81]

During the last ten years, several monitoring programs were performed over a six month period, and confirmed the continuous input and diffusion of organotins in freshwater systems. In England, rivers appeared to be systematically contaminated by the three butyltins, with a TBT range between 10 and $200\,\mathrm{ng}\,l^{-1}$ (as Sn).[43] The presence of such species were mainly attributed to the fluvial traffic, TBT concentrations being three to ten times higher in harbours than in the waterways.

A French survey of four rivers in the Rhine and Meuse drainage basins was undertaken for a period of 7 months.[79] This clearly demonstrated that waters and sediments from the Sarre, Moselle, Meuse and Rhine rivers were systematically contaminated by butyltins, with a similar average concentration of TBT of $20\,\mathrm{ng}\,l^{-1}$ (as Sn) in waters and of $10\,\mu\mathrm{g}\,\mathrm{kg}^{-1}$ (as Sn) in suspended matters for both water systems. Important accidental contamination could also occasionally be detected in the Rhine river during the summer months, with concentrations reaching $65\,\mu g\,l^{-1}$ (as Sn) in water. Phenyltins were also detected in waters, TPhT concentration being about $20–40\,ng\,l^{-1}$ (as Sn). Between 20 and $40\,ng\,l^{-1}$ (as Sn) of TeBT were also present in suspended matter in the Rhine. On average, a mean concentration of about $20–40\,ng\,l^{-1}$ (as Sn) can be defined as the background level of all the organotin species in such waters, which suggests continuous high-level inputs into the fresh waterways.

Organotins can be also present in rain water. Mono- and di-methyl- and monobutyl-tins were detected in Florida[15] at low $ng\,l^{-1}$ (as Sn) levels. This contamination seems quite low and seldom takes place. Agricultural spraying is probably the major cause of atmospheric inputs of organotins. Similar trends have been reported in Dutch rain water, raising again questions with regard to its sources.[82]

Urban wastes are also an important and continuous source of contamination of fresh aquatic environments.[12,83] This fact is not surprising considering the large array of domestic appliances in most households. Wastewater and sludge have been reported to contain both butyl- and octyl-tins.[75-77,81,84-87] It is likely that these species originate from continuous leaching from PVC pipes, where they are employed as stabilisers.[88] In the incoming effluent of treatment plants, butyltin concentrations may vary between a few $ng\,l^{-1}$ (as Sn) to over one hundred $\mu g\,l^{-1}$ (as Sn). These species are always present, indicating continuous inputs from households or PVC draining pipes.

Organotins are widely used in the home and humans are in contact with these compounds throughout the day. This use is usually as fungicides, and they can be found in a wide array of common commercial products, such as clothing and textiles (occurrence of dioctytltins, di- and tri-butyltin, trioctyltins) at very high concentrations, up to several thousands of $\mu g\,kg^{-1}$ as organotin cations.[89] These observations are, therefore, quite consistent with the ubiquitous occurrence of organotin compounds in municipal wastewaters. Disubstituted organotin species (dibutyl- and dioctyl-tins) can also be found at similar levels in sanitary towels, tampons, and maternity sanitary pads, raising new concerns regarding their use.[89]

As a result, urban wastewater generally contains higher TBT levels than those of potentially contaminated industrial wastes.[77] The major proportion of organotin compounds coming into the wastewater treatment plant is removed from the effluent by adsorption on sludge.[87,90] Organotin levels generally found under these conditions in sludges, currently reach concentrations ranging from 100 to $500\,\mu g\,kg^{-1}$ (as Sn). If the water treatment processes are efficient in removing the organotin from the water, they then concentrate it in the activated sludges, creating a new concentrated potential source of contamination. This fact should be taken into consideration when managing the disposal of the sludge. This point is quite similar to that occurring in the management of dredged harbour sediments. The usual practices of spreading activated sludges over cultivated fields may result in the direct transfer of the organotins to edible crops and also be a direct method of introduction of organotins into soils. Sewage sludge can also be contaminated by TeBT.[81]

Another important direct input of organotins into soils is the extensive use of direct agricultural spraying of organotin fungicides.[76,90] TBT and TPhT are directly introduced into the environment by such spraying, with evaporation and precipitation contributing to their dispersion. In the Netherlands, TPhT-based fungicide treatment of potatoes results in high TPhT concentrations in waterways located close to the fields. A total of 2.6 tons of organotin are thus directly dispersed each year into Dutch freshwaters.[91] These inputs may contaminate the groundwaters through drainage from agricultural soils.[13,76] The quoted studies show that agricultural practices, using either insecticide or fungicide sprays, are a potential source of contamination of plants cultivated for human consumption.

The use of organotin-based pesticides also leads to contamination of fruit and vegetables. A series of measurements have begun to address this issue, but these are mainly focused on total tin determinations. The levels recorded were high, ranging from 0.03 to 0.13 mg kg^{-1} (expressed as total tin) and were found in some vegetables, with the highest concentrations detected in lentils, potatoes, chicory and carrots.[92] It cannot, however, be assumed that these quantities are solely due to organotin fungicides, since only total tin was measured The presence of tricyclohexyltin hydroxide was studied in some fruits (apples, kiwis and pears) and, again, high concentrations of up to 0.06 mg kg^{-1}, expressed again as total tin, were detected. This suggests important direct inputs due to organotins, but this needs to be confirmed by appropriate analytical methods. These results were only expressed with regard to 4% of the whole fruit.[93] In both cases, the studies reveal a worrying situation regarding the use of tin fungicides in fruits and vegetables. A recent study illustrates that these concerns are valid, since concentrations up to 3 μg kg^{-1} of Fentin were detected in commercial potatoes[94] using a valid analytical approach. The three phenyl species were also found in pecan leaves. Ten days after spraying a TPhT-based solution, leaf samples contained between 11 and 50 mg kg^{-1} of organotins.[95] In this study, the dephenyltation process seemed to be rapid, since 66% of TPhT was degraded to mono- and especially di-phenyltin. Concentrations remained, however, very high, particularly for the highly toxic TPhT. Currently, it is difficult to assess the real impact and risk for human health, since available data are very scarce, but more information is urgently needed.

Significant direct discharge of organotins can also originate from the direct release of wastewater or cooling water from power stations, where TBT is used extensively to prevent bacterial growth in the cooling systems. Concentrations recorded in such case are high, ranging from 1 to 100 μg l^{-1} (as Sn).[96,97] Many other industrial sources contribute to the inputs of organotin compounds into the environment. Industrial sites using organotins for wood treatment processes can represent potential sources of pollution by accidental or deliberate release into the environment.[16]

Finally, the leaching or erosion of plastics discarded in the environment may also release the mono- and di-substituted moieties contained within the plastic.[98] Leachates from municipal dumping sites may also include mono-, di- or trialkyltins. This phenomenon has been recognised as an important source of organotin contamination in the United States.[29]

4   Increasing Awareness of Biological Effects of Organotin Compounds

As in most cases of severe environmental damage, biologists were the first to become aware of the toxicity of organotins released into water. It is beyond the scope of this chapter to review the successive evidence of biological impacts of organotins. It is, however, worth considering the facts in their time frames, and to discuss their implications.

The first adverse biological effects of organotins were detected in marine systems, with the Bay of Arcachon frequently quoted as the classical example after the severe decline of the oyster fishery. Shell malformations and the failure of juveniles to reach maturity almost wiped out this industry in the 1980s.[99] Next, the evidence of male characteristics and sex induction in female snails (the 'imposex' phenomenon) on the dogwhelk, *Nucella lapillus*, was observed, and a very low level of 1 ng l^{-1} of TBT was recognised as the cause of the phenomenon. These findings confirmed the extremely high toxicity of TBT in marine ecosystems, at concentrations as low as 1 ng l^{-1}.[100] These deleterious effects were then observed in a wide variety of marine organisms and later, in the common whelk, *Buccinum undatum*.[66] Populations of deposit-feeding bivalves, *Scrobicularia plana*, were also wiped out of the Bay of Arcachon.[101] Historical reconstruction, based on the temporal evolution of the failing dynamic of these populations in this Bay, has demonstrated that populations of biological organisms started to respond and decline as soon as TBT was introduced into the bay at the end of the 1970s.[38] The general impact of TBT and other organotin species on organisms has been fully reviewed, first by Bryan *et al.* in 1991[102] and later by Fent in 1996.[12]

At present, several other biological indicators are raising concern over the effects of organotins in the open ocean. There are increasing reports in the literature on the deleterious effects of tin compounds on large marine mammals, all over the world. Traces of TBT have been found in toothed whales, like the dolphin, and in members of the seal family, in places as far apart as the United States, South East Asia, the Adriatic and the Black Sea. Organotins have also been found in sperm whales beached on the Dutch coast.[63] These observations are consistent with global marine contamination.

The second trend is the growing evidence of the impact of organotin compounds on large biological systems. Important progress has been made at the biochemical level, and organotins are now considered to be endrocrine disrupters. These concepts started in the 1990s, and are now well established. A large array of recent literature, released in 2000, confirms these findings. This evolution of concepts and the improvement of knowledge on these issues have been reviewed by Matthiessen and Gibbs in 1998[103] and by Alzieu in 2000.[104] For marine organisms, TBTO has been demonstrated to be both toxic and genotoxic to embryo-larval stages of the common marine mussel, *Mytilus edulis*, and the ragworm, *Platynereis dumerilis*.[105] It has been shown to reduce the sperm count in guppies, even at very low levels.[106] The endocrine disruption effect of organotin compounds is suspected to be one of the causal factors for the decline of Japanese abalone stocks.[107] This endocrine disruption effect has also be been demonstrated on the apoptosis in PC12 cells.[108] Finally, the cyctotoxicity of tributyltin has been demonstrated in rat hippocampal slice culture. Results suggest that TBT provokes an apoptosis-like neuronal cell death.[109]

Questions also arise when one is concerned with the level of organotin compounds in shellfish and seafoods. In a report released by the IVM in 1999,[63] based on the literature available throughout the world, the authors demonstrated that, based on the tolerable daily intake (TDI) of TBT[110] and the average seafood consumption in different countries, tolerable average residue levels could be derived. The results, focusing on TBT, showed that these levels are exceeded in many samples from Italy and Japan, and in a number of samples from Canada, France, Korea, Taiwan, Thailand and USA. The major outcome of this study states that for the majority of countries, including many countries with a high seafood consumption, no data on organotin levels in seafood were found at all, which illustrates the need for improved and homogeneous data in these cases.

Three main categories of countries, and TDI levels, could be illustrated. In a group of 22 countries which had limited data sets on organotin levels in seafood, more than one third of these countries had exceeded, in one or more sample, the tolerable average residue level of TBT. This concerns Canada, France, Italy, Japan, Korea, Poland (DBT + TBT), Taiwan, Thailand and the USA. In these countries, the claims that levels of organotin compounds in seafood pose a negligible risk to human health has to be rejected, considering the current data. Poor quality results were found for Italy. Indeed, in the Italian samples and data sets, the average TBT residue level in mussels exceeded the tolerable average residue level by a factor of 2.5. Over 90% of the mussel samples contained TBT levels that exceeded the tolerable average residue level, the maximum by a factor of 10. Finally, in Japan the average TBT residue level in seafood was almost as high as the tolerable average residue level for this country. Almost 30% of the samples, taken from various species purchased from traditional fish markets, exceeded the TBT tolerable average residue level with a maximum factor of 5.8.

There is now a converging set of literature illustrating the importance of organotin contamination at all levels of the marine trophic chain. An increasing number of publications in the literature also demonstrate that concern should also be shown, and action taken, in freshwater systems. Indeed, literature data illustrate that aquatic fauna (*i.e.* shellfish, mollusc and fishes) can be contaminated by organotins as well.[43,85,111–113] Fairly high concentrations ranging between 10 and 200 μg kg^{-1} (as Sn) of the three butyl- and mono- and tri-phenyltins were found in fish from the Rhine.[44]

In Swiss harbours, the three butylated and three phenylated species were detected in mussels and fish tissues, with a concentration range of 1–10 mg kg^{-1} (as Sn), representing a bioaccumulation factor of 1 to 4 $\times$ 10^4 in the living organisms.[85] Similar ratios and trends were observed in fish from the Rhine, confirming the high potential for bioaccumulation in biota in freshwater systems.[44] Detailed mechanisms of bioaccumulation and pathways of TBT have been studied recently by Looser,[114] and illustrate that uptake of TBT by freshwater larvae (*Chironomus riparius*), living in freshwater sediments, is very efficient, and proceeds both from the organotins present in the pore water and also from TBT bound to particles in the solid substrate.

The presence of organotins in meat has also been detected in animals bred for human consumption. MBT, DBT and TBT have all been found in chicken livers, the mode of entry of organotins being from therapeutic products.[115]

Little is known regarding the occurrence of organotins in plants, although a few data sets have revealed that bioaccumulation factors can be very high in this area. The bioaccumulation may vary from 2 to 4×10^4, for different types of algae and for terrestrial plants.[18,43] This raises the question of the fate of organotin compounds that are indirectly introduced to agriculture *via* treatment of soils with sewage sludges. These questions should also be further investigated.

5 Critical Considerations with Respect to Determination and Environmental Monitoring Strategies

Environmental data always rely upon a complete and continuous analytical chain, which is only as strong as its weakest link. The sequence starts from sample collection, through sample storage, sample preparation and finally detection. These issues are most critical when dealing with organotin determination in the environment. It is beyond the scope of this chapter to review the different techniques developed for the determination of organometallic compounds, and more specifically organotins, in the environment. They have been critically reviewed recently by Ritsema and Donard[116] and Donard *et al.*[117] Nonetheless, it is worth examining some of the critical issues associated with the whole analytical chain for organotin compounds.

Great progress has been made in analytical methodologies and instrumentation in recent years, which has given rise to an increased awareness of organotin related issues in the international community. Much experience has been gained in this field by European research groups, who have combined their efforts to produce a wide array of reference materials for the quality control of organotin determinations in the environment.

5.1 Sample Preparation

5.1.1 Sample Treatment

Most of the analytical methodologies for organotin determination employ an extraction of these compounds from the matrix, followed by derivatisation, in the case of chromatographic separation, and a purification step. Sample treatment can be considered as one of the most critical steps, being prone to errors from many sources. Contamination of samples can arise from glassware, reagents, the laboratory and even the operator. In the case of organotin analysis, plastic labware, and in particular PVC (containing di- and mono-butyltin), but also plastic gloves (sometimes containing trioctyltin), must be avoided. However, the crucial point is in the choice of the most suitable extraction and derivatisation reagents. It is trivial to mention that low extraction efficiencies and/or low derivatisation yields lead directly to an underestimation of the real concentration.

Furthermore, changes in speciation during treatment, due to degradation, must be taken into account, optimising not only the choice of the most suitable reagents, but also the experimental conditions, such as temperature, light and time of processing.

5.1.2   Extraction

Extraction of the analytes from the matrix is the initial and crucial step of the determination procedure. Extraction should fulfil the following requirements:

- separate the analytes from the matrix without alteration to the chemical identity,[118] eliminating or reducing the interferences from other components
- concentrate the analytes to a detectable concentration level

Poor extraction yields, losses of analytes and contamination directly affect the quality of the results. Furthermore, changes in the organotin speciation occurring during this step lead, not only to wrong information about the contamination levels, but also to incorrect conclusions regarding the ongoing degradation phenomena. Therefore, recoveries should be carefully tested for each organotin compound. In the absence of reference materials, the most commonly applied approach is to assess the recovery by a spiking procedure. The recovery of the spike can be evaluated by analysing the matrix before and after the spike addition. This approach, however, suffers from some drawbacks, and these have been critically discussed.[119]

To date, the most valid approach for sound environmental speciation data relies on the integration of certified reference materials (CRMs) in the whole quality control procedure. It is worth mentioning, however, that the use of CRMs leads to a global evaluation of the performance of the overall analytical method, rather than an evaluation of the extraction efficiency only. Differences between the certified values and the one found can usually be attributed to an array of processes taking place during the extraction procedure, such as low yields from derivatisation or losses during the clean-up step. Critical comments on the evaluation of the extraction efficiencies in organometallic speciation are given elsewhere.[120]

A large number of materials have been used for the extraction of organotins from water, sediment and biological samples – acids, organic solvents with and without complexing agents (e.g. tropolone) and mixtures of them. Extraction after solubilisation with tetramethylammonium hydroxide (TMAH) or after enzymatic digestion (generally using lipase and protease) has been proposed for organotin determinations in solid samples. The different extraction methods available for organotin analysis have been reviewed in the literature.[121,122]

5.1.3 Derivatisation

The derivatisation step must also be considered as one of the most critical steps in organotin determination. The low volatility of organotins makes a derivatisation step essential, when a gas chromatographic based technique is to be used. Further, this derivatisation procedure will provide a reduction of possible interferences during the subsequent analytical steps, and particularly at the detection stage.

Low yields from derivatisation, and/or degradation phenomena (especially in the case of phenyltins), can greatly affect the quality of the results. A validation, or at least a careful study, of the procedure being used in the laboratory for the particular matrix, with its particular interferences, is necessary.[123] This validation is, however, almost always hindered by a lack of commercially available derivatised standards. Recently, the synthesis of derivatised standards for ethylation and Grignard derivatisation was carried out at the Free University of Amsterdam[124] within the framework of the Standards, Measurements and Testing Programme. These materials permit the establishment of optimisation and validation studies.

Most traditional derivatisation procedures for organotins were based on the Grignard reaction, in a solvent medium. These methods were long and tedious, with a large number of steps during preparation. Grignard derivatisation reactions offer the advantage of being usable for the determination of most organotin compounds present in the environment (methylated, butylated and phenylated species), in a large variety of matrices (water, sediments, biota). Grignard derivatisation is generally performed by methylation, ethylation, propylation, butylation, pentylation and hexylation. Among the different Grignard derivatisation reactions, hexylation and pentylation offer the advantage of providing derivatised compounds with relatively low volatility. These allow pre-concentration steps during sample pre-treatment, without taking special precautions. Condensation problems in the interface, due to the low volatility, have been described when coupling GC to AAS.[125] The use of propyl, ethyl and, above all, methyl Grignard reagents offers the advantage of a higher reactivity, even though the higher volatility of the products can lead to volatilisation losses during the pre-concentration steps. Methylation has an additional disadvantage in that it obviously cannot be applied to the determination of methylated species.

To date, most workers determining organotin compounds in environmental samples have turned to susbstitution reactions that take place in water, in order to simplify and improve the efficiency of the sample preparation procedure. The most common derivatisation reactions used for organotin analysis in water are hydride generation with $NaBH_4$ or ethylation with $NaBEt_4$.[126]

The performance and suitability of these techniques for organotin determination in environmental matrices have been recently critically evaluated.[127] Both hydride generation and ethylation with $NaBEt_4$ are particularly suitable for aqueous samples, since they both offer the advantage of being directly applicable to the samples. Simultaneous, *in situ*, derivatisation and extraction is possible, reducing the number of analytical steps, and thus, potential sources of error. Ethylation, in contrast to hydride generation, provides high derivatisation yields, not only for butyltin, but also for phenyltin compounds. In the case of solid samples, such as sediments and biological samples, hydride generation is often hindered by the presence of severe interferences from some inorganic species present in the matrix. The main advantage of $NaBEt_4$, its stability in water, is in such cases no longer exploitable. It is worth mentioning that $NaBEt_4$ in the presence of strong acids, often used in the extraction of organotin compounds from solid samples, is not stable and decomposes. In this case, the organotins have to be extracted into an organic solvent, prior to the derivatisation reaction. Some precautions must also be taken in the case of ethylation with $NaBEt_4$.

Independent of the applied derivatisation technique, it is necessary to check the yields for the chosen experimental conditions. The lack of commercially available derivatised organotin standards, however, hampers the quantitative evaluation of the derivatisation yields. The different possibilities and limitations of these substitution procedures, during organotin sample preparation, have recently been discussed.[128]

5.2 Sample Collection

5.2.1 Sampling

Considerable attention is generally paid to the analysis, in order to improve the reliability of the measurements, whilst sampling is, generally, scarcely considered. However, the contribution to the total uncertainty associated with the final result arising from the analysis can be estimated at no more than 15–20%, whereas that from sampling can contribute up to more than 99%.

This means that, in order to reduce the total uncertainty and correctly address the quality of a data set, more attention must be paid to the field sampling stage. Before sampling, a clear strategy should be planned. This strategy should be able to answer at least five questions related to:

- the correct choice of samples
- the number of samples to be collected
- how to collect them
- the sampling period
- the validity of the sampling site selected

5.2.2   Choice of Samples

After its direct release into water, TBT is accumulated by marine organisms and readily adsorbs on to suspended matter, sediments being the final sink. Thus, to understand the fate and distribution of TBT in the ecosystem under investigation it is necessary to collect water, sediment and biological samples at each sampling station, unless the aim of the sampling is different.

With regard to water samples, many workers perform the analysis, after filtration at $0.45\,\mu$m, to determine the organotin concentration in the dissolved phase. As the toxicity values of organotin compounds are generally referred to their concentration in this phase, such a procedure is widely accepted to provide information about the contamination levels. However, the strong tendency of TBT to be adsorbed on to suspended matter can lead to an underestimate of the true contamination of the site.[129] Depending on the particular site, the concentration of adsorbed TBT can be equal to, and in some case higher than, that in the dissolved phase. On the other hand, analyses of non-filtered samples provide results that are not easily comparable. This is due to the variety of the extraction techniques used for organotin analysis, and their respective performances with regard to extraction efficiencies from organotins on the particles. This point is worth mentioning since the stability of organotins may be affected by the presence of particles in the samples, even if the sample has been acidified. Nonetheless, there is a general lack of agreement between the monitoring authorities of different countries on these issues. This results in problems when making an inter-comparison of the levels in water and, even more important, a heterogeneous approach in setting up water quality criteria. The tremendous efforts that have gone into improving the analytical methodologies over the last decade should now be targeted at a homogenisation of the sampling strategies for urgent comparisons of the data, so that sound Water Quality Criteria can be established. Suspended matter collection, based on large volume filtration or centrifugation, should be taken into account. Indeed the suspended particulates are the primary source of entry of organotins to filter-feeding organisms.

The same careful approach needs to be taken on sediment collection, integrating the main geochemical parameters for meaningful comparative assessments. The sieving of the sediment samples at $60\,\mu$m may simplify the sediment collection procedure, in order to compare results obtained on sediments from other sites.[128] For surface samples, only the undisturbed surface layer (0–2 cm) should be sampled. Box corers, or some other low-disturbance sampler, should be used for sediment sampling. It should also be emphasised that sites for continuous sediment recording should be carefully selected, in order to have a better and global estimate of organotin half-lives in sediments.

Finally, biological monitoring, making use of the mussel watch programmes developed in different countries for the surveillance of organic pesticides, should also be considered here for the organotin compounds.

5.2.3   Number of Samples

The number of samples taken will depend on the site location and hydrodynamics of the area. The conclusions of an interagency workshop on aquatic sampling and analysis of organotins, held in 1986 in the USA, recommended the collection of three samples from each station and the analysis of two of them. If the results differ by more than 15%, the third sample should be analysed, and all of the results used to calculate the mean.[130]

5.2.4   Sample Collection

Contamination from the hulls of the sampling boats, treated with organotin-based anti-fouling paints, could lead to an overestimate of the organotin concentrations in water. For this reason, the use of rubber boats for sampling is strongly recommended.

Monobutyltin (MBT) and dibutyltin (DBT), the primary degradation products of TBT, are widely used as stabilisers of rigid PVC. Release of these compounds from PVC has been reported.[131] Thus, PVC containers should not be used as sampling bottles, in order to avoid possible changes in the organotin speciation due to the release of DBT and/or MBT from the walls of the containers. Losses of organotins, due to the adsorption on the container walls, are minimised by using polycarbonate bottles.[132]

PTFE sheets, glass plates and stainless steel mesh frames have been used for organotin sampling, but risks of adsorptive losses have not been studied.[133] A rotating drum polycarbonate sampler has been used to minimise losses.[132]

5.2.5   Spatial Variability

Significant differences of TBT and MBT concentrations usually occur in the water column. TBT shows a great tendency to accumulate in the surface microlayer, reaching much higher concentrations (one or two orders of magnitude) than in the water column.[134,135] Thus, superficial water samples should be collected at about 50 cm under the water surface. Organotin concentrations usually decrease as the water depth increases.[133-135] However, higher concentrations of organotins can be found at the water-sediment interface. The primary degradation product of TBT in sediments seems to be MBT[134,136] and, because of its hydrophilicity, it may enter the water column a short time after its formation. Moreover, sediment re-suspension phenomena can lead to a re-solubilisation of adsorbed organotins. Recent studies by Amouroux et al.[39] strengthen the above point. Diffusion of organotin compounds through volatilisation processes should lead to significant enrichments just above the sediment–water interface. Hence, deep water samples should be collected from a depth sufficiently above the bottom to avoid a direct influence from re-suspension, or de-gassing, from the sediment.

5.2.6   Temporal Variability

Seasonal variations in organotin concentrations often occur, due to the increase in the anthropogenic inputs during the summer (due to leisure boats), and to the different rates of degradation processes between the hot and cold seasons. Higher concentrations of organotins are usually determined during spring and summer.[133,135] Strong seasonal fluctuations in TBT concentrations have been reported in the south and southwest of England.[137] Sampling carried out at different states of the tide can lead to an over- or under-estimate of the organotin concentration in the dissolved phase, due to the dilution effect of the tide. Large differences in organotin concentrations in water samples collected before and after a tidal cycle have been reported.[138] Slack tide is the time recommended by the US EPA.

5.2.7   Sample Storage

Storage of samples is one of the most critical aspects of the whole analytical procedure, during which physico-chemical alterations occurring in the samples could affect the quality of the final determination steps. This is particularly true for organotins in biological and sediment samples, where degradation phenomena are well documented. Data on the stability of organotin compounds in water,[139] sediment[139-141] and mussel[142] samples have been reported.

TBT has been proven to be stable over a 4 month period in samples of seawater, filtered and acidified to pH 2, and stored in the dark both at $+4\,°C$ and at $+20\,°C$. The degradation products, DBT and MBT, were shown to be stable only at $+4\,°C$.[139] No significant variations in the TBT and DBT concentrations were observed in sediment extract samples stored at $+4\,°C$ or at $+20\,°C$. A significant degradation of MBT was, however, noted.[141] To minimise the risks of degradation, it is preferable to deep-freeze the samples immediately after sampling. Storage at $-20\,°C$ in the dark should ensure long-term organotin stability.

The stability of organotin compounds in freeze-dried mussel samples, stored under different conditions of light and temperature, was investigated over a 4 year period. The samples were stored at four different temperatures in the dark: $-20\,°C$, $+4\,°C$, room temperature and $+40\,°C$. At room temperature, the samples were also stored in daylight. TBT appeared to be stable for 6 months, in samples stored at $-20\,°C$ and $+4\,°C$. Samples stored at $+40\,°C$ and at room temperature displayed significant degradation, particularly if they were stored in the light. At room temperature, the concentration of TBT, after 1 year of storage, was about 30% of the original content, decreasing from about $3\,mg\,kg^{-1}$ to less than $1\,mg\,kg^{-1}$ (the worst stability was obviously obtained at $+40\,°C$). In the same samples a simultaneous increase of MBT concentration was observed, reflecting the known TBT degradation pattern. More pronounced degradation phenomena were observed for TPhT. A significant decrease of TPhT concentration was observed also in samples stored at $+4\,°C$, while in samples stored at room temperature in the light TPhT was completely degraded. The conclusion of this study was that long-term stability of butyl- and phenyl-tins, in freeze-dried mussel tissue, can be achieved only by storing samples at $-20\,°C$ in the dark. Storage of samples at room temperature must be avoided. Complete details of this study are reported elsewhere.[142,143]

6   Environmental Policies and Regulations

6.1   Policy Evolution

Since the 1960s, organotin compounds have been used in PVC, and as broad spectrum biocides. As discussed above, biology was the primary indicator and raised concerns after the decline of the French oyster farming activities in the 1980s.[144]

Later, the endrocrine disrupting effect of TBT, causing the imposex phenomenon in the snail *Nucella lappilus*, was observed right along the coast of the UK, and on many other European coastlines. Similar regulations banning TBT from anti-fouling paints then followed in many countries, such as the UK, the USA, Switzerland, the Netherlands, New Zealand and Japan and was implemented throughout the rest of Europe by 1989. In most cases, an environmental water quality standard was set at $2\,ng\,l^{-1}$ (as Sn). The EEC directive 89/677/EC set a similar limit, although this recommendation did not receive much attention and TBT anti-fouling paints were still available on the market despite the restrictions on their use.

Most of the action at this time was focused on coastal and enclosed harbours, and there was very little concern about open ocean systems, and little or no regard to freshwater environments. At the end of the 1980s, no further measures were considered necessary by the Marine Environmental Protection Committee (MEPC) of the International Maritime Organisation (IMO). These views changed in the early 1990s. With evidence of the increasing endrocrine disrupting effects of TBT on marine life, the MEPC proclaimed a worldwide recommendation of the IMO, to ban TBT in the 1990s. These views were further confirmed following the finding of imposex at the bottom of the sea correlated with the tracks of shipping routes and, most specifically, in south-east Asian waters.[145] TBT-treated ships were banned from Japan, and these results further supported the complete ban on the use of TBT.[146] At present TBT and triphenyltin, which is also used as a biocide in anti-fouling paints, have been included in the US EPA Environmental Endocrine Disrupters list, and, very recently, in the European Priority Pollutant Lists (EU Directive 76/464). These developments have led the MEPC to reach a unanimous agreement to impose a global ban on the application of TBT by January 2003 and on its presence by the 1 January 2008.

6.2   Industrial Responses

The high efficacy of TBT was rapidly realised by the industrial paint manufacturers and hence, given the prospect of a ban, a wide array of solutions have been, or are being, actively developed. Due to economic considerations, however, owners of large sea-going vessels try to minimise the time spent in dry-dock.

Thus, at present, ship-owners prefer anti-fouling paints that extend, as much as possible, the interval between dry-docking operations.

A wide array of alternatives to TBT-based paints have been developed. They are, however, generally less effective. Copper-based paints have been known for a long time, and are now regaining attention, despite being less effective than TBT-based paints. However, copper, and its derived formulations are only effective against marine fauna. In order to prevent weed growth, herbicides such as Diuron and Irgarol are often added, and these generate new threats to the environment. Other alternatives use non-stick coatings, which are not particularly efficient, or direct cleaning of the ships by mechanical underwater removal.

7 Conclusions

The fate and impacts of organotin compounds in the environment provide an excellent example of an environmental case that is far from being resolved. The historical evolution of the environmental case is illustrated in Figure 3.

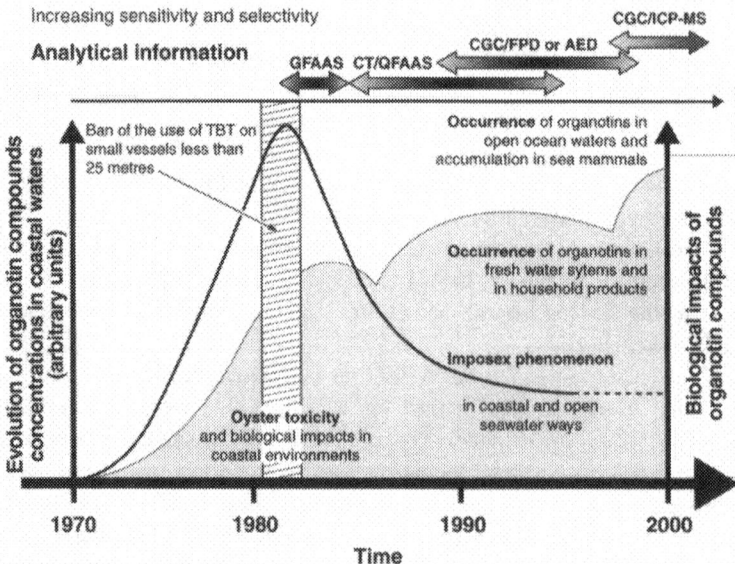

Figure 3 *Historical perspective of organotin compounds concentrations in the environment and the growing awareness of their occurrence and impacts of organotin compounds in the environment. Biological and instrumental responses. (GFAAS Graphite Furnace Atomic Absorption Spectrometry, CT/QFAAS Cryo Trapping Quartz Furnace Atomic Absorption Spectrometry, Capillary Gas Chromatography Flame Photometric Detection, Atomic Emission Detection, Capillary Gas Chromatography Inductive Coupled Plasma Mass Spectrometry)*

Typically, as for many major environmental crises, the impact was revealed when macroscopic biological disorders were observed. The fallout from the oyster farming activities triggered international interest in the issue. This initial concern concluded that the problem was confined to harbours and coastal environments. Further concerns were raised when the endocrine disrupting properties of these compounds were established, with the development of imposex on snail populations throughout the world, particularly in populations living below major shipping lanes. The presence of these compounds in large marine organisms at the top of the oceanic food chain, and now the direct evidence (made possible by the development of powerful selective and ultra-sensitive analytical techniques) of the occurrence of trace amounts of these substances in the open ocean, lead us to think that maritime inputs of organotin, to the global oceans, are far from insignificant.

Although coastal issues are well known, they are still likely to be under-estimated. Indeed, very little is known of the accumulation and impacts of organotin compounds in seafood products. These products are chiefly produced in coastal areas, and are therefore severely affected by the presence of organotin compounds, as revealed by the recent study produced by the IVM.[63]

If international regulations on the reduction or elimination of organotins from anti-fouling paints, scheduled for 2002, do take place, this will certainly contribute to a massive reduction in direct maritime inputs. This, however, will not directly eradicate the impact of these compounds on coastal areas. Despite the ban on the use of organotin-containing anti-fouling paints on small vessels, butyltin concentrations have not significantly declined in coastal waters, and concentration levels still exceed the environmental quality targets of most countries.

Indeed, sediments, which where earlier considered to be a sink, will now have to be considered as a source of the compounds, accumulated over time. The organotins will slowly diffuse back into the water column after biotic or abiotic trans-methylation reactions.[39] Despite the very low concentrations in the environment, nothing is known of the toxicity and uptake route of the methylated moieties on biological systems. Many questions are still unanswered regarding these serious issues. There is, now, evidence of important contamination of freshwater systems. These contaminations probably originated from the extensive use of organotin compounds in the large array of industrial products that can currently be found in many households. They will end up in sewage treatment plants, and will later be released into water-ways, or enter the food chain, *via* the amendment of cultivated soils with sewage sludge. The direct inputs of triorganotins as terrestrial fungicides should also be evaluated. Further, the fate, occurrence and biological impacts of these compounds, in the top trophic level of the food chain, including humans, is not well known. This needs to be thoroughly evaluated, due to the compounds' endocrine disruption properties at very low concentrations.

In spite of the fact that these questions have triggered the international efforts of the scientific community, and that a large array of analytical solutions have been developed, there is still an urgent need for the harmonisation of sampling and analytical procedures for organotins in the main compartments of the ecosystem. This is needed to allow meaningful comparability of the data. These efforts should be accompanied by a greater understanding of the biochemical actions of organotins at the biomolecular and cellular level.

We now have the evidence, first through biology and secondly from the direct confirmation of their presence, *via* the remarkable developments in instrumentation, that organotin compounds affect all compartments of the ecosystem. They should definitely be considered as global pollutants, in a similar way to now well established considerations for PCBs, mercury and dioxins. The efficiency of trisubstituted organotin species as endocrine disrupters, even at very low concentrations, should put them at the top of the list of the priority pollutants.

Acknowledgement

The authors would like to thank Christelle Benoit for her help in the preparation of this manuscript.

8 References

1. J.W. Ludgate, *Ocean '87, Conf. Rec.*, 1987, **4**, 1309.
2. S.J. Bushong, M.C. Ziegenfuss, M.A. Unger and L.W. Hall, *Environ. Toxicol. Chem.*, 1990, **9**, 359.
3. R.J. Maguire, R.J. Tkacz, Y.K. Chau, G.A. Bengert and P.T.S. Wong, *Chemosphere*, 1986, **15**, 253.
4. R.J. Maguire, *Water Poll. J. Can.*, 1991, **26**, 243.
5. R.J. Hugget, M.A. Unger, P.F. Seligman and A.O. Valkirs, *Environ. Sci. Technol.*, 1992, **26**, 232.
6. O.F.X. Donard and P. Michel, *Analusis*, 1992, **20**, M45.
7. C. Alzieu, *Mar. Environ. Res.*, 1991, **32**, 7.
8. R.J. Huggett, M.A. Unger, P.F. Seligman and A.O. Valkir, *Environ. Sci. Technol.*, 1992, **26**, 232.
9. S.J. De Mora and E. Pelletier, *Environ. Technol.*, 1997, **18**, 1169.
10. C. Alzieu, J. Sanjuan, J.P. Deltreil and M. Borel, *Mar. Pollut. Bull.*, 1986, **17**, 494.
11. D. Adelman, K.R. Hinga and M.E.Q. Pilson, *Environ. Sci. Technol.*, 1990, **24**, 1027.
12. K. Fent, *Crit. Rev. Toxicol.*, 1996, **26**, 1.
13. L. Ebdon, S.J. Hill and C. Rivas, *Trends Anal. Chem.*, 1998, **17**, 277.
14. S. Yamada, Y. Fujii, E. Mikami, N. Kawamura, J. Hayakawa, K. Aoki, M. Fukaya and C. Terao, *J. AOAC Int.*, 1993, **76**, 436.
15. R.J. Maguire, *Water Pollut. Res. J. Can.*, 1991, **26**, 243.
16. S.J. De Mora and E. Pelletier, *Environ. Technol.*, 1997, **18**, 1169.
17. X. Dauchy, Thèse d'Université, Spéciation des butylétains dans les sédiments par couplage HPLC-ICP-MS, Université de Pau, 1993.
18. S. Simon, Rapport de recherche, DEA Environnement et matériaux, Transfert des organoétains vers les plantes, Université de Pau, 2000.
19. K.A. Barnes, R.J. Fussell, J.R. Startin, H.J. Mobbs, R.James and S.L. Reynolds, *Rapid Comm. Mass Spectrom.*, 1997, **11**, 159.
20. B.W. Krugh and D. Miles, *Environ. Toxicol. Chem.*, 1996, **15**, 495.
21. M.A. Champ and P.E. Seligman, *Organotin: Environmental Fate and Effects*, Chapman and Hall, London, 1996.
22. R.J. Maguire, J.H. Carey and E.J. Hale, *J. Agric. Food. Chem.*, 1983, **31**, 1060.

23. R.B. Laughling Jr., H.E. Guard and W.M. Colman III, *Environ. Sci. Technol.*, 1986, **20**, 201.
24. O.F.X. Donard, S. Rapsomanikis and J.H. Weber, *Anal. Chem.*, 1986, **58**, 772.
25. L. Randall and J.H. Weber, *Sci. Total Environ.*, 1986, **57**, 191.
26. J.J. Cleary and A.R.D. Stebbing, *Mar. Pollut. Bull.*, 1987, **18**, 238.
27. M.A. Unger, W.G. MacIntyre and R.J. Huggett, *Environ. Toxicol. Chem.*, 1988, **7**, 907.
28. Ph. Quevauviller and O.F.X. Donard, in *Element Speciation in Bioinorganic Chemistry*, eds. S. Caroli, John Wiley & Sons, New York, 1996, 331.
29. D.R.J. Moore, D.G. Noble, S.L. Walker, D.M. Trotter, M.P. Wong and R.C. Pierce, Recommandations pour la qualité de l'eau au Canada, Direction Générale des Sciences et de l'évaluation des écosystèmes, Direction de la santé des écosystèmes, Ottawa, Ontario, Canada, Etude no 191, Série scientifique, 1992.
30. D. Adelman, K.R. Hinga and M.E.Q. Pilson, *Environ. Sci. Technol.*, 1990, **24**, 1027.
31. R.F. Lee, A.O. Valkirs and P.F. Seligman, *Environ. Sci. Technol.*, 1989, **23**, 1515.
32. D. Barug, *Chemosphere*, 1981, **10**, 1145.
33. O.F.X. Donard, F.T. Short and J.H. Weber, *Can. J. Fish. Aquat. Sci.*, 1987, **44**, 140.
34. R. François, F.T. Short and J.H. Weber, *Environ. Sci. Technol.*, 1989, **23**, 191.
35. P. Michel and B. Averty, *Environ. Sci. Technol.*, 1999, **33**, 2524.
36. M. Astruc, R. Lavigne, R. Pinel, F. Leguille, V. Desauziers, Ph. Quevauviller and O.F.X. Donard, *Speciation of Tin in Sediments of Arcachon Bay, France*, in *Metals Speciation and Recovery*, Lewis Publisher, Rome, Vol II, 1989, pp. 263–274.
37. Ph. Quevauviller, O.F.X. Donard, R. Ritsema and M. Ewald, *Sampling Organotin Variability in Estuarine and Coastal Environments: A Review*, in *Heavy Metals in the Hydrological Cycle*, Lisbon, 1988, p. 401.
38. J.M. Ruiz, G. Bachelet, P. Caumette and O.F.X. Donard, *Environ. Pollut.*, 1996, 93, **2**, 195.
39. D. Amouroux, E. Tessier and O.F.X. Donard, *Environ. Sci. Technol.*, 2000, **34**, 988.
40. C.G. Arnold, Triorganotin compounds in natural waters and sediments: Aqueous speciation and sorption mechanisms, PhD Thesis, Swiss Federal Institute of Technology, Zurich, Switzerland, 1998.
41. R.J. Maguire, R.J. Tkacz, Y.K. Chau, G.A. Bengert and P.T.S. Wong, *Chemosphere*, 1986, **15**, 253.
42. P.M. Sarradin, Thèse d'Université, Répartition et évolution du tributylétain dans les sédiments marins, Université de Pau, 1993.
43. M.E. Waite, K.E. Evans, J.E. Thain and M.J. Waldock, *Appl. Organomet. Chem.*, 1989, **3**, 383.
44. C. Carlier-Pinasseau, G. Lespes and M. Astruc, *Environ. Technol.*, 1997, **18**, 1179.
45. Ph. Quevauviller, R. Ritsema, R. Morabito, W.M.R. Dirkx, S. Chiavarini, J.M. Bayonne and O.F.X. Donard, *Appl. Organomet. Chem.*, 1994, **8**, 541.
46. J.A. Stab, W.P. Cofino, B. Van Hattum and U.A.Th. Brinkman, *Anal. Chim. Acta*, 1994, **286**, 335.
47. S. Kan-Atireklap, S. Tanabe and J. Sangansin, *Mar. Pollut. Bull.*, 1997, **34**, 894.
48. A. Steward and J. Thomson, 1997 in IFREMER RNO Report *Surveillance du milieu marin*, edition 1999.
49. H. Harino, M. Fukushima and S. Kaway, *Environ. Pollut.*, 1998, **105**, 1.
50. B.S. Tselentis, M. Maroulakou, J.F. Lascourreges, J. Szpunar, V. Smith and O.F.X. Donard, *Mar. Pollut. Bull.*, 1999, **38**, 146.
51. S. Diez, M. Abalos and J.M. Bayona, *Wat. Res.*, 2001, in press.
52. Ph. Quevauviller, O.F.X. Donard and H. Etcheber, *Environ. Pollut.*, 1994, **84**, 89.

53. Ph. Quevauviller, H. Etcheber, C. Raoux and O.F.X. Donard, *Oceanol. Acta*, 1991, **11**, 247.

54. P.M. Sarradin, A. Astruc, R. Sabrier and M. Astruc, *Mar. Pollut. Bull.*, 1994, **28**, 621.

55. P.M. Sarradin, A. Astruc, V. Desauziers, R. Pinel and M. Astruc, *Environ. Technol.*, 1991, **12**, 537.

56. Ph. Quevauviller and O.F.X. Donard, *Appl. Organomet. Chem.*, 1990, **4**, 353.

57. W.P. Ridley, L.J. Dizikes and J.M. Wood, *Science*, 1977, **197**, 329.

58. H.E. Guard, A.B. Cobet and W.M. Coleman III, *Science*, 1981, **213**, 770.

59. J.T. Byrd and M.O. Andreae, *Science*, 1982, **218**, 565.

60. L.E. Hallas, J.C. Means and J.J Cooney, *Science*, 1982, **215**, 1505.

61. C. Alzieu, J. Sanjuan, P. Michel, M. Borel and J.P. Dreno, *Mar. Pollut. Bull.*, 1989, **20**, 22.

62. R. Ritsema, *Appl. Organomet. Chem.*, 1994, **8**, 5.

63. A.C. Belfroit, M. Puperhart and F. Ariese, *Organotin Levels in Seafood in Relation to the Tolerable Daily Intake (TDI) for Humans*, Report E-99/12, Institut for Environmental Studies, Amsterdam, The Netherlands, 1999.

64. US Environmental Protection Agency, Report to Congress on effectiveness of existing laws and rules on reducing risks from organotin compounds, prepared by the Office of Prevention, Pesticides, and Toxic substances, Lynn Goldman, Assistant administrator, and the Office of Water, Robert Perciaseppe, Assisstant Administrator, 1997.

65. P.E. Gibbs, G.W. Bryan, P.L. Pascoe and G.R. Burt, *J. Mar. Biol. Ass. UK*, 1987, **67**, 503.

66. B.P. Mensink, *Imposex in the Common Whelk, Buccinum undatum*, eds. Ponsen and B.V. Looijen, Thesis, Wageningen University, The Netherlands, 1999.

67. C.C. ten Hallers-Tjabbes, J.F. Kemp and J.P. Boon, *Mar. Pollut. Bull.*, 1994, **28**, 311.

68. S. Hashimoto, M. Watanabe, Y. Noda, T. Hayashi, Y. Kurita, Y. Takasu and A. Otsuki, *Mar. Environ. Res.*, 1998, **45**, 169.

69. H. Yamada, K. Takayanagi, M. Tateishi, H.Tagata and K. Ikeda, *Environ. Pollut.*, 1997, **96**, 217.

70. S. Tanabe, M. Prudente, T. Mizuno, J. Hasegawa, H. Iwata and N. Miyazaki, *Environ. Sci. Technol.*, 1998, **32**, 193.

71. K.S. Guruge, H. Iwata, H. Tanaka and S. Tanabe, *Mar. Pollut. Bull.*, 1997, **44**, 191.

72. H. Tao, R.B. Rajendran, R. Naganawa, T. Nagazato, A. Miyakazi, M. Kunugi and A. Harashima, *J. Environ. Chem.*, 1999, **3**, 661.

73. H. Tao, R.B. Rajendan, C.R. Quetel, T. Nakazato, M. Tominaga and A. Miyazaki, *Anal. Chem.*, 1999, **71**, 4208.

74. R.J. Maguire, *Tributyltin in Canadian Waters*, INRE Contribution, Institut National de Recherche sur les Eaux, Burlington, Ontario, Canada, 1989.

75. Y.K. Chau, R.J. Maguire, M. Brown, F. Yang and S.P. Batchelor, *Water Qual. Res. J. Can.*, 1997, **32**, 453.

76. M. Muller, L. Renberg and G. Rippen, *Chemosphere*, 1989, **18**, 2015.

77. K. Becker Van Slooten, L. Merlini, A.-M. Stegmueller, F. DeAlencastr and J. Tarradellas, *Gaz Wasser Abswasser*, 1994, **2**, 104.

78. A.I. Sadiki, D.T. Williams, R. Carrier and B. Thomas *Chemosphere*, 1996, **32**, 2389.

79. C. Montigny and G. Lespes, *Analyse de composés organo-métalliques sur eaux et matières en suspension sur divers cours d'eau de Bassin Rhin-Meuse*, Rapport Agence de l'Eau Rhin-Meuse, 1997.

80. G. Lespes, V. Desauziers, C. Montigny and M. Potin-Gautier, *J. Chromatogr. A*, 1998, **826**, 67.

81. C. Bancon-Montigny, G. Lespes and M. Potin-Gautier, *J. Chromatogr. A*, 2000, **896**, 149.

82. F. Ariese, personal communication, 1999.

83. World Health Organisation, *Tributyltin Compounds*, United Nations Environment Program, World Health Organisation, Environmental Health Criteria 116, Geneva, Switerland, 1990, p. 273, ISBN 92-4-157116-0.

84. O.F.X. Donard, Ph. Quevauviller and A. Bruchet, *Wat. Res.*, 1993, **27**, 1085.

85. K. Fent, J. Hunn, D. Renggli and H. Siegrist, *Mar. Environ. Res.*, 1991, **32**, 223.

86. I. Heninger, M. Potin-Gautier, M. Astruc, L. Galvez and V. Viguier, *Chem. Spec. Bioavailab.*, 1998, **10**, 1.

87. C. Bancon-Montigny, G. Lespes and M. Potin-Gautier, *Analyst*, 1999, **124**, 1265.

88. Ph. Quevauviller and O.F.X. Donard, *Appl. Organomet. Chem.*, 1991, **5**, 125.

89. RIVM Report, *Health Risk Assessment for Organotins in Textiles*, eds. Alberts and Dannen, Report 613350 002, The Netherlands, 2000.

90. K. Fent, R. Fassbind and H. Siegrist, *Organotins in a Municipal Wastewater Treatment Plant*, First Conference of Ecotoxicology, Copenhagen, eds. H. Lokke, H. Tyle and F. Bio-Rasmussen, The Technical University of Denmark, Lyngby, Denmark, 1989, pp. 72–80.

91. E. Kortland and J. Stronkhorst, *An Issue of Substance: TBT in Marine Antifouling Paints*, National Institute for Coastal and Marine Management/RIKZ, 1998.

92. G.H. Biégo, M. Joyeux, P. Hartemann and G. Debry, *Arch. Environ. Contamin. Toxicol.*, 1999, **36**, 227.

93. W.A.C. Anderson, S.A. Thorpe, L.M. Owen, S.E. Anderson, H.M. Crews and S.L. Reynolds, *Food Add. Contam.*, 1998, **15**, 288.

94. H.H. Van den Broek, G.B.M. Hermes and C.E. Goewie, *Analyst*, 1988, **113**, 1237.

95. K. Kannan and R. F. *Environ. Toxicol. Chem.*, 1996, **15**, 1492.

96. G.P. Gabrielides, C. Alzieu, J.W. Readman, E. Bacci, O. Aboul Dahab and I. Salihoglu, *Mar. Pollut. Bull.*, 1990, **21**, 233.

97. K. Fent and J. Hunn, *Environ. Technol. Chem.*, 1995, **14**, 1123.

98. W. Wu, R.S. Roberts, Y.C. Chung, W.R. Ernst and S.C. Havlicek, *Arch. Environ. Contam. Toxicol.*, 1989, **18**, 839.

99. C. Alzieu, in ed. S.J. de Mora, Cambridge University Press, p. 94.

100. I.M. Davies, S.K. Bailey and M.J.C. Harding, *J. Mar. Sci.*, 1998, **55**, 34.

101. J.M. Ruiz, J. Szpunar and O.F.X. Donard, *Sci. Total Environ.*, 1997, **198**, 225.

102. P.E. Gibbs, G.W. Bryan and P.L. Pascoe, *Mar. Environ. Res.*, 1991, **1–4**, 79.

103. P. Matthiessen and P.E. Gibbs, *Environ. Toxicol. Chem.*, 1998, **1**, 37.

104. C. Alzieu, *Ecotoxicology*, 2000, **1–2**, 71.

105. A.N. Jah, J.A. Hagger and S.J. Hill, *Environ. Mol. Mutagenesis*, 2000, **4**, 343.

106. E. Haubruge, F. Petit and M.J.G. Gage, *Biol. Sci.*, 2000, **1459**, 2333.

107. T. Horoguchi, N. Takiguchi, H.S. Cho, M. Kojima, M. Kaya, H. Shiraishi, M. Morita, H. Hirose and M. Shimizu, *Mar. Environ. Res.*, 2000, **50**, 223.

108. O. Yamanoshita, M. Kurasaki, T. Saito, K. Takahasi, H. Sasaki, T. Hosokawa, M. Okabe, J. Mochida and T. Iwakuma, *Biochem. Biophys. Res. Comm.*, 2000, **272**, 557.

109. S. Mizuhashi, Y. Ikegaya and N. Matsuki, *Environ. Toxicol. Pharmacol.*, 2000, **8**, 205.

110. A.H. Penninks, *Food Add. Contam.*, 1993, **10**, 351.

111. K. Becker Van Slooten and C. Studer, *Pollution des eaux par les organoétains: situation encore inchangée malgré l'interdiction*, Bulletin de l'Office Fédéral de l'environnement des forêts et du paysage, Protection de l'environnement en Suisse, 1993, p. 11.

112. J.A. Stab, M. Frenay, I.L. Freriks, U.A.Th. Brinkman and W.P. Cofino, *Environ. Toxicol. Chem.*, 1995, **14**, 2023.

113. C. Carlier-Pinasseau, A. Astruc, G. Lespes and M. Astruc, *J. Chromatogr. A*, 1996, **750**, 317.

114. P.W. Looser, *Bioaccumulation of triorganotin compounds by a sediment-dwelling organism* (*Chironomus riparius*): *assessment of bioavailability, uptake and elimination processes*, dissertation submitted to the Swiss Federal Institute of Technology Zürich for the degree of Doctor of Natural Sciences, Diss. ETH No. 13 661, Zürich, 2000.

115. K. Kannan, S. Tanabe and R. Tatsukawa, *Bull. Environ. Contamin. Toxicol.*, 1995, **55**, 510.

116. R. Ritsema and O.F.X. Donard, *Organometallic Compounds Determination in the Environment by Hyphenated Techniques*, in *Sample Handling and Trace Analysis of Pollutants*, D. Barcelo, ed., Elsevier, Amsterdam, 2000, p. 1003.

117. O.F.X. Donard, E. Krupp, C. Peycheran, D. Amouroux and R. Ritsema, *Trends in Speciation Analysis for Routine and New Environmental Issues*, in *Elemental Speciation – New Approaches for Trace Element Analysis*, J.A. Caruso, K.L. Sutton and K.L. Ackley, eds., Elsevier, Amsterdam, 2000, p. 451.

118. R. Morabito, *Fresenius' J. Anal. Chem.*, 1995, **351**, 378.

119. R. Morabito, *Microchem. J.*, 1995, **51**, 198.

120. Ph. Quevauviller and R. Morabito, *Trends Anal. Chem.*, 2000, **19**, 86.

121. W.M.R. Dirkx, R. Lobinski and F.C. Adams, *Anal. Chim. Acta*, 1994, **286**, 309.

122. R. Morabito, H. Muntau, W. Cofino and Ph. Quevauviller, *J. Environ. Monit.*, 1999, **1**, 75.

123. R. Ritsema, F.M. Martin and Ph. Quevauviller, *Hydride Generation for Speciation Analysis Using GC/AAS*, in *Quality Assurance for Environmental Analysis*, Ph. Quevauviller, ed., Elsevier, Amsterdam, 1995, p. 490.

124. C. Pellegrino, P. Massanisso and R. Morabito, *Trends Anal. Chem.*, 2000, **19**, 97.

125. W.M.R. Dirkx, R. Lobinski and F.C. Adams, *Speciation Analysis of Organotin by GC-AAS and GC-AES after Extraction and Derivatization*, in *Quality Assurance for Environmental Analysis*, Ph. Quevauviller, ed., Elsevier, Amsterdam, 1995, p. 360.

126. M.B. de la Calle-Guntiñàs, R. Scerbo, S. Chiavarini, Ph. Quevauviller and R. Morabito, *Appl. Organomet. Chem.*, 1997, **11**, 693.

127. R. Morabito, P. Massanisso and Ph. Quevauviller, *Trends Anal. Chem.*, 2000, **19**, 97.

128. O.F.X. Donard, Ph. Quevauviller, R. Ritsema and M. Ewald, in *Heavy Metals in the Hydrological Cycle*, Lisbon, pp. 401–412.

129. D.R. Young, P. Schatzberg, F.E. Brinckman, M.A. Champ, S.E. Holm and R.B. Landy, *Summary Report – Interagency Workshop on Aquatic Sampling and Analysis for Organotin Compounds*, Proceedings, Ocean 86, Washington DC, 1986, pp. 1135–1140.

130. Ph. Quevauviller, A. Bruchet and O.F.X. Donard, *Appl. Organomet. Chem.*, 1991, **5**, 125.

131. C.A. Dooley and V. Homer, *Organotin Compounds in the Marine Environment: Uptake and Sorption Behavior*, Naval Ocean System Center Technical Report No. 917, San Diego, CA, 1983.

132. G. Batley, *Distribution and Fate of Tributyltin in the Marine Environment, Tributyltin: Case Study of an Environmental Contaminant*, Cambridge Environmental Chemistry Series, 1996.

133. J.J. Cleary and A.R.D. Stebbing, *Organotins in the Water Column – Enhancement in the Surface Microlayer*, Proceedings, Ocean 87, An International Workplace, Hali-

fax, 1987, pp. 1405–1410.

134. S. Chiavarini, C. Cremisini and R. Morabito, *Distribution and Fate of TBT and its Degradation Products in the La Spezia Gulf*, FAO/UNEP/IAEA MAP Technical Report Series No. 59, 1991, pp. 179–187.

135. A.M. Caricchia, S. Chiavarini, C. Cremisini, M. Fantini and R. Morabito, *Sci. Total Environ.*, 1992, **121**, 133.

136. P.M. Stang, R.F. Lee and P.F. Seligman, *Environ. Sci. Technol.*, 1992, **26**, 1382.

137. L. Ebdon, K.E. Evans and S.J. Hill *Sci. Total Environ.*, 1988, **68**, 207.

138. C. Clavell, P.F. Seligman and P.M. Stang, *Automated Analysis of Organotin Compounds: a Method for Monitoring Butyltins in the Marine Environment*, Proceedings, Ocean 86, Washington DC, 1986, pp. 1152–1154.

139. Ph. Quevauviller and O.F.X. Donard, *Fresenius' J. Anal. Chem.*, 1991, **339**, 6.

140. Ph. Quevauviller, R. Morabito, L. Ebdon, W. Cofino, H. Muntau and M.J. Campbell, *The Certification of the Contents (Mass Fractions) of Monobutyltin, Dibutyltin and Tributyltin in Mussel Tissue (CRM477)*, EUR Report 17921 EN, Brussels, 1997.

141. W.C. Li, S. Zhang, Y.K. Chau and A.S.Y. Chau, *Preservation of Organics. Part IV Stability of Bis(tri-n-butyltin) Oxide in Sediment Extracts*, NWRI Contribution No. 90-139, 1, 1990.

142. W.C. Li, S. Zhang, Y.K. Chau and A.S.Y. Chau, *Preservation of Organics. Part V Stability of Butyltin Species in Sediment Extracts*, NWRI Contribution No. 90-146, 1, 1990.

143. A.M. Caricchia, S. Chiavarini, C. Cremisini, R. Morabito and R. Scerbo, *Anal. Chim. Acta*, 1994, **286**, 329.

144. C. Alzieu, M. Héral, Y. Thibaud, M.J. Dardignac and M. Feuillet, *Rev. Trav. Inst. Peches Maritimes*, 1982, **45**, 101.

145. C. Swenn, N. Ruttanadakul, S. Ardseungnern, H.R. Singh, B.P. Mensink and C.C. ten Hallers-Tjabbes, *Environ. Technol.*, 1997, **18**, 1245.

146. C.C. ten Hallers-Tjabbes, *Environ. Technol.*, 1997, **18**, 1265.

CHAPTER 9

An Environmental Case History of the Platinum Group Metals

SEBASTIEN RAUCH AND GREGORY M. MORRISON

1 Introduction

During recent years the use of the platinum group metals (PGMs) has increased dramatically, mainly as a catalyst for the removal of pollutants in automobile exhaust. The increasing use of platinum and more recently palladium and rhodium has raised questions concerning their release into the environment and the related ecological and human health risks.

The potential problems caused by the presence of PGMs in the environment can only be assessed through a clear understanding of their environmental pathways, transformations and speciation.[1] In early years, research was dedicated to the difficult task of the determination of total concentrations, but increasingly research groups are trying to provide this clear understanding that can only be achieved through developments in speciation analysis.

This article gives an insight into platinum group metal research with a particular focus on speciation.

2 Natural Occurrence

Platinum was discovered in South America by Ulloa in 1735 and by Wood in 1741. Palladium and rhodium were found in a platinum ore in 1803. The PGMs are concentrated in the Earth's core and mantle. PGM abundance in the Earth's crust is low (Table 1) and their repartition inhomogeneous. It is only under certain conditions that platinum group metals occur at higher concentration, allowing their recovery. PGMs are found in two kinds of deposits: primary deposits, in which they are usually associated with other elements, for example nickel and copper, in igneous rocks, and secondary deposits formed by erosion and relocation of PGMs in a pure metallic form.[2]

It follows, therefore, that the background levels of the PGMs in the environment are very low and difficult to measure. Background levels in air samples[3] are below $0.05 \, \text{pg m}^{-3}$. Background PGM levels have been estimated to be in the pg g^{-1} range in road sediments[4] and soil.[5] Background platinum and rhodium concentrations in grass[6] are below $0.03 \, \text{ng g}^{-1}$.

3 Uses of Platinum Group Metals

Before its chemical recognition in 1735, platinum was used in jewellery by pre-Colombian Indians and indeed this use continues today. It was only later, during the 19th century, that platinum started to be used in dentistry and for applications in electrical and chemical engineering. The invention of ammonia oxidation, with platinum as a catalyst, enabled the large-scale production of fertilisers in the late 1940s. At the same time, platinum started to be used as a catalyst by the petrochemical industry.

In the mid-1960s platinum was found to have therapeutic properties against cancer[7] (see box). Recently, the interest for platinum interactions with the human body led to the development of platinum-based drugs with reduced side effects and increased efficiency. Speciation techniques have been a particularly useful tool to test new drugs and understand their mechanism of action.[8,9]

With the introduction of more stringent air pollution control regulations in the USA, car companies developed technologies for the reduction in the emission of exhaust pollutants, particularly carbon monoxide and hydrocarbons. The first successful approach to this problem was the oxidation catalytic converter, in which platinum was used for the oxidation of carbon monoxide and hydrocarbon. Basically, a catalytic converter is a device fitted directly to the exhaust system of a vehicle to reduce the amount of harmful gases emitted. The introduction of the autocatalyst in the USA was followed by their introduction in Japan (1976) and Canada (1976). Since 1993 autocatalysts have been mandatory on all new cars in Europe.

Previously, palladium was mainly used by the electrical industry and for dentistry. Recently, palladium has been incorporated into automotive catalysts and its use is increasing. Palladium has similar catalytic properties to platinum and a lower market price. Consequently, palladium has been used in catalytic

---

Cancer therapy

Intensive clinical research for more than 20 years has shown that several platinum-based drugs interact with DNA and can prevent cell division. Therefore, these drugs have been found to be efficient against several types of cancer by slowing tumour growth.[7]

The principal drugs used are cisplatin and carboplatin. Cisplatin was initially used,[4] but side effects, *e.g.* nausea and neurotoxicity, led to the development of new drugs. Carboplatin was found to have reduced side effects but shares a cross resistance with cisplatin.[10] Third generation drugs have been developed with the hope of increased efficiency, lowered side effects and cross resistance.[11]

converters as a substitute for platinum or to reduce its amount.

Until the 1980s the use for rhodium was very low. It was found that rhodium can catalyse the reduction of nitrogen oxides. Rhodium was, therefore, added to platinum in a new generation of autocatalysts, or three-way converters, which enable the oxidation of carbon monoxide and hydrocarbons and the reduction of nitrogen oxides. Modern catalytic converters, therefore, use a trimetallic combination of palladium, platinum, and rhodium.[12]

There is now no doubt that catalytic converters represent the major input of platinum group metals entering the environment[13,14] and the increasing use of the PGMs in autocatalysts has led to increasing concentrations in the urban environment.

4 Platinum Group Metal Research

After the introduction of catalytic converters in the US in the mid 1970s, several research groups reported on platinum concentrations in the environment. Hill and Mayer[15] were the first to report noble metal emissions from autocatalysts, and background levels of platinum and palladium were determined in human samples.[3] However, noble metals occurred at concentrations too low to be determined by available techniques in environmental samples. Moreover, platinum was thought to be inert and, therefore, interest in this research was relatively low.

Research on platinum in the environment went through a revival in the mid-1980s. With the development of sensitive analytical techniques, including electrochemical and spectroscopic techniques, it became possible to determine platinum at environmental concentrations. After the introduction of autocatalysts in Germany in 1986, several German research groups started to analyse platinum in vehicle emissions[16] and to investigate the fate of the emitted platinum. At the same time in Australia, platinum concentrations were estimated in blood,[17] and diet was found to be an important source of platinum for man.[18] In Sweden, platinum was found to be increasing in road sediments[19] and the first attempt at speciation was performed through sequential extraction.[20]

The further development of analytical techniques now enables chromatographic speciation with coupling to sensitive detection techniques such as inductively coupled plasma-mass spectrometry, but the analysis of PGMs at environmental concentrations remains difficult, as does their speciation.

An increasing number of research groups in Europe are now contributing to the understanding of platinum group metals in the environment. This interest resulted in the European Commission funding the CEPLACA project (see box).

However, while we are starting to understand the pathways and transformations of platinum in the environment, little is known about palladium and rhodium.

CEPLACA

The European Commission is funding a project on the 'Assessment of Environmental Contamination Risk by Platinum, Palladium and Rhodium from Automobile Catalyst' (CEPLACA) which started in December 1997 and involves academic departments, research institutes and car companies in five European countries. CEPLACA has three main objectives: the study of platinum group metal emissions, the assessment of environmental contamination and the study of PGM bioaccumulation, as well as the development of analytical techniques. The results will be used for environmental and health risk assessments.

The CEPLACA project should provide an improved understanding of environmental contamination by PGMs. Moreover, the CEPLACA project can be considered strategic, since the results can be implemented by governmental institutions for new regulations as well as directly by the involved car companies to avoid environmental contamination and the loss of precious resources.

---

5 The Analytical Challenge

The accurate determination of PGMs in the environment has been a difficult task owing to their low concentrations in most samples. Several techniques have a proven sensitivity for PGMs, the most commonly used being inductively coupled plasma-mass spectrometry (ICP-MS) and cathodic stripping voltammetry (CSV).

The determination of platinum group metals by ICP-MS suffers from molecular ion interferences.[4,21-23] Non-spectral interference can be avoided through the use of an internal standard. Spectral interference can be eliminated when sufficiently high spectral resolution is used.[4] However, only a few instruments can achieve the required resolution for platinum, and for palladium and rhodium only some of the interferences can be eliminated by employing high resolution ICP-MS. Another approach is to estimate the contribution of the interference and correct mathematically.[23]

Platinum and rhodium can be analysed by CSV based on the catalytic properties of PGMs.[24] The technique is very sensitive but relatively slow. Moreover, it does not offer the possibility of analysing palladium at low concentrations. Combined results from ICP-MS and CSV for the analysis of PGMs in environmental samples are perhaps the closest we can come to a true result today.[4]

It should also be stressed that the assessment of platinum concentration and its effect on the environment can only be achieved with improved quality control. So far there is a lack of reference materials certified for platinum, palladium and rhodium. The European Commission funded PACEPAC project aims at the certification of platinum group metal concentrations in tunnel dust and should be a good basis for improved analysis of PGMs in environmental samples.[25]

6 Speciation of Platinum Emission

It has now been demonstrated that catalytic converters (see box) emit significant amounts of PGMs to the environment.[13,14] Reported emission rates vary, but are believed to lie in the ng km^{-1} range. While measurements have usually been performed under good engine conditions, it has been demonstrated that emission rates can be significantly increased under normal, on-road engine conditions.[26]

PGM emission may be an abrasive mechanism, leading to the release of washcoat particles on to which PGMs are bound. PGMs are mostly emitted in a metallic form with a small soluble fraction.[16] Noble metals are used as catalysts and should, therefore, remain in a metallic form. However, platinum in fresh autocatalyst is not only present in a metallic form, but also as oxides, as chlorides, and bound to hydrocarbons. The chloride species probably originate from chloroplatinic acid, which is used as a precursor in the converter fabrication process. Rhodium is present in a metallic form or as an oxide.[27]

PGM species other than those normally found in a fresh autocatalyst might form during ageing of the converter. High temperatures can result in sintering of noble metal particles leading to particle sizes of up to several μm.[28] Moreover, particular conditions in the autocatalyst may explain the transformation of PGMs into a soluble form. The study of PGM speciation in vehicle exhaust is of interest for the understanding of the catalytic mechanism and deactivation and as a source for PGMs into the environment. Catalysts tend to be deactivated by the action of elements present in fuel (Pb, S) or lubricants (P, Ca, Zn).[12] Some of these elements might react with PGMs in the converter. In a rich mixture, sulfur was found to bind strongly to noble metals[29] and is known to increase the solubility of crushed catalyst.[30] Moreover, PGMs might react with other catalyst components. Platinum can form alloys[31] and consequently, platinum and rhodium occur together in the automobile exhaust.[32] Since platinum is emitted on the autocatalyst particle, cerium is found together with platinum.[33]

High exhaust temperature may help reactions to occur in the converter.

The three-way catalytic (TWC) converter
TWC converters are autocatalysts that enable the removal of hydrocarbons, carbon monoxide and nitrogen oxides from the exhaust of cars. The main active components in three-way catalytic converters are platinum, palladium and rhodium. Noble metals are deposited on an aluminium oxide washcoat coated on a honeycomb ceramic support. Moreover, the auto-catalyst contains additives such as Ce, Zr, La or Ba to reinforce the catalyst activity, stabilise the support and inhibit the sintering of noble metals.

TWC converters operate under stoichiometric air-to-fuel ratio conditions. The stoichiometry of the exhaust is monitored by an oxygen sensor which controls the fuel injection system in order to keep the exhaust at the right conditions.[12]

Moreover, the nanocrystalline structure of PGMs favours these reactions by increasing the exposed surface.

7 Platinum Speciation in the Environment

As a result of autocatalyst emissions, platinum group metals are found in the environment at concentrations significantly higher than the background level (Table 1).

PGM concentrations in road sediments have been increasing since the intro-duction of catalytic converters and this trend has been demonstrated for road sediments collected in Sweden (Figure 1). Pt in road dust occurs in the metallic state with a thin film of platinum at the surface available for transformation.[1]

Road sediment speciation by sequential extraction shows that platinum is predominantly in an inorganic form, with a significant exchangeable fraction which may lead to the release of dissolved platinum during storm events.[19] Laser ablation-ICP-MS of road sediments suggest that platinum is bound to Ce in road sediments indicating that platinum is still attached to autocatalyst particles on the road surface. Palladium and rhodium show a lower degree of association with Ce.[4]

Platinum in urban gullypots (roadside catch-basins) occurs in an organic form, possibly as a result of bacterial action.[20] Platinum might have been

Table 1 *Concentration in environmental compartments*

	Concentration					
Compartment	*Pt*	*Rh*	*Pd*	*Unit*	*Note*	*Reference*
Earth crust	0.4	0.06	0.4	$ng\,g^{-1}$		34
Urban air particles	0.02–5.1			$pg\,m^{-3}$		35
Urban road dust	189–416	56–136	74–104	$ng\,g^{-1}$	$<63\,\mu m$	4
	164–388	56–136	140		$63–250\,\mu m$	
Roadside soil	25.3–253	4.8–39.7	1.2–12.5	$ng\,g^{-1}$	Topsoil, road distance 0.1 m	36
Runoff water	15	—	—	$ng\,l^{-1}$	median	37
River water	0.22–0.64	—	—	$ng\,g^{-1}$		38
Urban river sediment	4.8–15	2.5	—	$ng\,g^{-1}$		4
Roadside grass	10.6	1.54		$ng\,g^{-1}$	mean	6
Human serum	0.9	—	50.2	$ng\,l^{-1}$	mean	39
Human urine	1.02	11.7	9.5	$ng\,l^{-1}$	median	21

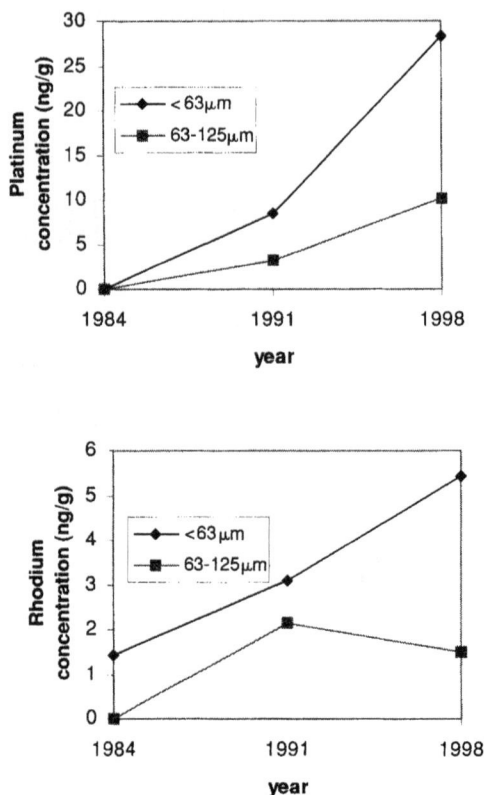

Figure 1   *Temporal trends for platinum and rhodium in road sediments*[4]

transformed into a methyl derivative. Methylation of platinum follows a complex series of reactions and requires both Pt(II) and Pt(IV).[1] Methylation of palladium follows a simpler process.[40] Another explanation for the occurrence of organic species is through binding to organic complexing agents. Platinum also deposits on roadside soil where it accumulates in the top layer.[41] Platinum is subject to complex transformations, possibly solubilisation followed by re-mobilisation through complexing agents in soil.[42] Therefore, platinum is relatively immobile in soil.

The PGMs accumulated in roadside soil might accumulate in plants. Platinum can be remobilised by complexing agents such as EDTA and plants have similar complexones.[43] Platinum has been found in plants close to traffic,[5] although a six month experiment did not show a significant platinum accumulation from soil.[44] One explanation for platinum in plants close to traffic is the long-time accumulation and another explanation might be adsorption of particles deposited, possibly related to the higher solubility of air particles compared with road sediments. Platinum in plants was found to bind to high molecular weight proteins under natural conditions.[45] In grass treated with a platinum-containing solution with uptake exclusively from the roots, platinum was found to bind mostly to low molecular weight proteins.[46]

Transport to receiving water through stormwater runoff is a possible pathway for metals in the urban environment (Figure 2). Platinum is known to be transported through runoff.[37] In the river, platinum accumulates in sediments while soluble platinum remains low.

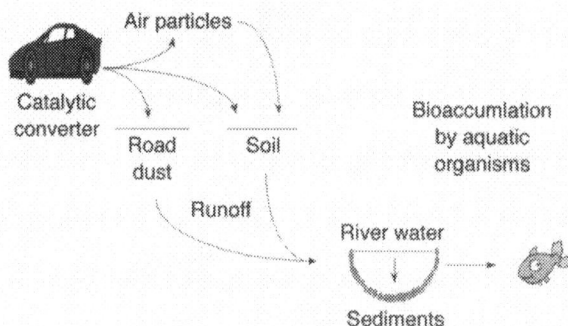

Figure 2 *Possible route of PGMs to receiving waters*

It was recently suggested that platinum is relatively immobile in sediment, since it may remain on a catalytic converter particle as it does in road dust. Palladium and rhodium may have a different behaviour and it was suggested that they are relatively mobile in urban rivers, thereby posing a greater risk.[4]

The uptake of platinum by the freshwater isopod *Asellus aquaticus* has been studied in detail. Asellids live on the sediments and are directly exposed to platinum in urban rivers. Platinum content was found to be very variable. Two different uptake mechanisms have been suggested: binding to metallothionein-type proteins and accumulation in intracellular granules. Granules can explain the accumulation of high concentrations of metals without any apparent toxic effect. Moreover, granules can be associated with the exoskeleton and may be excreted each time the exoskeleton is lost (every 2–3 weeks), providing an efficient depuration mechanism. Exposure experiments show that platinum accumulation is species dependent[47] (Figure 3). Further studies of the biological accumulation of other PGMs is needed and is currently being undertaken. A speciation approach is very important to provide a clearer understanding of the accumulation of these metals in the environment and Figure 3 shows that different PGM species do indeed have very different bioaccumulation patterns.

8   Platinum Group Metals in Man

Reported levels for the PGMs in humans is sparse, mostly due to difficulty of analysis and low concentrations. Recently, with the introduction of catalytic

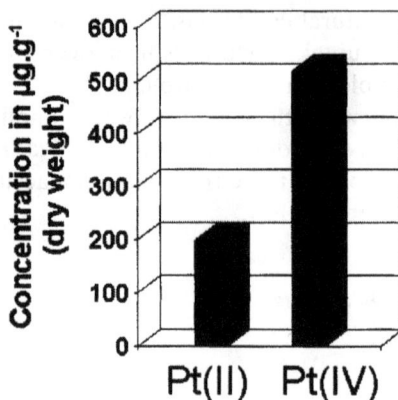

Figure 3   *Accumulation of different platinum species by the freshwater isopod* Asellus aquaticus[47]

converters, but also the use of platinum-based drugs and occupational exposure in the manufacture of autocatalysts, there has been a growing interest in the concentration of PGMs in body fluids. PGM levels in blood and urine for non-occupationally and medically exposed populations are shown in Table 1.

Platinum-based drugs research has led to a better understanding of platinum accumulation and metabolism in humans. Several speciation techniques have been used for this purpose, including size exclusion chromatography–ICP-MS,[9] HPLC–ICP-MS[8] and two-dimensional electrophoresis.[48]

Platinum is known to bind to proteins in blood[49] and although only Pt(II) was found to bind to proteins, Pt(IV) can be reduced to Pt(II) before binding.[50] Proteins in blood are known to play a crucial role in the transport of metals to target organisms. Platinum is transported to the liver and kidneys where it accumulates before being excreted.[3] In the liver and kidneys, platinum binds to

Health risks

Environmental concentrations of platinum have not been shown to directly affect ecosystems or human health. However, platinum salts are known to be potent sensitisers.[54] Recently, the importance of speciation has been stressed for occupational exposure to platinum salts.[55] Moreover, platinum was found to react with DNA[7] and is widely used in cancer therapy, where the speciation is vital. Reaction of platinum with DNA might be a potential hazard. Platinum-based drugs also show side effects including nephrotoxicity and suppression of bone marrow formation.

Even less is known about the environmental toxicity of palladium and rhodium. Palladium is known to have allergenic effects when in an ionic form.[56] Cytotoxic and mutagenic effects of palladium and rhodium were found to be lower than that of platinum.[57] However, the high reported levels of palladium and rhodium found in urine[21,22] are a concern.

metallothioneins (MTs). MTs are low molecular weight proteins known to regulate metal concentration in organisms. Pt(II) was found to bind to and induce the synthesis of MTs,[51] while Pt(IV) is reduced to Pt(II) before binding.[52] Platinum is stored and released progressively during a long period as shown by the long-term excretion, monitored in the urine of patients treated with cis-platin.[53]

Highway workers exposed to emissions from cars did not show a significantly higher platinum group metal concentration in urine when compared with non-occupationally exposed individuals.[22]

Diet has been considered the major source of platinum for man,[18] although the high concentrations reported have led to some doubts being raised about these results.

PGM levels in human blood and urine have been determined for non-occupationally exposed and non-treated individuals.[21,22,39] The level of platinum was found to be lower than that of palladium and rhodium. This result is surprising owing to the fact that their environmental level is lower than that of platinum. It may, however, be explained by different mobility in the environment as well as different uptake and metabolism.

9   Future Research

Environmental PGM contamination is a relatively new problem. Research has focused on obtaining accurate total platinum concentrations in environmental matrices, although recently the environmental concentrations of palladium and rhodium have also been determined. Environmental PGM concentrations are low and the analysis remains difficult. PGM speciation is an even more difficult task.

As a result, there is no clear understanding of PGM behaviour under environmental conditions. There is now a need to (1) understand the speciation of PGMs in catalytic converters, (2) determine the mobility of PGMs in the catalytic converter particles on the road surface, (3) assess the possibility for PGMs to be methylated under environmental conditions and (4) study the ecotoxicology of platinum group metals in both aquatic and terrestrial environments.

In the meantime, PGM concentrations in urban environments will continue to increase as long as there is no viable substitute for PGM based-autocatalysts. Recent findings of PGMs in Greenland snow is also of concern,[58] although concentrations are very low, as it shows that non-directly polluted environments may also be affected by automobile emissions.

10   References

1. S. Rauch and G.M. Morrison, *Routes for Bioaccumulation and Transformation of Platinum in the Urban Environment*, in *Anthropogenic Platinum-Group-Element Emissions and Their Impact on Man and Environment*, F. Alt and F. Zereini, eds., Springer Verlag, Berlin, 1999.

2. H. Renner and G. Schmuckler, *Platinum-Group Metals*, in *Metals and Their Compounds in the Environment*, E. Merian, ed., VCH, 1991 pp. 1135–1151.
3. D.E. Johnson, J.B. Tillery and R.J. Prevost, *Environ. Health Persp.*, 1975, **12**, 27.
4. S. Rauch, G.M. Morrison, M. Motelica-Heino and O. Donard, *Evaluation of Speciation, Transport and Ecological Risk of Palladium, Platinum and Rhodium in Urban Stormwater Systems*, 8th International Conference on Urban Storm Drainage Proceedings, The Institution of Engineers, Sydney, Australia, 1999, **1**, 202.
5. F. Alt, H.R. Eschnauer, B. Mergler, J. Messerschmidt and G. Tölg, *Fresenius' J. Anal. Chem.*, 1997, **357**, 1013.
6. E. Helmers and N. Mergel, *Fresenius' J. Anal. Chem.*, 1998, **362**, 522.
7. B. Rosenberg, L. Van Camp and T. Krigas, *Nature*, 1965, **205**, 698.
8. W.R.L. Cairns, L. Ebdon and S.J. Hill, *Fresenius' J. Anal. Chem.*, 1996, **355**, 202.
9. J. Szpunar, A. Makarov, T. Pieper, B.K. Keppler and R. Lobinski, *Anal. Chim. Acta*, 1999, **387**, 135.
10. C.F.J. Barnard, M.J. Cleare and P.C. Hyde, *Chem. Br.*, 1986, **22**, 1001.
11. L.R. Kelland and M.J. McKeage, *Drugs Aging*, 1994, **5**, 85.
12. P. Degobert, *Automobiles and Pollution*, Technip, Paris, France, 1995.
13. M.A. Palacios, M. Moldovan, M. Gomez, G. Morrison, S. Rauch, C. McLeod, S. Caroli, P. Schramel, S. Lustig, J. Laserna, P. Lucena, J.C. Saenz, M. Luna, J. Santamaría and U. Wass, *Catalytic Converters of Gasoline and Diesel Engines as Main Source of Platinum, Palladium and Rhodium into the Environment*, Poster presentation at the 5th European Winter Conference on Plasma Spectrochemistry, Pau, France, 1999.
14. M.A. Palacios, M. Moldovan and M. Gomez, in *Anthropogenic Platinum-Group-Element Emissions and Their Impact on Man and Environment*, F. Alt and F. Zereini, eds., Springer Verlag, Berlin, 1999.
15. R.F. Hill and W.J. Mayer, *IEEE Trans. Nucl. Sci.*, 1987, NS24(6), 2549.
16. V.R. Schölg, G. Indlekofer and P. Oelhafen, *Angew. Chem.*, 1987, **99**, 312.
17. O. Nygren, G.T. Vaughan, T.M. Florence, G.M.P. Morrison, L.M. Warner and L.S. Dale, *Anal. Chem.*, 1990, **62**, 1637.
18. G.T. Vaughan and T.M. Florence, *Sci. Total Environ.*, 1992, **111**, 47.
19. C. Wei and G.M. Morrison, *Sci. Total Environ.*, 1994, **146/147**, 169.
20. C. Wei and G.M. Morrison, *Anal. Chim. Acta*, 1994, **284**, 587.
21. M. Krachler, A. Alimonti, F. Petrucci, K.J. Irgolic, F. Forastiere and S. Caroli, *Anal. Chim. Acta*, 1998, **363**, 1.
22. J. Bergerow, M. Turfeld and L. Dunemann, *Fresenius' J. Anal. Chem.*, 1997, **359**, 427.
23. M. Parent, H. Vanhoe, L. Moens and R. Dams, *Talanta*, 1997, **44**, 221.
24. C. León, H. Emons, P. Ostapczuk and K. Hoppstock, *Anal. Chim. Acta*, 1997, **356**, 99.
25. S. Lustig, P. Schramel and P. Quevauviller, *PACEPAC: A Road Dust Certified Reference Material for Quality Assurance in the Analysis of Pt, Pd and Rh in Environmental Samples*, in *Anthropogenic Platinum-Group-Element Emissions and Their Impact on Man and Environment*, F. Alt and F. Zereini, eds., Springer Verlag, Berlin, 1999.
26. E. Helmers, *Environ. Sci. Pollut. Res.*, 1997, **4**, 100.
27. A.J. Oakes and J.C. Vickerman, *Surf. Interface Anal.*, 1996, **24**, 695.
28. B. Stenbom, G. Smedler, P.H. Nilsson, S. Lundgren and G. Wirkmark, *Thermal Deactivation of a Three-way Catalyst: Changes in Structural and Performance Properties*, SAE Technical Paper Series, 1990, 900273.
29. D.D. Beck and J.W. Sommers, *Impact of Sulfur on Three-way Catalysts: Comparison of Commercially Produced Pd and Pt–Rh Monolyth*, in *Catalysis and Automotive Pollu-*

tion Control III, A. Frennet and J.M. Bastin, eds., Elsevier, Amsterdam, 1995.

30. D. Nachtigall, H. Kock, S. Artelt, K. Levsen, G. Wünsch, T. Rühle and R. Schlögl, *Fresenius' J. Anal. Chem.*, 1996, **354**, 742.
31. V. Pitchon, F. Garin and O. Maire, *Appl. Catal. A*, 1997, **149**, 245.
32. C. Lüdke, E. Hoffmann, J. Skole and S. Artelt, *Fresenius' J. Anal. Chem.*, 1996, **355**, 261.
33. P.J. Silva and K.A. Prather, *Sci. Technol.*, 1997, **31**, 3074.
34. K.H. Wedepohl, *Geochim. Cosmochim. Acta*, 1995, **59**, 1217.
35. F. Alt, A. Bambauer, K. Hoppstock, B. Mergler and G. Tölg, *Fresenius' J. Anal. Chem.*, 1993, **346**, 693.
36. M. Cubelic, R. Pecoroni, J. Schäfer, J.D. Eckard, Z. Berner and D. Stüben, *Z. Umweltchem. Okotox.*, 1997, **9**, 249.
37. D. Lashka, T. Striebel, J. Daub and M. Nachtwey, *Umweltwiss. Schadst.-Forsch.*, 1996, **8**, 124.
38. D. Lashka and M. Nachtwey, *Untersuchungen zum Eintrag von Platin aus Kfz-Katalysatoren in die Umwelt*, Bayerische Landesanstalt für Wasserforschung, Munich, 1993.
39. J. Begerow, M. Turfeld, and L. Dunemann, *J. Anal. At. Spectrom.*, 1997, **12**, 1095.
40. W.M. Scovell, *J. Am. Chem. Soc.* 1974, **96**, 3451.
41. F. Zereini, B. Skerstupp, F. Alt, E. Helmers and H. Urban, *Sci. Total Environ.*, 1997, **206**, 137.
42. S. Lustig, S. Zang, B. Michalke, P. Schramel and W. Beck, *Sci. Total Environ.*, 1996, **188**, 195.
43. S. Lustig, S. Zang, W. Beck and P. Schramel, *Mikrochim Acta*, 1998, **129**, 189.
44. S. Lustig, S. Zang, B. Michalke, P. Schramel and W. Beck, *Fresenius' J. Anal. Chem.*, 1997, **357**, 1157.
45. D. Klueppel, N. Jakubowski, J. Messerschmidt, D. Stuewer and D. Klockow, *J. Anal. At. Spectrom.*, 1998, **13**, 255.
46. F. Alt, J. Messerschmidt and G. Weber, *Anal. Chim. Acta*, 1998, **359**, 65.
47. S. Rauch and G.M. Morrison, *Sci. Total Environ.*, 1999, **235**, 261.
48. S. Lustig, J. De Kimpe, R. Cornelis and P. Schramel, *Fresenius' J. Anal. Chem.*, 1999, **363**, 484.
49. T.J. Einhäuser, M. Galanski and B.K. Keppler, *J. Anal. At. Spectrom.*, 1996, **11**, 747.
50. D. Nachtigall, S. Artelt and G. Wunsch, *J. Chromatogr. A*, 1997, **775**, 197.
51. B.L. Zhang, W.Y. Sun and W.X. Tang, *J. Inorg. Biochem.*, 1997, **65**, 295.
52. W. Zhong, Q. Zhang, Y. Yan, S. Yue, B. Zhang and W. Tang, *J. Inorg. Biochem.*, 1997, **66**, 179.
53. R. Schierl, B. Rohrer, J. Hohnloser and H. Hohnloser, *Cancer Chemother. Pharmacol.*, 1995, **36**, 75.
54. R. Merget, G. Schlutze-Werninghaus, T. Muthorst, G. Friedrich and J. Meyer-Sydow, *Clin. Allergy*, 1988, **18**, 569.
55. P.J. Linnett and E.G. Hughes, *Occup. Environ. Med.*, 1999, **56**, 191.
56. J.C. Wataha and C.T. Hanks, *J. Oral Rehabilitation*, 1996, **23**, 309.
57. J. Bünger, J. Stork, and K. Stahlder, *Int. Archives Occup. Environ. Health*, 1996, **69**, 33.
58. C. Barbante, T. Bellomi, G. Mezzadri, P. Cescon, G. Scarponi, C. Morel, S. Jay, K. van de Velde, C. Ferrari and C.F. Boutron, *J. Anal. At. Spectrom.*, 1997, **12**, 925.

CHAPTER 10

Speciation in the Frame of Environmental Biomonitoring – Challenges for Analytical and Environmental Sciences

HENDRIK EMONS

1 Environmental Biomonitoring

Public awareness and scientific understanding of various environmental issues have dramatically improved, particularly during the last two decades. It is now widely accepted that human activities which influence the chemical composition of the environment have to be systematically controlled. At present, most regular environmental monitoring is focused upon the observation of local emission sources, and the operation of more or less closely controlled networks for air and water monitoring in countries around the world. Modern environmental observation has, however, to provide more effect-related information about the state of our environment and its changes with time. Therefore, it cannot only be based on investigations of abiotic environmental samples such as air, water, sediment or soil. Environmental studies and controls have to take greater heed of the situation of the biosphere. This includes the transfer of contaminants, mainly of anthropogenic origin, into plants, animals and finally into human beings.

Therefore, biomonitoring plays an increasing role in modern environmental observation programs.[1-3] Here, selected biological organisms, called bioindicators, are used for the monitoring of pollutants either by observation of phenomenological effects (loss of needles, discolouring of leaves *etc.*) or by the measurement of chemical compounds taken up by the specimens. The latter approach is based on the chemical analysis of appropriate bioindicators and should answer the following questions:

- Which pollutants appear where, when and at what concentration?
- How mobile and stable are pollutants in ecosystems?

- Which transformations occur with anthropogenic emissions in the biosphere?
- Where are pollutants accumulated or finally deposited, and in which chemical form?
- Which short- and long-term effects do they have with respect to mankind and the environment?

Obviously, these problems require interdisciplinary research activities with major contributions from analytical chemistry. They cannot be studied by determining only total element concentrations in environmental samples. Since the properties of the compounds, such as transport, distribution behaviour, bioavailability, chemical reactivity and toxicity all depend on their oxidation state, binding form, and binding partners, speciation is obviously of great importance.

In the following, several aspects of trace element speciation for the purpose of environmental bio-monitoring will be outlined and challenges will be discussed.

2 Bioindicator Matrices

Specimens which have been used for environmental bio-monitoring represent a very broad variety with respect to matrix composition. For active monitoring different plant species, mainly grass cultures, are exposed at selected locations for a limited time. Passive monitoring is performed with the help of environmentally representative plants as well as animals that are naturally present in the ecosystem of interest. Typical examples are shown in Figure 1.

For a scientifically sound description of the environmental situation, well-composed sets of bio-monitoring specimens have to be selected, dependent upon the ecosystem of interest and the available biological populations. From an analytical point of view, such samples are highly complex and differ with regard to concentration levels of trace elements, matrix influence on sample preparation

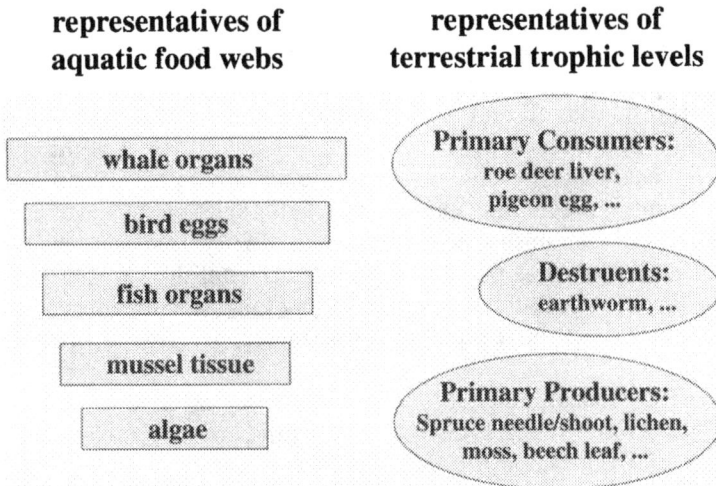

Figure 1 *Examples of specimens used for passive biomonitoring*

and analyte determination (see Section 3), as well as chemical stability. In the past, the criterion of accumulation of pollutants by the respective organism played an important role in its application. The tremendous progress in analytical chemistry, with respect to the determination of many compounds at the trace and even ultra-trace level, has decreased the importance of this aspect. Nowadays, the level of knowledge about biological properties and ecological functions of the available specimens appears to be the limiting factor which decides utility.

Several of the bioindicators listed in Figure 1 have previously been used only for total element monitoring.[2,4] At present, the conceptual shift to assessment-oriented monitoring is combined with an increasing interest in the investigation and control of unchanged chemical species and organic molecules. This objective introduces new requirements into the analytical strategy, for instance in the investigation of fresh material. Such samples are often difficult to prepare for speciation analysis and critical steps of complete procedures will be outlined in the following section.

3   Total Procedures for Speciation

In the development of procedures for trace element speciation with such delicate sample material such as bioindicators, one has always to consider the main purpose of the environmental monitoring, which is to obtain information about the original species pattern in the environment. Therefore, it is not sufficient to take into account only the operations in the analytical laboratory. Rather one has to design the complete process starting with sampling and ending with an assessment of results, as illustrated in Figure 2. Parts of the whole procedure, which are currently underdeveloped in most of the speciation schemes from the point of view of retaining and exploiting species information, are highlighted in Figure 2 by a black background. Critical aspects of the individual steps are now discussed together with an approach developed for arsenic speciation for the biomonitoring of aquatic ecosystems.[5]

Species transformation and loss can already occur during sampling at the sampling site. Degradation depends upon the chemical nature of the species, it may be influenced by enzyme activity and is usually more critical in animal organs than for plant samples. A key parameter for reaction rates is temperature. Therefore, one can diminish species transformations by decreasing the temperature as far and as early as possible. This was accomplished for our As species biomonitoring, by shock-freezing of the samples (algae, common mussels, fish muscles, seabird eggs) above liquid nitrogen immediately after collection, at the sampling site.[3]

The majority of samples collected for biomonitoring have to be treated by mechanical operations such as dissection of organs, grinding, mixing and taking aliquots. Most of this sample processing, even the separation of mussel tissue and shell, can be performed at very low temperatures.[6] Working at near liquid nitrogen temperature offers also the opportunity to grind and homogenize fresh

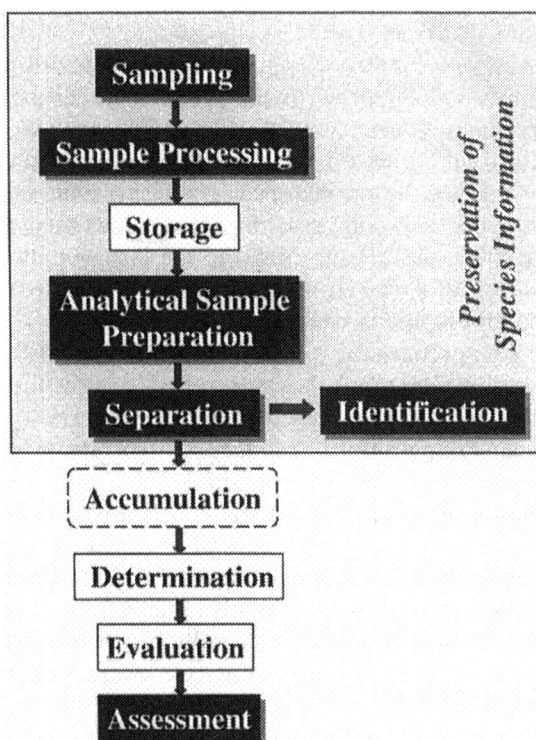

Figure 2 *Steps of total analytical procedures for speciation*

biological material with very different mechanical properties, resulting for instance from varying water and fat contents. We processed our samples to a fine powder with particle sizes below 200 μm. One should avoid freeze-drying because of possible species transformations and loss.[7] In the case of As speciation in the bioindicators listed above, however, no differences between the total As content in fresh and freeze-dried materials were found.[5]

Presently, one of the most critical steps from the aspect of retaining the species information is the analytical sample preparation. The necessary transfer of the species of interest, into liquid and/or gas phases, introduces a major uncertainty about unwanted changes of the speciation pattern. Most of the current preparation protocols include a step for the extraction of the metal(loid) species from the biomatrices.[8,9] For instance, our As speciation procedure is based on the extraction of the finely ground powder with methanol/water mixtures, under conditions optimized from an approach reported by Kuehnelt *et al.*[10] One can control the extraction yield on the basis of the quantity of extracted metal atoms from the biological sample and we were able to obtain yields of more than 95% with optimized procedures. The preservation of the original molecular structure and the stoichiometry and completeness of chemical reactions for derivatization (*e.g.* alkylation) cannot, at present, be controlled.

Species separation is currently most commonly achieved by chromatographic techniques based on dissolved or gaseous species.[8,9,11-13] If one takes into account the very complex matrix composition of biomonitoring samples, even in the extraction solution, the large variations of analyte concentrations dependent upon the specimens selected, sampling time and location, the influence of interactions at the solid/liquid or solid/gas interface, mobile phase composition and temperature changes, the original species pattern is difficult to estimate. Due to this complexity, control experiments with pure standard solutions allow only a first guess at possible species transformations. For instance, chemical changes seem to occur easily with As(III) in case of various ion-exchange conditions during liquid chromatographic separation of arsenic species.[14]

Currently, the main approach for species identification involves the comparison of chromatographic retention behaviour of standards of pure compounds and the sample extract. Unfortunately, during recent years, the progress in method development for powerful techniques of structural analysis, such as NMR and mass spectrometry, was not comparable at the ultra-trace level, to the achievements in quantification of known analytes. In the case of As speciation in bioindicators from aquatic ecosystems, several chromatographic signals for unidentified As species have been obtained by HPLC–ICP-MS, in addition to peaks for As(V), dimethylarsinic acid, arsenobetaine, tetramethylarsonium, arsenocholine, and various arsenosugars.[5] Generally, the restricted availability of pure and stable standard compounds represents a severe limitation for many speciation developments.

The methods for species determination have been significantly extended and improved during the last few years.[8,9,11,12,15,16] Powerful detection techniques such as ICP-MS, GF-AAS and AFS offer sufficient quantification capability for many biomonitoring purposes, if the dilution factor during the whole analytical procedure (Figure 2) is minimized. Other techniques, such as ICP-AES, may require considerable pre-concentration procedures.[17] The use of efficient sample introduction devices can decrease the limits of detection – in the case of As speciation by HPLC–ICP-MS, from 11 pg As(V) (cross-flow nebulizer) to 0.7 pg As(V) (hydraulic high pressure nebulizer).[5] In real sample solutions, memory effects and species instability often influence the applicability of these devices and one has to design the optimum instrumentation set-up while considering the specific sample and the problem of interest.[14]

The evaluation of analytical data is, in itself, a multi-step process. The basic statistical treatment of raw analytical data from speciation is often limited by the small number of repetitions performed for the relatively long, and still delicate, total procedure. The more challenging problem seems to be the calculation of reasonable ratios for obtaining concentration values from the measured, absolute, species content. For example, the widespread use of analyte concentrations, related to dry mass (weight) of the sample, is of limited value for environmentally relevant comparisons of specimens with different water content. Relating the analyte content to fresh mass, or even to the volume of sampled material, would allow a more meaningful estimation of bioconcentration factors. Moreover, variable specimen parameters which influence the uptake and distribution of the species of interest, such as the exposed surface of leaves in the case of air monitoring, or the lipophilicity of a fish organ, which defines its storage capacity for lipophilic organometallic species such as methylmercury, have to be taken into account. The latter is often considered by relating the measured analyte quantities to the 'fat content' of the sample,[18] despite the problem that this is only an operationally defined determination. Moreover, the contribution of fat analysis to the total uncertainty of the concentration value may be significant.

Overall, the assessment of speciation data with respect to the description of the state of our environment is presently insufficiently developed. At this stage, one has to combine the analytical information with knowledge of environmentally relevant effect levels, doses and metabolic processes within the studied organism and ecosystem. An important limitation for the definition and selection of speciation development for biomonitoring is actually the present lack of toxicological and ecological knowledge.

4 Future Needs

From a critical evaluation of current procedures for trace element speciation in fresh biomonitoring samples, one can conclude that there are three main areas which have to be preferentially developed:

- sampling procedures which do not alter the original species pattern
- structural identification of species at the ultra-trace level, particularly in their natural surroundings, *e.g.* tissues
- quality assurance for the whole speciation procedure with the emphasis on trueness control

On the basis of differentiated data about critical loads and dangerous effects of metal(loid) species from the consideration of human health, as well as ecotoxicology, the definition of target species for environmental biomonitoring has to be made more specific. Until now the main elements of interest for biomonitoring speciation have been Hg, As, and Sn. Other elements such as Al, Cr, Pb, Sb, Se, V, Co, Rh, Ni, Pd, Pt, Cu, and Ag may form the focus of such activities over the next few years. A starting point for decisions about the necessity of speciation is the concentration level of the trace element in the selected biomonitoring sample. Examples of specimens collected and investigated, in the frame of environmental specimen banking and monitoring programs in Germany, are shown in Figure 3.

For the future assessment of environmental situations, one will consider not only background levels of trace elements and their natural variations dependent on the specific specimen, ecosystem type and its location, but also species patterns within the biomonitoring samples. For the design of complete procedures, which allow the preservation of original species information, fundamental chemical investigations are necessary with respect to equilibrium and kinetic parameters in complex systems, such as biological cell compartments, as well as with respect to metabolic reactions under stress conditions, which can take place during the sampling step.

The validation of speciation procedures is of the utmost importance in the field

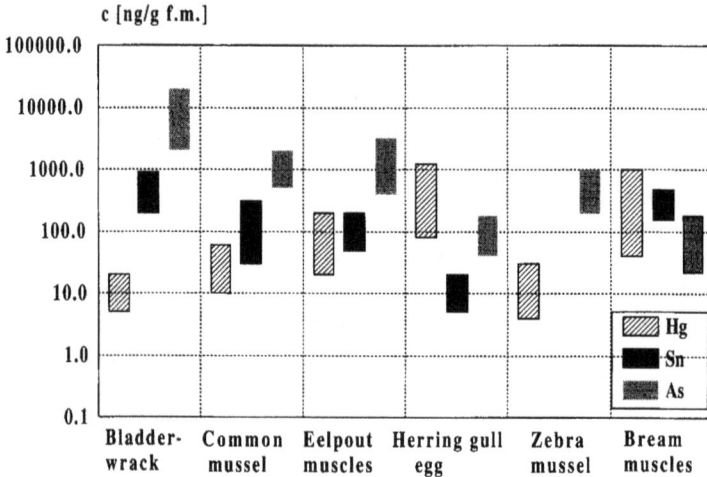

Figure 3 *Concentration ranges of Hg, Sn, As in selected biomonitoring samples from aquatic ecosystems in Germany*

of environmental biomonitoring because of the social and economic impact of such environmental information and resulting decisions. Therefore, one has not only to control possible species transformations during the whole procedure, but it should also be mandatory to perform a careful budget of species amounts. Such mass balance studies are essential, especially for steps that include phase transformations, such as extraction. From the point of view of further method developments and their validation, the availability of matrix-matched, certified reference materials is very important. Unfortunately, there is no realistic hope of a commercial reservoir for the whole range of biological matrices necessary for biomonitoring programs, which also cover the large variations in species composition and concentration levels at differently polluted sampling sites. More activity in the area of the preparation and characterization of environmental reference materials for speciation should be strongly encouraged.

In general, speciation of trace elements for biomonitoring purposes represents a typical interdisciplinary field of research and development, with major challenges for analytical and environmental sciences. It's future will be based on progress in many areas and the definition of new research needs and projects that require the combination of chemical, physical, biological, and medical knowledge.

Acknowledgement

The cooperation of many co-workers at the ESB Jülich for the development and application of speciation for biomonitoring is gratefully acknowledged. I thank Mrs. G. Nork for the careful technical preparation of this manuscript.

5 References

1. J.H. Phillips and P.S. Rainbow, *Biomonitoring of Trace Aquatic Contaminants*, Elsevier, London, 1993.
2. B. Markert (ed.), *Plants as Biomonitors*, VCH, New York, 1993.
3. H. Emons, J.D. Schladot and M.J. Schwuger, *Chemosphere*, 1997, **34**, 1875.
4. E. Kubin, H. Lippo, J. Karhu and J. Poikolainen, *Chemosphere*, 1997, **34**, 1939.
5. K. Falk, PhD Thesis, University of Essen, 1999.
6. D. Koglin, F. Backhaus and J.D. Schladot, *Chemosphere*, 1997, **34**, 2041.
7. S. Shawky, H. Emons and H.W. Dürbeck, *Anal. Comm.* 1996, **33**, 107.
8. Ph. Quevauviller, E.A. Maier and B. Griepink (eds.), *Techniques and Instrumentation in Analytical Chemistry – Volume 17: Quality Assurance for Environmental Analysis*, Elsevier, Amsterdam. 1995.
9. Ph. Quevauviller, *Method Performance Studies for Speciation Analysis*, Royal Society of Chemistry, Cambridge, 1998.
10. D. Kuehnelt, W. Goessler and K.J. Irgolic, *Appl. Organomet. Chem.*, 1997, **11**, 289.
11. L. Dunemann and J. Begerow, *Kopplungstechniken zur Elementspeziesanalytik*, VCH, Weinheim, 1995.
12. D. Pérez-Bendito and S. Rubio, *Speciation in Environmental Analysis*, in *Wilson & Wilson's Comprehensive Analytical Chemistry – Volume XXXII: Environmental Analytical Chemistry*, ed. S.G. Weber, Elsevier, Amsterdam, 1999, 649.
13. A.M. Ure and C.M. Davidson (eds.), *Chemical Speciation in the Environment*, Blackie Academic & Professional, London, 1995.
14. K. Falk and H. Emons, *J. Anal. Atom. Spectrom.*, 2000, **15**, 643.
15. S. Caroli, *Chemical Analysis Vol. 135: Element Speciation in Bioinorganic Chemistry*, John Wiley & Sons, New York, 1996.
16. R. Lobinski and Z. Marczenko, *Wilson & Wilson's Comprehensive Analytical Chemistry – Volume XXX: Spectrochemical Trace Analysis for Metals and Metalloids*, ed. S.G. Weber, Elsevier, Amsterdam, 1996.
17. M.K. Donais, R. Saraswati, E. Mackey, R. Demiralp, B. Porter, M. Vangel, M. Levenson, V. Mandic, S. Azemard, M. Horvat, K. May, H. Emons and S. Wise, *Fresenius' J. Anal. Chem.*, 1997, **358**, 424.
18. K.-W. Schramm, P. Marth, A. Wolf, K. Hahn, K. Oxynos, J. Schmitzer and A. Kettrup, *UWSF – Z. Umweltchem. Okotox.*, 1999, **11**, 277.

CHAPTER 11

Arsenic Speciation in the Environment

PÉTER FODOR

1 General

For many centuries, the word 'arsenic' had only one meaning for most people –
'poison'. This situation has only begun to change in the late 20th century, due, in
part, to the introduction of arsenic-containing medicines. The differing toxicities
of arsenic-containing compounds became apparent when organoarsenic compounds, which are the predominant arsenic compounds in aquatic systems, were
shown to be nearly 100 times less toxic than inorganic ones. Furthermore,
arsenobetaine (AsB), which is the major arsenic compound in marine animals
was shown to be metabolically inert and non-toxic in a number of studies[1] (see
also Chapter 16).

It is now well-known that determination of the total concentration of an
element does not give enough information on its mobility, ecotoxicity, bioavailability *etc.*, and that physico-chemical characteristics of the various forms of the
element in question influence strongly its behaviour in the environment. For
example, arsenic present in soil will interact with the soil biota according to its
physico-chemical form (oxidation state, organic–inorganic bonds *etc.*). The bioavailability of arsenic to plants, its adsorption on to soil particles and leaching to
groundwater systems may also be very different for different species. It is clear
that the biogeochemical cycling of arsenic depends predominantly on the
physico-chemical forms (species) in which it is present.

2 Arsenic Sources

Arsenic is ubiquitous in the environment, since it is present, even if in small
quantities, in all rock, soil, dust, ash, water and air. A number of books and
review articles contain material on the origin and sources of arsenic in the
environment.[2-6] Natural phenomena, such as weathering of minerals and biological and volcanic activity, along with anthropogenic inputs are primarily
responsible for the emission of arsenic into the atmosphere, soil, groundwater

systems and food chain.

Arsenic in rocks is usually found as sulfide minerals (*e.g.* FeAsS), in iron deposits and in sedimentary iron ores. The average arsenic concentrations in most rocks range from 0.5 to 2.5 $mg kg^{-1}$, though higher concentrations are found in some argillaceous sediments and phosphorites. Very high arsenic concentrations can be found in some reducing marine sediments which may contain as much as $3000 mg kg^{-1}$.[7] Other important geological sources of arsenic are active volcanoes (*e.g.* Mexico, Italy, Japan) where volcanic activity can transport vast amounts of material from the Earth's core into the biosphere in the form of magma and volcanic dust. Studies have shown arsenic concentrations an order of magnitude higher in soils close to these active volcanoes compared with concentrations in reference soil samples.[8]

In spite of the high loading of arsenic in the biosphere from natural sources, input from human activities are not negligible. Chilvers and Peterson[9] have estimated the ratio of natural to anthropogenic sources of arsenic at 60:40. Today, the use of arsenic-containing insecticides and herbicides in agriculture has completely ceased in several countries, but residual arsenic content is added to the amount of arsenic that keeps cycling in the biosphere and thus contaminating soil, crop, surface and groundwaters *etc.* In addition to input from the above-mentioned agricultural chemicals, other industries also contribute to the anthropogenic burden, since arsenic compounds are used in the pottery industry, in glass making, as wood preservatives, in the production of anti-fouling paints for ships, in the electronics industry in the production of semiconductors and are alloyed with selenium to coat photocopier drums, as feed additives *etc.* Some arsenic-containing warfare agents used during Word War II (*e.g.* Adamsite, Lewisite) were dumped into certain areas of the Baltic Sea, and can be expected to constitute a threat to these areas.[10] The use of mine waters for irrigation may also pose a threat to alluvial aquifers and surface and groundwaters (*e.g.* in Japan and China).

Contamination caused by industrial activities has not decreased in the last few decades: on the contrary, emissions of arsenic by the smelting industry and from industrial and animal wastes, along with dusts from the burning of fossil fuels, have actually increased.[11] Dusts usually contain arsenic as As_2O_3, and the solubility of this compound, and hence its possible leaching into drinking water systems, depends on several parameters but primarily on the pH of the leachant. The solubility of As_2O_3 in pure water at 25 °C is 2.16 g in 100 g water, and in environments containing hydrochloric acid the solubility will increase due to the formation of chloro-complexes. In basic environments (*e.g.* in some soils) the solubility of As_2O_3 is also considerably higher than at neutral pH. The arsenic present in the soil as a result of, for example, treatments by herbicides (sodium arsenite, monosodium and disodium methanearsonate, sodium cacodylate *etc.*) may also be converted into inorganic or organic arsenic compounds such as arsine (AsH_3), methylarsine (CH_3AsH_2), dimethylarsine [$(CH_3)_2AsH$], trimethylarsine [$(CH_3)_3As$] *etc.* as a result of microbial activity.[12]

In Figure 1 the structures of some arsenic compounds commonly encountered in nature are shown.

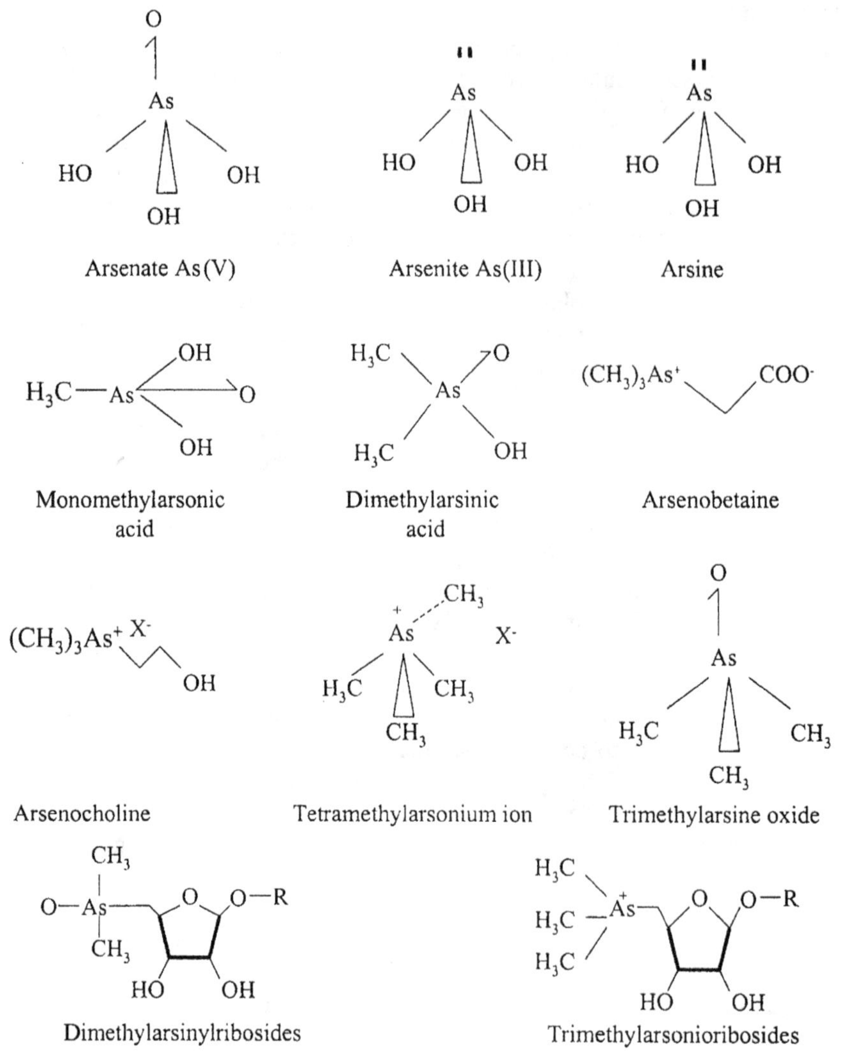

Figure 1   *Arsenic-containing compounds often found in the environment*

3   Arsenic in Soils and Sediments

3.1   Soils

Arsenic may be bound to solids through sorption by metal (hydr)oxide minerals or by structural incorporation into minerals. The adsorption of arsenic species present in the soil, their redistribution into other environmental compartments such as groundwater or air and their assimilation by plants are governed by complex processes and are determined by numerous parameters of the milieu.[12]

The form of arsenic which can be found in the soil depends primarily on the type and amount of the sorbent component of the soil, on the pH and redox conditions (E_h/pH) and on the characteristics of microbial activity. In general, the water-soluble arsenic content is directly proportional to the total arsenic content and inversely proportional to iron, manganese, aluminium and calcium concentrations, since these elements interact easily with arsenic.[13] Under oxidising conditions the predominant species is As(V), which exists as oxyanions of arsenic acid ($H_2AsO_4^-$, AsO_4^{2-}, AsO_4^{3-}). Under mildly reducing conditions arsenite [As(III)] is thermodynamically stable and exists as arsenious acid (H_3AsO_3, $H_2AsO_3^-$, $HAsO_3^{2-}$).[14] As(III) is more soluble, more mobile and so more toxic than arsenate [As(V)], but conversion of As(III) into As(V) can easily happen in oxidative environments, such as the As(V) → As(III) conversion in reductive environments.[15] Generally, E_h/pH conditions determine the As(V)/As(III) ratio, but microbial action can also promote oxidation/reduction and methylation.[16] Methylated arsenic compounds such as monomethylarsonic acid (MMA), dimethylarsinic acid (DMA) and trimethylarsine oxide (TMAO) are also present in soils, either as a result of microbial activity or as fingerprints of human activities (agriculture, industry). The result of microbial activity may also be the formation of volatile arsines, resulting in the transfer of arsenic from soil to the atmosphere.[12]

3.2 Sediments

Processes that influence arsenic distribution in sediments are similar to those influencing the cycling of arsenic in aquatic environments. It is interesting therefore that arsenic concentrations in freshwater, estuarine and oceanic sediments are so much higher (ppm) than those of the overlying waters (ppb).[6] Arsenic distribution in sediments is influenced by similar factors as in the case of soils – equilibrium between As(III) and As(V) is pH and redox dependent[17] and biotransformation may lead both to methylation and demethylation of arsenic species in sediments, depending upon the conditions which govern the microbial activity.

4   Microbial Activity

The interaction of micro-organisms with arsenic-containing compounds can be classified into two groups:[6] interaction without the formation of an arsenic–carbon bond and microbial alkylation of arsenic. Several fungi, yeast and bacteria strains are able to transform the different arsenic species. These reactions are basically reduction, oxidation, methylation and degradation reactions. From the biological point of view, some of these processes may be considered as protection mechanisms of the micro-organisms against the toxic effect of arsenic. By these means, some organisms are able to tolerate high arsenic concentrations [up to As(V) concentrations of 48000 mg l^{-1}][18] and to adapt themselves to living in arsenic-polluted environments. Various mechanisms of protection have been described, for example a plasmid-mediated resistance to arsenate, which is a result of a highly specific arsenate efflux pump that eliminates intracellular arsenate, or a chromosomally determined resistance to arsenate that results in a reduced accumulation of arsenate with the help of a special phosphate transport system. Oxidation of arsenite to arsenate by an inducible enzyme system is also classified as a mechanism of resistance, since arsenite accumulated in the cells is converted into a much less toxic form.[6,19] Volatilisation of inorganic arsenic compounds by methylation is also a common way of eliminating toxic compounds. The high rates of arsine evolution in soils lead to the conclusion that volatilisation is an important cause of loss of arsenic from soils.[12,20]

Investigating the behaviour of *Scopulariopsis brevicaulis* in contact with thirty arsenic containing compounds, Challenger[21] has summarised that altering concentrations, variable composition of medium and time in most cases resulted in methylation of the arsenic compounds. The garlic-like odour of the reaction products, the different arsines [*e.g.* $(CH_3)_3As$, $(C_2H_5)(CH_3)_2As$], is a very intense and characteristic indicator of reaction. The odour threshold for $(CH_3)_3As$ seems to be $0.002 \, \mu g \, kg^{-1}$ in dilute aqueous solutions.[22]

The methylation capacity of around 20 yeast and fungi strains was reviewed by Cullen and Reimer.[6] *Candidata humicola*, *Glicoladium roseum* and species of *Penicillium* were mentioned as species isolated from raw sewage.[23] This was the first time that modern analytical techniques (gas chromatography and mass spectrometry) were used in the identification of $(CH_3)_3As$, and a progressive enrichment technique was used to isolate the three fungi strains from the sewage. It is worth mentioning here that from the numerous arsenic compounds examined only arsenobetaine could resist the transformation activity of *Candidata humicola*.[24] A further interesting observation is that even a wood-rotting fungus has the capacity to metabolise arsenic-containing compounds. This is curious since arsenic-containing wood preservatives are often applied in the timber industry.

Several schemes of fungal methylation of arsenic compounds have been described, and modelling of the biochemical processes involved in the methylation of arsenic species has also been mentioned. In 1945, Challenger had assumed that bacteria were unable to methylate arsenic.[21] The first publication on arsine production by bacteria was published in 1971.[25] Since then, more than 20 such non-methanogenic bacteria strains have been identified which are able to methylate arsenic.[26] These strains were isolated from the environment under aerobic conditions; the methylation products were various dimethyl and trimethyl derivatives of arsenic. The experiments have proved that natural bacterial activity (*Escherichia coli*, *Staphylococcus aureus etc.*) can easily transform the oxides of arsenic into arsine. When $(CH_3)_3AsO$ was added to fresh river water, to sewage sludge, to rumen fluid and to sea-sediment, $(CH_3)_3As$ was detected by GC and by odour.[27]

The degradation of organic arsenicals by micro-organisms has been studied in soils, sediment and culture media. Studies have shown that methylated organic arsenicals (TMAO, sodium cacodylate and MMA) were converted into less methylated compounds in culture media by sediment micro-organisms.[28]

Degradation of monosodium methylarsonate (MSMA) has also been reported in soils.[29] Hanaoka and co-workers have studied the degradation of AsB in marine sediments[30] and found that this compound is decomposed completely to As(V) through the steps of TMAO, DMA and MMA.

5 Arsenic in the Atmosphere

The first evidence that volatile arsenic compounds might exist in the environment was reported in 1912, when Gosio identified a toxic gas ('Gosio gas'), responsible for a number of deaths and also for the garlic-like odour in the buildings in which it was produced. Studies have proved that the poisonous gas was produced by moulds growing on wallpaper coloured with arsenic-containing pigments, or those pasted up with adhesives containing arsenic. Gosio identified the gas incorrectly as diethylarsine, Challenger later correctly identifying it as trimethylarsine.[6] Since Challenger's classic studies on the biomethylation of arsenic by *Scopulariopsis brevicaulis*,[21] several other strains of micro-organisms capable of producing a full range of volatile arsenicals have been identified. Transfer of methyl groups to metals and metalloids can take place not only under anaerobic conditions but also in aerobic environments and in fossil substances such as gas or oil. Besides methylation another possible volatilisation pathway is the formation of arsenic hydrides.[31] Inorganic airborne arsenic species may originate from non-biological processes, both natural and anthropogenic. The presence of volatile alkylarsenic compounds in the atmosphere usually requires biological processes, but the synthesis of these products without microbial activity has been illustrated.[32]

In reductive environments, such as domestic waste deposits and sewage gases,[33-35] arsine, monomethyl-, dimethyl- and trimethyl-arsine and often unidentified arsenic species have been detected. Carbon dioxide, methane and hydrogen are typical products of such reductive environments, and this makes the detection of small amounts of volatile arsenic compounds more difficult.

Anthropogenic emissions, such as those from smelting and fossil fuel combustion, introduce considerable amounts of arsenic into the atmosphere, usually in the form of oxides. Most of the arsenic in the atmosphere can be found in the form of particulate matter: less than 10% is present in the vapour phase or on particles smaller than $0.2 \mu m$.[6] Analysis of airborne particulate matter has revealed that it is often significantly enriched in arsenic compared with the continental crust, which is probably due to precipitation (adsorption/complexation) of volatile arsenicals on particle surfaces.[36,37] In the case of fly ash, concentrations of arsenic depend on original concentrations in the burned fossil fuel and combustion conditions during burning. Arsenic in fly ash exists predominantly as As(V), but As(III) can also be found in aqueous extracts of fly ashes, usually at low concentrations. Exceptions may be the acidic fly ashes that have been found to have ratios of As(III)/As$_{total}$ of 30–50%.[38] Possible oxidation of As(III) to As(V) by atmospheric oxygen has to be considered, especially at high pH.[14]

The majority of fly ash currently produced by industry is disposed of in landfills and surface impoundments, which raises questions concerning soil and groundwater contamination by elements potentially leached from the ash. The investigation of these complex environmental systems is the subject of an increasing number of publications. Jackson and Miller[39] have studied the speciation of arsenic and selenium in soil solutions from fly ash, poultry litter and sewage sludge amended soils, and have found that the mixing of fly ash with either organic waste affected both the solubility and the speciation of arsenic in the resulting soil system. Poultry litter increased solubility and appeared to be itself, a significant source of soluble arsenic.

6 Arsenic in Natural Waters

Maintaining, and where necessary, improving the environmental state of natural waters are of prime importance, since these sources provide drinking water to mankind. Reliable information on the arsenic content and species distribution of arsenic in waters is therefore vital. Currently, there are no limit values for arsenic species in drinking water anywhere in the world. In 1958, the World Health Organisation set the maximum permissible arsenic level in drinking water as $0.2 \, \text{mg} \, \text{l}^{-1}$; this was lowered to $0.05 \, \text{mg} \, \text{l}^{-1}$ in 1963 and today is $0.01 \, \text{mg} \, \text{l}^{-1}$. The present limit value, which has been reduced twenty-fold, is indicative of the importance of arsenic intake *via* drinking water. Many publications deal with the identification of arsenic species in drinking water. In some cases no arsenic species were detected in the water samples investigated.[40] When arsenic was present, As(V) was found as the dominant form,[41,42] while As(III) and As(V) were detected in similar proportions in drinking water contaminated by geological causes.[43]

In natural water systems the oxidation state of arsenic ($+5$, $+3$, 0, -3) is determined mainly by the E_h/pH conditions,[44] but usually As(V) and As(III) are found in natural water systems. These forms are thermodynamically unstable under natural conditions; limit values for drinking water are possibly given as total arsenic for this reason. Beside E_h/pH conditions, several other parameters, such as temperature, light conditions, materials in contact with water *etc.*, may influence the oxidation state of inorganic arsenic compounds during sampling and storage.[43,45,46]

In geological areas where the arsenic content is much higher in the groundwater than the permissible value, detailed studies have been carried out to identify the species distribution in these waters. Such an area, for example, is the now well-known Blackfoot Disease (BFD) area in Taiwan where the average concentration of dissolved arsenic in the wells is $671 \pm 149 \, \mu\text{g} \, \text{l}^{-1}$.[47] The first symptoms of the disease are the spotted discoloration of the skin on the feet that turns from brown to black as the disease develops until amputation of the affected extremities, the final resort to save the BFD victims, becomes necessary. The predominant arsenic species in well waters of the BFD area is As(III) with an average As(III)/As(V) ratio of 2.6. Concentrations of MMAs and DMAs in these waters were below the limits of detection($< 1 \, \mu\text{g} \, \text{l}^{-1}$).[47] Arsenic speciation was investigated in 200 tube wells in Bangladesh, and scientists have found also that the inorganic arsenic species is dominant.[48] More than 40 million people now live in this area contaminated naturally with arsenic. Similar problems have to be faced in some parts of India. Hungary also has some problems due to the relatively high arsenic content present in its drinking water resources. The Hungarian arsenic database, registered by MÁFI (Hungarian State Institute for Geosciences) contains arsenic concentration data of more than 10 000 water samples. Geologists have concluded that the origin of arsenic in these Hungarian waters is not detrital arsenopyrite, as may be initially thought, but goes back to the Quaternary period. Arsenic precipitates, either in the form of arsenates with iron and manganese colloids, or spherical adsorption in the marsh and washland reservoirs are responsible.[49] In these waters arsenic can be found primarily as As(III), but As(V) may also be important, while organic forms of arsenic can be found in trace amounts only, if at all.[46,49]

On investigation of the As(III)/As(V) distribution of geothermal springs in the Massif Central of France, the Island of Dominica in the Leeward Islands of the Caribbean, the Valles Caldera of New Mexico (USA) and South-west Bulgaria, arsenic speciation gave different results.[50] Thermodynamic calculations in this study were tested for modelling purposes. In water samples from the Island of Dominica, As(III) was the major component in the oxidised waters of the acid sulfate type, and also in fluids from the deep levels of geothermal reservoirs, as a result of sulfur–arsenic oxidation kinetics. Bicarbonate-rich spring waters from New Mexico contain more As(III) than As(V), but As(V) was found to be the predominant arsenic species in the acid sulfate springs. Carbon dioxide rich French spring water samples showed great variability in the As(III)/As(V) distribution, while in the Bulgarian warm alkali waters As(III) was the predominant form.[50] Turkish geothermal waters had more than $1000 \mu g \, l^{-1}$ of arsenic in total, and Na_2HAsO_4 was found to be the dominant form in these samples.[51]

Some theoretical considerations of arsenic stability and solubility in freshwater environments were made by Wagemann.[52] The influence of 14 metals on the possible concentration of total dissolved arsenic and various arsenic species was investigated. He found that from the 14 elements investigated Ba, Cr, Fe and Cu could significantly influence the concentration of arsenic species at different pHs. A review devoted to the cycle of arsenic in natural waters also explains the presence and distribution of arsenic species according to As–S, E_h/pH diagrams.[44] Living organisms may influence the accumulation and transformation of arsenic species in water systems, for example, arsenite–arsenate oxidation is often catalysed by them, but processes such as methylation are not possible without their interaction. The pH dependence of the arsenate/arsenite content in open freshwater environments is also influenced by the absorbed sunlight resulting in differing photosynthetic processes in these environments. These processes show cyclic behaviour, which makes the evaluation of changes of kinetic processes possible.[53] Distribution of As(III) and total arsenic content was investigated in the oxygenated and anoxic waters of the Oslofjord (Norway), and it was found that there exists a correlation between the arsenic and phosphate content of deep waters. Whilst As(III) is present in small quantities in the deeper layers of the fjord, there are higher concentrations during spring, summer and autumn in the upper layers, and in the anoxic layer. The total arsenic content in the fjord is influenced by the biochemical cycle, whereas the As(III) content is influenced by the prevailing anoxic conditions and the release of assimilated reduced arsenic.[54] Methylated forms of arsenic were rarely detected in natural waters and these are usually in the pentavalent form. Trivalent methylarsenicals were first found and measured in 1994 in natural waters in Japan.[55]

Compared with the great number of experiments conducted in the marine environment, relatively few experiments have been performed in freshwater. Freshwater fish and other organisms seem to accumulate much less arsenic than their counterparts in seawater. The permissible limit values defined by the European Union for freshwater fish are significantly lower than those applied for sea-fish. Experiments on arsenic transformation in different freshwater food chains (*e.g.* green algae → shrimp → shellfish,[56] freshwater algae → zooplankton grazer → carnivorous guppy[57]) have revealed that total arsenic accumulation in each species *via* food decreased by one order of magnitude or more towards higher organisms. The concentration of methylated arsenic relative to total arsenic accumulated was found to be higher at the higher tropic level, mostly as a result of biomethylation. Both methylation and demethylation of the arsenicals in these systems were observed.[56]

7   Terrestrial Organisms

7.1   Plants

Elemental analysis of terrestrial plants has conventionally used ashing or acid digestion techniques as sample preparation, and therefore the results refer almost exclusively to total concentrations of the elements in question. As far as arsenic is concerned, little is known about the fate of arsenic in the terrestrial environment. The average concentration of arsenic in terrestrial plants is found to lie between 0.2 and 0.4 μg g^{-1}, but certain plants such as *Chrysanthemum leucanthemum*, *Pinus halopensis*, *Agrostis tenuis*, *Catharantus roseus etc.*, may accumulate significant amounts of arsenic in different organs (leaves, roots).

Biomethylation of inorganic arsenic by terrestrial organisms to simple methylated compounds (MMA, DMA) is known to occur in fungi, bacteria, mushrooms, freshwater algae and higher plants. Freshwater algae, when in contact with arsenic from contamination (*e.g.* mining or agricultural activities), often accumulate arsenic by transforming the original oxidation state of the species, rather than by methylating arsenic-containing compounds and/or integrating them into their lipids.[58-60] Mushrooms seem to be good candidates for speciation studies in terrestrial organisms because they can accumulate remarkable amounts of arsenic, *Sarcosphaera coronaria* may contain arsenic up to 2000 mg kg^{-1} (dry weight). Aqueous and methanol/water extracts of mushrooms have been analysed and have shown the presence of As(III), As(V), MMA, DMA and AsB.[61,62] The various mushroom species may accumulate arsenic in differing forms, some contain only MMA, others mainly As(III) and As(V). In *Lac-*

caria amethystina DMA was the major arsenic compound, while *Sarcodon imbricatum* and two *Aspergillus* species were found to contain primarily AsB. On analysing extracts of the mushroom *Amanita muscaria*, besides arsenite, arsenate, dimethylarsinic acid, arsenobetaine and tetramethylarsonium cation, Kuehnelt et al.[63,64] also detected the presence of arsenocholine for the first time in a terrestrial sample. In the author's opinion, the presence of arsenocholine and arsenobetaine, prominent compounds of the marine arsenic cycle, in the terrestrial environment strongly indicates that the terrestrial arsenic cycle is similar to the marine arsenic cycle. Nissen and Benson[59] have observed that higher plants are capable of the reduction and methylation of arsenic. In a phosphate and nitrate deficient environment, tomato, maize and melon were able to reduce As(V) and to methylate arsenite to MMA (22%) and to DMA (76%).

The main difficulty in the investigation of soil to plant transfer of arsenic involves the evaluation of the extraction results, since it is not known how realistically the applied extraction methods model the real uptake by plants. The need to find better extraction methodologies, which better correlate with interactions between soil and plant, motivates several laboratories, and as a result in the case of rice it has been found that a solution of hydrochloric acid, between 0.1 and 1 M is effective.[65] Generally, increased concentrations of arsenic in the soil resulted in increased arsenic concentration in the rice. In the case of other cultures, the ratio depends strongly on the soil type.

The main source of anthropogenic arsenic contamination for terrestrial plants is arsenic-containing chemicals, such as monosodium methylarsonate (MSMA), which are widely used as herbicides. In the USA alone, the estimated amount of MSMA used in 1977 was 1.5×10^6 kg.[66]

7.2 Animals, Humans

Arsenic speciation in terrestrial animals has been determined primarily in different mammalian species (including humans) in order to gain more information on arsenic metabolism, since arsenic is known to be toxic and a potent human carcinogen. Amongst terrestrial mammalian species, rats, dogs, monkeys, mice, rabbits and hamsters have been investigated. Data on arsenic compounds in animals other than mammals (such as arsenic species in ants, *Formica* sp.[64]) or in earthworms[67] are rare. Exposed to inorganic arsenic, mammals and microorganisms have the ability to methylate arsenic to less reactive and more readily excretable metabolites, but there are considerable differences between both species and individuals. In most mammals, inorganic arsenic is methylated to MMA and DMA and these species are excreted in the urine between the second and fourth days. This is more rapid than the excretion of inorganic arsenic, especially As(III), which is highly reactive with tissue components.[68] Absorbed As(V) is reduced to trivalent arsenic before the methyl groups are attached. DMA is considered to be the final product of the methylation process in most mammalian species: compared with humans, very little MMA is produced and excreted in the urine. If rodents and humans are exposed to DMA, about 5% of the dose is excreted in the urine as trimethylarsine oxide. The rate of methylation and its mechanism are very much dependent upon the species, populations and individuals. Marmosets and chimpanzees for example, do not seem to methylate inorganic arsenic at all,[69] whilst a native Andean women was found to excrete arsenic mainly as inorganic arsenic and DMA.[70]

8   Marine Environment

8.1   Seawater

In contrast to the limited data available on terrestrial biota, arsenic speciation in marine systems enjoys a much richer position. Since marine organisms (fish, crabs, mussels *etc.*) are extensively eaten by humans, the arsenic content of these organisms is well documented and is discussed elsewhere in this book. Here we focus primarily on the distribution of arsenic species in seawater and sediment.

Arsenic is typically found in the open ocean and coastal seawater at concentrations of $1-3\,\mu g\,l^{-1}$, but diverse values have been measured in surface and deep waters and in the seawaters of certain parts of the world.[71] Arsenic in seawater is predominantly present in two redox states, As(III) and As(V), and in two methylated forms, MMA and DMA. Studies indicate that MMA and DMA in seawater result from the biotransformation of arsenate by phytoplankton.[72] In addition to As(III), As(V), MMA and DMA, other organoarsenic compounds (many of them unidentified), are usually present within organisms, and may therefore be present at low concentrations in the water column. The presence of arsenobetaine, arsenocholine, MMA and DMA has also been shown in estuarine waters.[73] Arsenic can enter into seawater systems from the air, by atmospheric deposition, and from rivers, as dissolved species or as part of suspended particulate matter. These compounds may than be taken up by biota and/or settle and be adsorbed into the sediment. Estimates predict a slow increase of accumulated arsenic in sediments, thus contaminating seawater systems continuously.[9] The soluble Fe(III) content of waters has to be mentioned since it promotes As–Fe co-precipitation if the ionic strength and pH conditions dictate, and this process significantly decreases the arsenic flux in seawaters.[74]

8.2   Marine Sediment

Sediments existing at the bottom of the water column usually reflect the quality of the water system above and can therefore be used to detect the presence of contaminants that do not remain soluble in water. Sediments may act as sources or sinks for pollution, depending on the physico-chemical conditions (such as pH, redox conditions) of the material and the presence of organic chelators and living organisms. There have been few studies on the arsenic species present in marine sediments, largely because the extraction methods developed and used so far cannot usually assure the integrity of the species. The average total arsenic content in sea-sediments is $3-15\,\mu g\,g^{-1}$. Although arsenic concentrations in deep-sea sediments may be high (up to $450\,mg\,kg^{-1}$), sediment-bound arsenic species are not usually considered to be available to biota.[71] Studies by Maher showed that there were no detectable methylated arsenic compounds in sea-sediments.[75] In his opinion there are no methylation processes taking place in these sediments, and the possible presence of methylated species must be explained by the presence of arsenic-based pesticides. Other studies agree with these results, and have found, almost exclusively, inorganic arsenic species in sea-sediments, with the dominance of As(V).[76] Although studies investigating the forms of arsenic in interstitial waters (porewaters) of sediments report results similar to those for seawater, inorganic arsenic predominates and between 1 and 4% of the total arsenic was reported as methylated arsenic species, possibly as a result of the biomethylation processes.[77,78]

8.3 Marine Algae

The speciation of marine algae is an important issue in countries where they form a substantial part of everyday human nourishment (*e.g.* Japan). According to Francesconi and Edmonds, marine algae contain the greatest number of arsenic species among marine samples.[71] Studies have shown that half of the total arsenic content of marine algae is arsenate and arsenite, while the other half is given by different dimethylarsenoribosides (DMAR).[79] Species of algae exist, however, where DMARs were found to be the predominant forms of arsenic. No studies have been carried out so far to evaluate the toxicity of arsenosugars and arsenolipids.[80]

8.4 Marine Animals

The major arsenic compound present in marine animals is arsenobetaine, a form that is not considered toxic to humans. The tetramethylarsonium ion is also commonly found in marine organisms,[81,82] but no reliable toxicity data for this compound are known at present. Arsenocholine and trimethylarsine oxide are also found, but generally occur as very minor constituents.

The variety of arsenic-containing compounds in marine biota seems to be almost unlimited. Research continues to reveal numerous unidentified organoarsenic compounds and as analytical techniques improve and new marine species are investigated, more will surely be found. Transformation of arsenic in the different natural systems is still not clear; the search for key intermediates in the proposed biogeochemical pathways is the subject of study of several research groups.[83]

9 References

1. J.S. Edmonds and K.A. Francesconi, *Mar. Pollut. Bull.*, 1993, **26**, 665.
2. T. Tanaka, *Appl. Organomet. Chem.*, 1988, **2**, 283.
3. F.N. Riedeh and T. Eikmann, *Wirs. Umwelt.*, 1986, **3–4**, 108.
4. Huang Yan-Chu, in *Arsenic in the Environment. Part I: Cycling and Characterisation*, ed. J.O. Nriagu, Vol. 26, John Wiley & Sons Inc., New York, 1994, p. 17.

5. J.M. Ballantyne and J.N. Moore, *Geochim. Cosmochim. Acta*, 1988, **52**, 475.
6. W.R. Cullen and K.J. Reimer, *Chem. Rev.*, 1989, **89**, 713.
7. I. Thorton, in *Environmental Geochemistry and Health*, ed. I.D. Appleton, R. Fuge and G.H. McCall, Geological Society Special Publication, No. 113, 1996, p. 154.
8. A.P. Vingradov, *The Geochemistry of Rare and Dipersed Chemical Elements in Soils*, 2nd Edn., New York, 1959, p. 65.
9. D.C. Chilvers and P.J. Peterson, in *Lead, Mercury, Cadmium and Arsenic in the Environment*, ed. T.C. Hutchinson and K.M. Meema, Wiley, New York, 1987, p. 279.
10. J. Henriksson, A. Johannisson, P.-A. Bergqvist and L. Norrgren, *Arch. Environ. Contam. Toxicol.*, 1996, **30**, 213.
11. G.R. Sandberg and I.K. Allen, *ACS Symp. Ser.*, 1975, **7**, 124.
12. S. Gao and R.G. Burau, *J. Environ. Qual.*, 1997, **26**, 253.
13. E.A. Woolson, J.H. Axley and P.C. Kearney, *Soil Sci. Soc. Am. Proc.*, 1973, **37**, 254.
14. B. A. Manning and S. Goldberg, *Environ. Sci. Technol.*, 1997, **31**, 2005.
15. E.A. Rochette, G.C.Li and S.E. Fendorf, *Soil. Sci. Soc. Am. J.*, 1998, **62**, 1530.
16. R.Pongratz, *Sci. Total Environ.*, 1998, **224**, 133.
17. H. Xu, B. Allard and A. Grimvall, *Water Air Soil Pollut.*, 1991, **57–58**, 269.
18. E.W.B. Da Costa, *Appl. Microbiol.*, 1972, **23**, 46.
19. C. Cervantes, G. Ji, J.L. Ramirez and S. Silver, *Microbiol. Rev.*, 1994, **15**, 355.
20. E.A. Woolson, *Weed Sci.*, 1977, **25**, 412.
21. F. Challenger, *Chem. Rev.*, 1945, **36**, 315.
22. H. Yamauchi and Y. Yamamura, *Chem. Abstr.*, 1982, **96**, 33494z.
23. D.P. Cox and M. Alexander, *Bull. Environ. Contam. Toxicol.*, 1973, **9**, 84.
24. W.R. Cullen, A.E. Erdman, B.C. McBride and A.W. Pickett, *J. Microbiol. Meth.*, 1983, **1**, 297.
25. B.C. McBride and R.S. Wolfe, *Biochemistry*, 1971, **10**, 4312.
26. M. Shariatpanahi, A.C. Andreson and A.A. Abdelghani, *Trace Subst. Environ. Health*, 1982, **16**, 170.
27. A.W. Pickett, B.C. McBride and W.R. Cullen, *Appl. Organomet. Chem.*, 1988, **2**, 479.
28. K. Hanaoka, S. Hasegawa, N. Kawabe, S. Tagawa and T. Kaise, *Appl. Organomet. Chem.*, 1990, **4**, 239.
29. K.H. Akkari, R.E. Frans and T.L. Lavy, *Weed Sci.*, 1986, **34**, 78.
30. K. Hanaoka, K. Uchida, S. Tagawa and T. Kaise, *Appl. Organomet. Chem.*, 1995, **4**, 573.
31. O.F.X. Donard and J.H. Weber, *Nature (London)*, 1988, **332**, 339.
32. W. Delgardo-Morales, M.S. Mohan and R.A. Zingaro, *Int. J. Environ. Anal. Chem.*, 1994, **54**, 203.
33. J. Feldmann, R. Grümping and A.V. Hirner, *Fresenius' J. Anal. Chem.*, 1994, **350**, 228.
34. J. Feldmann and A.V. Hirner, *Int. J. Environ. Anal. Chem.*, 1995, **60**, 339.
35. E.B. Wickenheiser, K. Michalke, C. Drescher, A.V. Hirner and R. Hensel, *Fresenius' J. Anal. Chem.*, 1998, **362**, 498.
36. D.L. Johnson and R.S. Braman, *Chemosphere*, 1975, **6**, 333.
37. D. El-Mogazi, D.J. Lisk and L.H. Weinstein, *Sci. Total Environ.*, 1988, **74**, 1.
38. B.P. Jackson and W.P. Miller, *J. Anal. At. Spectrom.*, 1998, **13**, 1107.
39. B.P. Jackson and W.P. Miller, *Environ. Sci. Technol.*, 1999, **33**, 270.
40. M.L Magnuson, J. T. Creed and C.A. Brockhoff, *J. Anal. At. Spectrom.*, 1997, **12**, 689.
41. S. Saverwyns, X. Zhang, F. Vanhaecke, R. Cornelis, L. Moens and R. Dams, *J. Anal. At. Spectrom.*, 1997, **12**, 1047.
42. P. Thomas and K. Sniatecki, *J. Anal. At. Spectrom.*, 1995, **10**, 615.
43. G.E.M. Hall, J.C. Pelchat and G. Gauthier, *J. Anal. At. Spectrom.*, 1999, **14**, 205.

44. F. Ferguson and J. Gavis, *Water Res.*, 1972, **6**, 1259.
45. Zs. Jókai, J. Hegoczky and P. Fodor, *Microchem. J.*, 1998, **59**, 117.
46. A. Bartha, I. Csalapovits, I. Horváth, U. Siewers and J. Stummeyer, Proceedings of the 42nd Hungarian Spectrochemical Meeting, Veszprém, Hungary, 1999, p. 16.
47. S.L. Chen, S.R. Dzeng, M.H. Yang, K.H. Chlu, G.M. Shieh and C.M. Wai, *Environ. Sci. Technol.*, 1994, **28**, 877.
48. S.A. Talukder, J. Zheng and W. Kosmus, *Proceedings of the International Symposium on Trace Elements in Humans*, Athens, Greece, 1997, p. 373.
49. Z. Mester and P. Fodor, *J. Chromatogr. A*, 1996, **756**, 292.
50. A. Criaud and C. Fouillac, *Chem. Geol.*, 1989, **76**, 259.
51. E. Büyüktuncel, S. Bektas, B. Salih, M.M. Evirgen and O. Genc, *Fresenius' Environ. Bull.*, 1997, **6**, 494.
52. R. Wagemann, *Water Res.*, 1978, **12**, 139.
53. C.C. Fuller and J.A. Davis, *Nature*, 1989, **340**, 52.
54. M.I. Abdullah, Z. Shiyu and K. Mosgren, *Mar. Pollut. Bull.*, 1995, **31**, 116.
55. H. Hasewaga, Y. Sohrin, M. Matsui, M. Hojo and M. Kawashima, *Anal. Chem.*, 1994, **66**, 3247.
56. T. Kuriowa, A. Ohki, K. Naka and S. Maeda, *Appl. Organomet. Chem.*, 1994, **8**, 325.
57. S. Maeda, A. Ohki, T. Tokuda and M. Ohmine, *Appl. Organomet. Chem.*, 1990, **4**, 251.
58. S. Maeda, K. Kumeda, M. Maeda, S. Higashi and T. Takeshita, *Appl. Organomet. Chem.*, 1987, **1**, 363.
59. P. Nissen and A.A. Benson, *Physiol. Plants*, 1982, **54**, 446.
60. A.A. Benson and P. Nissen, in *Biochemistry and Metabolism of Plant Lipids*, ed. J.F.G.M. Wintermans and P.J.C. Kuiper, Elsevier, Amsterdam, 1982, p. 121.
61. S. Londesborough, J. Mattusch and R. Wennrich, *Fresenius' J. Anal. Chem.*, 1999, **363**, 577.
62. A.R. Byrne, Z. Slejkovec, T. Stijve, L. Fay, W. Göessler, J. Gailer and K.J. Irgolic, *Appl. Organomet. Chem.*, 1995, **9**, 305.
63. D. Kuehnelt, W. Göessler and K.J. Irgolic, *Appl. Organometal. Chem.*, 1997, **11**, 459.
64. D. Kuehnelt, W. Göessler and K.J. Irgolic, *Appl. Organometal. Chem.*, 1997, **11**, 289.
65. T. Kato, T. Izawa, and I. Okimura, *Aichi-ken Nogyo Shikenjo Kenkyu Hokoku*, 1977, **9**, 157.
66. J.W. Mason, A.C. Anderson, P.M. Smith, A.A. Abdelghani and A.J. Englande, *Bull. Environ. Contam. Toxicol.*, 1979, **22**, 612.
67. A. Geiszinger, W. Göessler, D. Kuehnelt, K. Francesconi and W. Kosmus, *Environ. Sci. Technol.*, 1998, **32**, 2238.
68. M. Vahter, *Sci. Progr.*, 1999, **82**, 69.
69. M. Vather, G. Concha, B. Nermell, R. Nilsson, F. Dulout and A.T. Natarajan, *Eur. J. Pharmacol. Environ. Toxicol. Pharmacol. Section*, 1995, **293**, 455.
70. D.L. Johnson and M.E.Q. Pilson, *J. Mar. Res.*, 1972, **30**, 140.
71. K.A. Francesconi and J.S. Edmonds, *Croat. Chem. Acta*, 1998, **71**, 343.
72. M.H. Florêncio, M.F. Duarte, S. Facchetti, M.L. Gomes, W. Göessler, K.J. Irgolic, H.A. van't Klooster, R. Montanarella, R. Ritsema, L.F. Vilas Boas and A.M.M. de Bettencourt, *Analusis*, 1997, **25**, 226.
73. O. Andeae, J.T. Byrd and P.N. Froelich, *J. Environ. Sci. Technol.*, 1983, **17**, 731.
74. W.A. Maher, *Chem. Geol.*, 1984, **47**, 333.
75. M.O. Andreae, *Limnol. Oceanogr.*, 1979, **24**, 440.
76. L. Ebdon, A.P. Walton, G.E. Millward and M. Whitfield, *Appl. Organomet. Chem.*, 1987, **1**, 427.
77. K.J. Reimer and J.A.J. Thompson, *Biogeochemistry*, 1988, **6**, 211.

78. K.A. Francesconi and J.S. Edmonds, *Oceanogr. Mar. Biol. Annu. Rev.*, 1993, **31**, 111.
79. K. Shiomi, in *Arsenic in the Environment. Part II: Human Health and Ecosystem Effects*, ed. J.O. Nriagu, Vol. 27, John Wiley & Sons Inc., New York, 1994, p. 261.
80. K. Shiomi, Y. Kakehashi, H. Yamanaka and T. Kikuchi, *Appl. Organomet. Chem.*, 1987, **1**, 177.
81. M. Morita and Y. Shibata, *Anal. Sci.*, 1987, **3**, 575.
82. K.A. Francesconi, W. Göessler, S. Panutrakul and K.J. Irgolic, *Sci. Total Environ.*, 1998, **221**, 139.

CHAPTER 12

Rapid Tests – A Convenient Tool for Sample Screening with Regard to Element Speciation

MARTINA UNGER-HEUMANN

1 Introduction

The demand for fast and simple analytical methods is growing. This is due to the fact that the number of analyses to be done has been increasing for many years and, so far, this trend shows no sign of abating. In the field of environmental monitoring and industrial in-process control, information is often needed on-site and within a few minutes. Conventional laboratory methods are often unable to fulfil these requirements. Using a rapid spot test is one way of achieving the desired aims, and not only save time but also money. Rapid tests are cost-effective with low purchase costs, low time consumption and low running costs, which all decrease the cost per analysis. The information provided is in many cases sufficient for purpose and this contributes to their growing importance.

2 Definition of Rapid Test

As there are so many different expressions used for a rapid test, *e.g.* screening test, test kit, field test, spot test *etc.*, it seems to be necessary to explain first what a rapid test is and what are the main features.[1]

In general, a rapid test is a specially adapted form of a well-known analytical method. The main objective is always to provide a test that is fast, highly portable and easy to use. A rapid test may be, for example, a test strip or a sensor which can be taken anywhere and which gives an immediate result if brought into contact with the sample to be analysed. A test kit always contains all that is necessary to run the analysis. All reagents are pre-mixed and ready to use. Thus, a rapid test consists of two important components: the chemical reaction and the evaluation system. The chemistry and the evaluation system have to fit each other and the quality of the analysis and the measurement result depend on how they are combined. Figure 1 illustrates the efficiency of the different methods used in rapid tests with regard to precision, sensitivity and speed.

The chemistry selected is mainly based on classical colour reactions, which are known to be very specific. Colour reactions due to complexation of metals, redox reactions using a suitable indicator, enzymatic reactions and others, like the Griess reaction (for nitrite and with some modifications also for nitrate)[2] and the indophenol blue reaction (for ammonium),[3] are very common as rapid tests. As the main focus is to have an easy-to-use and fast test system, manufacturers concentrate on a meaningful combination of the necessary reagents, on the stabilisation of reagent mixtures and solutions, and on the reduction of interferences by the addition of masking agents.

The chemistry used on the nitrate test strip is discussed in some detail in order to illustrate these features. Such test strips are handled annually by thousands of people, but few are aware of why their use is as easy and as fast as it is. Such a test strip consists of a plastic foil made of PVC or polyester (which is more environmentally friendly since it is biologically degradable) and a reaction zone sealed on to the foil. This reaction zone is made of a special paper which is impregnated with the colour reagent, in this case a reducing agent, the Griess reagent [sulfanilic acid and *N*-(1-naphthyl)ethylenediamine], and also with an acidic buffer,

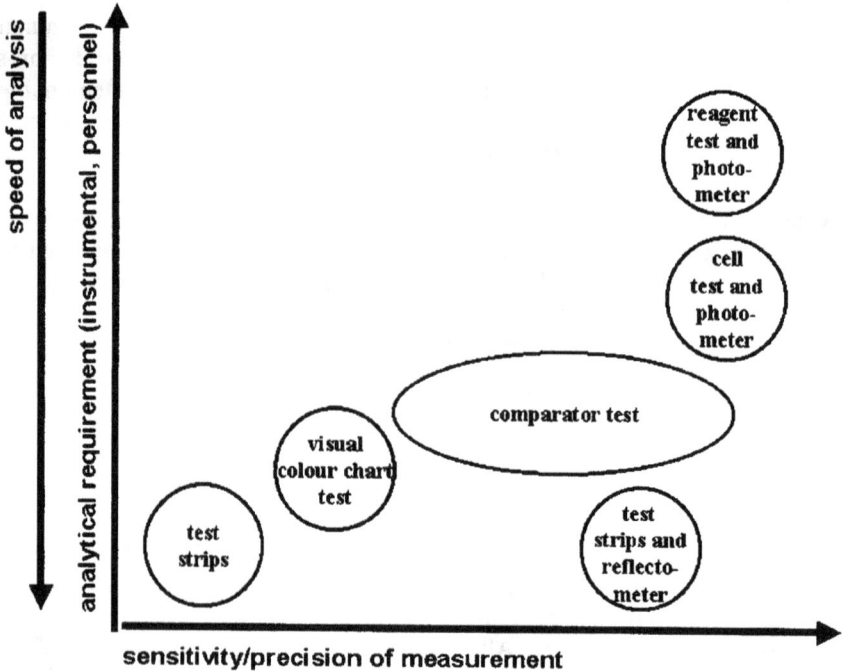

Figure 1 *Efficiency of the different methods that are mainly used in rapid tests*

since the reaction needs an acidic medium. Without this buffer, it would be necessary to adjust the pH of the sample in an additional handling step, which is less convenient. For correct functioning, it must be ensured that the chemicals do not react with each other prior to the actual analysis, and that the glue, which is used to fix the reaction zone on the plastic foil, is chemically inert.

Returning to the other component of a rapid test, the evaluation system, the methods used can be divided into chemical and physicochemical methods on the one hand and biochemical and biological methods on the other. While chemical and physicochemical methods are most frequently used for the determination of inorganic compounds and elements, biochemical and biological methods are primarily used for the determination of organic compounds. Within the chemical and physicochemical methods employed, the following methods are mainly used for the evaluation of rapid tests:

Visual. Due to a colour change, a qualitative result or, in connection with a colour chart, a semiquantitative estimation can be achieved, *e.g.* litmus paper or pH test strips.

Colorimetric. In this case the reaction also produces a colour change that is measured using a colorimeter or comparator. This procedure usually allows colour compensation in faintly coloured samples, and is sensitive enough to be used in the ppb-range and provide semiquantitative results.

Titrimetric. These are the well-known laboratory methods, *e.g.* the determination of the total hardness by complexometric titration with EDTA. The titrant is not added by means of a burette but by means of a simple plastic syringe or by just counting drops. Titrimetric methods provide quantitative results with slightly reduced precision compared with the respective laboratory method.

Reflectometry. Small pocket-size reflectometers are used for the quantitative evaluation of test strips. This method has been used in clinical diagnostics for many years, *e.g.* for home monitoring of the blood sugar content of diabetics, although it is still new for the determination of inorganic species in environmental and food samples.[4,5] Such a reflectometric system is shown in Figure 2.

Photometry. Photometric test kits are often used together with a portable filter photometer. The calibration data for the different determinations may be stored in the photometer software, thus relieving the operator of the need to calibrate the instrument. Currently, some manufacturers of UV/VIS spectrophotometers provide special software programmes that enable their customers to use pre-prepared, easy-to-handle reagents for rapid tests.

Within the biochemical and biological methods the following test principles can be differentiated:

Biotests. These only give information about the toxicity of environmental samples in biological systems without identifying the substances concerned. As a consequence such a test cannot differentiate whether the toxic effect comes, for example, from mercury, methylmercury or a chlorinated hydrocarbon. An example of biotests is the luminescent bacteria test,[6] which is also available in a rapid test format.

Enzymatic assays. Enzymatic reactions are very specific and are often used in food analysis for the determination of ingredients.[7] Enzymatic inhibition tests

are useful for the screening of pesticides and have most promise for speciation.
Immunoassays. Most of the recent developments in rapid testing have been made
in the field of immunoassays, *e.g.* ELISA.[8] Immunoassays allow for specific and
sensitive determinations of organic compounds.

3 Reasons for the Use of Rapid Tests

The main reason for the use of rapid tests is undoubtedly their speed. They
provide an analytical result in a short time. In addition, the ease of handling
usually associated with these tests means fewer possibilities for error. Another
factor becoming increasingly important is the economic aspect. Rapid tests do
not require high investment in either equipment or in highly skilled staff. Such
cost-orientated thinking will become even more essential for industry as well as
for governmental authorities in the future.

Rapid tests provide results on-site, which is sometimes absolutely necessary. In
effect, the sample does not go to the laboratory but the laboratory comes to the
sample.

There are several circumstances where rapid tests have clear advantages
compared with analytical laboratory methods.

Screening. Rapid tests are, as already mentioned, highly portable due to their
small size and are therefore ideal for sample screening in the field. It is possible to
get on-site and gain a quick overview of a situation without spending much time
and wasting money. They can be used to identify critical samples. Only these
critical samples would require further analysis in a laboratory using validated
analytical procedures.

Process monitoring. Process monitoring often requires nothing more than
monitoring limit values. Normally, there is no need for overkill in terms of
precision. A fault has to be recognised as soon as possible so that countermeasures can be taken. More frequent on-site measurements can provide more
useful data than fewer samples being analysed with high precision.

Highly perishable samples. If preservation of the sample is not possible, a rapid
test done on-site provides a more accurate result, with respect to the time of
sampling, than a highly sophisticated laboratory method carried out many hours
later or on the next day.

Critical situations, health and safety protection. In cases where decisions have
to be made in the shortest possible time frame, the conventional way of carrying
out analysis is often unsatisfactory due to the fact that by the time the analysis
result is available the problem may have disappeared or, in the worst case, has
run out of control. Here rapid tests can provide immediate answers to the cause,
and corrective action can be made.

Setting an analysis strategy. In unknown samples it might be very helpful to
start an analysis by roughly estimating the concentration range of the component to be determined using a rapid test. With the help of this result, a specific
analysis strategy concerning sample preparation, dilution, *etc.* can be set without
losing time carrying out several trials or, even more importantly, contaminating

trace analysis instrumentation with high levels of certain elements.

Check of plausibility. If a mix-up of samples is suspected in a laboratory, rapid tests allow fast clarification of the problem.

4 Rapid Tests for Element Speciation

In general, there are many more rapid tests available for the speciation of anions than for cations. The chemical reactions that are used for the determination of metal cations normally require the presence of the free hydrated metal ion. Therefore, organometallic compounds cannot be detected; *e.g.* rapid tests for the determination of lead are not able to detect tetraethyl-lead. As most cations are determined by forming coloured complexes, it depends upon the stability of the existing complexes whether or not they are detected. In the following, rapid tests are described which are able to detect element species.[9]

4.1 Nitrogen

There is a wide variety of rapid tests which can clearly differentiate between nitrogen species such as NH_4^+, N_2H_4, NO_2^- and NO_3^-.

Ammonium tests are based on either the reaction with Nessler's reagent or the indophenol blue reaction which has been described by Berthelot.[3]

Hydrazine reacts with 4-(dimethylamino)benzaldehyde or similar aromatic aldehydes to form coloured hydrazones.

The well-known Griess reaction[2] is mostly used for the determination of *nitrite*. In the presence of an acidic buffer, nitrite is converted into nitrous acid which diazotises sulfanilic acid. Coupling the diazonium salt with *N*-(1-naphthyl)ethylendiamine produces a red–violet azo dye.

By integrating a reducing agent into the Griess reaction it detects *nitrate*, too. Other chemical reactions used for nitrate tests are reactions with aromatic compounds in the presence of strong acids, *e.g.* sulfuric acid or phosphoric acid, to form yellow or red nitro compounds.

In drinking water, wastewater and soil analysis, it is important to differentiate between ammonium, nitrite and nitrate as they effect the environment in different ways.[10] A typical application is described later.

4.2 Phosphorus

Although phosphorus as well as phosphate test kits are found in the manufacturer's catalogues and brochures, all tests detect *orthophosphate* only. Polyphosphates, organic phosphates and other phosphorus species are not detected. The chemistry used is mainly based on two reactions: the phosphomolybdenum blue reaction (PMB) and the vanadate–molybdate reaction (VM). In the PMB reaction, orthophosphate and molybdate form, in the presence of sulfuric acid, molybdophosphoric acid, which is reduced by a suitable reducing agent to phosphomolybdenum blue. For the VM method ammonium vanadate and ammonium heptamolybdate are used. In the presence of sulfuric acid they form together with PO_4^{3-} yellow–orange molybdovanadophosphoric acid.

4.3 Sulfur

The following sulfur species can be determined by rapid tests: S^{2-}, SO_3^{2-} and SO_4^{2-}.

Hydrogen sulfide reacts with *N,N'*-dimethyl-1,4-phenylendiamine and an oxidising agent to form Methylene Blue. While this chemical reaction was described in the 19th century it was used for the colorimetric detection of sulfide for the first time in 1901.[11]

Sulfite is determined either by a titrimetric or a colorimetric method. In the first case sulfite is titrated with potassium iodate against zinc iodide starch solution. The colorimetry uses the reaction of sulfite with potassium hexacyanoferrate(II), zinc sulfate and sodiumnitroprusside to form a red complex.

There are three different methods used to detect *sulfate*. In each of them, sulfate ions react with barium ions to produce sparingly soluble barium sulfate. In one test, the resulting turbidity is determined photometrically. In the other cases, sulfate is determined indirectly by colour reactions. In one case, sulfate reacts with the red thorin–barium complex to form barium sulfate, and as a consequence the colour turns yellow due to the released thorin. In another case, sulfate reacts with barium iodate, again to form barium sulfate – the released iodate forms a brownish-red complex with tannin.

Sulfite plays an important role in food processing, since it is often used as an antioxidant to preserve products made from fruit and vegetables such as wine, potato products *etc.* More details about this application are given later.

4.4 Chlorine

In the literature, rapid tests for chloride and chlorine are found.

Chloride is determined either by mercurimetric titration against diphenylcarbazone or by an indirect colour reaction. In the latter, chloride reacts with mercury(II) thiocyanate to form mercury(II) chloride and chloromercurate(II). The released thiocyanate reacts with iron(III) nitrate to form red iron(III) thiocyanate, the concentration of which is measured visually or photometrically. A speciality is the chloride test strip since it uses a kind of titration done on the different reaction pads of the strip. The chloride strip consists of five reaction zones, each of which is impregnated with different amounts of red–brown silver chromate. Chloride reacts with the silver ions and thus decolorises the silver chromate. Depending upon the concentration of chloride present in the sample, only the reaction zone with the lowest silver chromate content and zones with higher concentrations are decolorised. This principle allows a very quick, rough estimation of the chloride content.

In *chlorine* tests, the DPD method is normally applied. Elemental chlorine, as well as hypochlorite, oxidises dialkyl-*p*-phenylenediamine to form a red–violet product with a semiquinoid structure. Most of the chlorine tests allow the determination of 'free' (elementary, *i.e.* Cl_2) chlorine as well as total chlorine content (including chloramines).

In water analysis the differentiation between chlorine and chloride is very important. While the chloride content influences corrosion, the chlorine content must be controlled with regard to disinfection issues, *e.g.* in swimming pool water and industrial process water.

4.5 Chromium

The chemical reaction used in test kits for the determination of chromium is based on a redox reaction between chromate and diphenylcarbazide. This means that only Cr^{6+} (chromate and dichromate) is detected. As the toxicity of chromium depends on its oxidation state, it is necessary to differentiate between Cr^{6+} and Cr^{3+} in order to get meaningful information. Cr^{6+} is much more toxic than Cr^{3+} and is also suspected to be carcinogenic for man. Thus, the determination of the Cr^{6+} content in the working place, in blood and in urine plays an important role in the control of occupational diseases.

4.6 Iron

In rapid tests, iron is detected by complex-building reactions. Complexing agents such as 2,2'-bipyridine and 1,10-phenanthroline react with Fe^{2+} only. Some iron test kits contain an additional reducing agent, *e.g.* ascorbic acid, which allows both the measurement of Fe^{2+} alone and, additionally, the sum of Fe^{2+} and Fe^{3+}.

5   Fields of Application

Rapid tests are mainly used in environmental[12] and food analysis.[13]

5.1   Environmental Analysis

Water. The most important application, especially for photometric test kits, is in wastewater analysis. The use of photometric test kits is, in many countries, officially accepted for the control of wastewater quality. Modern systems consisting of portable battery-powered photometers and convenient cell tests, in which all the reagents needed for the analysis are contained in pre-dosed measurement cells, provide good analytical quality assurance (AQA). Besides wastewater analysis, rapid tests are also used to analyse many other types of water, *e.g.* drinking water, surface water and process water.

Soil. The determination of plant nutrients, *e.g.* nitrogen, phosphorus and potassium in soil can easily be performed with test strips. The need for farmers and growers to know how much fertiliser is needed for their land is obvious, and these results are required quickly, since the value changes as crops are planted

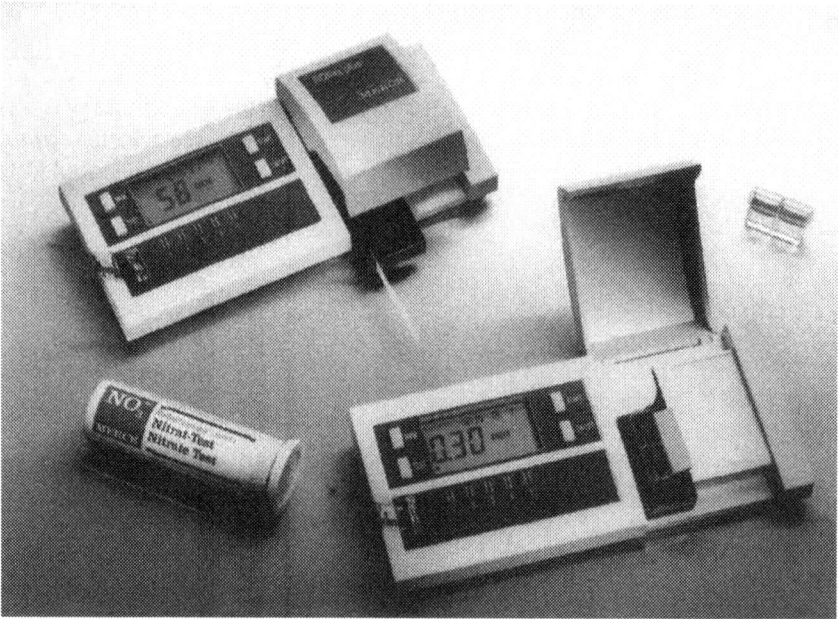

Figure 2 *The Reflectoquant® system (Merck KGaA) for the quantitative evaluation of test strips*

Table 1 *Comparison of costs and time for soil analysis (laboratory method vs. reflectometric test kit method)*

	Laboratory method	*Reflectoquant® system**
Result	after 2–4 weeks	after 1 day
Costs	35 DM/sample	approx. 10 DM/sample

Number of analyses: 651 soil samples.
Coefficient of correlation: $r = 0.92$.
*Trademark of Merck KGaA, Darmstadt.

and start to grow. Conventional laboratory analysis simply takes too long.

In a large scale trial organised by the Technical University of Munich, (Bayerische Landesanstalt für Landtechnik), the nitrate content of 651 soil samples was determined.[14] Each of the samples taken over a period of 10 days was divided. One part was analysed on-site using the Reflectoquant® system, a reflectometric rapid test system of Merck KGaA (Figure 2), the other part was analysed in a specialist laboratory according to standard methods. The performance is compared in Table 1. The correlation between both methods is excellent if one takes into account that storage conditions, sample preparation and analysis times were all different. In this example the advantages of using an on-site test kit are obvious.

Such a rapid test can of course also be used to determine the concentration of other plant nutrients in soil, *e.g.* ammonium, phosphate and potassium.

5.2 Food Analysis

In food processing there are different control points where food ingredients have to be analysed.

Control of raw material. A potato processing factory, for example, accepts for the production of its brand products only potatoes with a nitrate content below $150\,mg\,kg^{-1}$. Due to over-fertilisation of the soil, potatoes, as well as other vegetables, may accumulate high nitrate contents. Potatoes with high nitrate levels are more perishable and therefore not acceptable. During the harvest season up to 20–50 contract farmers deliver the potatoes daily to the processing factory. Before they are allowed to unload the potatoes, each delivery has to be analysed on a random basis as rapidly as possible. Therefore, a fast, simple (which can be done by unskilled persons) and quantitative method which is specific for nitrate is required. It was decided to use nitrate test strips combined with quantitative reflectometric evaluation for this purpose.

Control of food additives and ingredients. Ascorbic acid and/or sulfite are added to food as antioxidants. The Food Law for sulfite added to food lays down limit values. Potato products are normally treated with sulfite solution in order to avoid brown coloration. Besides the measurement of the sulfite content in the product it is also necessary to control the sulfite content in the sulfite solution which is used for the sulfurisation process, because if the sulfite content is too low it is not guaranteed that the required product quality can be achieved. On the other hand, if the sulfite content is too high the allowed limit value in the product may be exceeded. A simple and rapid in-process control that clearly can differentiate between sulfite and sulfate allows producers to take countermeasures and avoids misproduction.

Disinfection and cleanliness control. Chlorine-containing disinfectants are still the most commonly used ones in the food industry. The effect of the disinfection is only guaranteed if the concentration of the active compound is high enough. On the other hand, it is also necessary that all disinfectant residues have been rinsed carefully out of the equipment after the cleaning process in order to avoid residues in the food product. Rapid tests can be used either to control the concentration for use or to check for residues.

6 Conclusion

Rapid tests are not a cheap replacement for laboratory methods. They should not be seen to be in competition with laboratory methods, but are a useful addition. There are special applications where rapid tests have advantages compared with more expensive and time-consuming laboratory methods. In view of the increasing number of analyses required and the subsequent costs involved, rapid tests may provide an answer to many analytical problems.

7 References

1. M. Unger-Heumann, *Fresenius' J. Anal. Chem.*, 1996, **354**, 803.

2. P. Griess, *Berichte*, 1879, **12**, 426.
3. P. Berthelot, *Réport Chim. Pure Appl.*, 1859, **1**, 284.
4. L. Glessner, *Environ Solut.*, 1995, **8**, 23.
5. G. Schwedt, *Umwelt Magazin*, 1994, **23**, 134.
6. DIN 38412 part 34, Determination of the inhibitory effect of waste water on the light emission of *Photobacterium phosphoreum* – luminescent bacteria waste water test using conserved bacteria (L34). Beuth, Berlin, Wien Zürich, 1991.
7. R. Gromes, B. Schnellbacher, T. Siegl and T. Vatter, *Dtsch. Lebensm. Rdsch.*, 1995, **91**, 171.
8. C. Keuchel and R. Niessner, *Fresenius' J. Anal. Chem.*, 1994, **350**, 538.
9. Merck, *Rapid Test Handbook*, Merck, Darmstadt, 1987.
10. K. Aurand *et al.*, Die Trinkwasser-Verordnung, Erich Schmidt, Berlin, 1987.
11. W.G. Lindsay, *School of Mines Quarterly*, 1901, **23**, 24.
12. L.A. Hütter, *Wasser und Wasseruntersuchung*, Diesterweg, Frankfurt, 1988.
13. W. Baltes *et al.*, *Schnellmethoden zur Beurteilung von Lebensmitteln und ihren Rohstoffen*, Behr's, Hamburg, 1995.
14. M. Schurig, B. Lehmann and G. Rödel, personal communication, 1993.

Food

CHAPTER 13

The Importance of Trace Element Speciation in Food Issues

HELEN M. CREWS

1 Introduction

This chapter does not aim to be a comprehensive review of speciation in food. It should be read in conjunction with those chapters dealing with trace element species in the environment, since the human food chain can be affected by environmental issues, and those dealing with health effects since ingestion of chemical species with food can be both beneficial and detrimental depending upon the element and its form. In addition, other chapters in this publication cover many of the broader and individual issues which formed the basis of many fruitful presentations and discussions during the period of the Speciation 21 project. Here, in this chapter, some of the generic issues are highlighted, and the use of, and need for, speciation studies are illustrated by briefly citing examples for specific element species of the essential elements Fe and Se, and the toxic elements As and Sn.

Historically, many European countries have conducted surveillance exercises for the total levels of trace elements, in particular for the toxic elements Pb, Cd, Hg and As, as well as some of the nutritionally essential elements, in individual or mixed food groups. Only recently has the idea of measuring specific element species, which may provide a better estimate of consumer risk, been addressed, as methods for determining species are developed and validated.

2 Bioavailability

For essential elements to exert an effect they must be available from food, or potentially available, both for absorption by the gut and for subsequent utilisation by the body; in other words they must be bioavailable. The degree of bioavailability is affected by the individual's physiological response and their diet.[1] Food is the primary source of essential elements for humans. The elements which are considered essential are the major elements (Na, Mg, P, Cl, K and Ca) and the minor elements (Cr, Mn, Fe, Co, Cu, Zn, Se, Mo and I). In addition there are the 'newer' trace elements, which are possibly essential, these being Li, B, F, Si, V, Ni, As, Sn and Pb.[2]

2.1   Iron

The chemical form or speciation of an element is one of the primary influences on its bioavailability. In this context, the determination of valency states, metal–ligand complexes and mineral compounds may be required. The nature of the chemical species and its subsequent effect(s) vary from element to element. For example, Fe is essential for good health. In the food industry, the many discrete forms of Fe have long been known, since their use as food fortificants in some countries demands that the organoleptic properties of each species are well understood and used appropriately. It has long been established that the absorption of heme-Fe, (found in red meat), is minimally influenced by the type of meal and it is absorbed as an intact complex into the mucosal cell.[3] As such, it represents a different dietary pool to non-heme Fe (found in fruit, vegetables and cereals), the absorption of which depends upon the type of dietary ligands present in the intestinal tract.[4,5] The bioavailability of both forms of Fe can be related to physiological requirements and body stores.[2,6]

2.2   Selenium

A second essential element, of increasing interest, is Se and in this publication Chapter 18 deals with the different aspects of Se speciation in relation not only to its essentiality, but also to its role in disease prevention. The use of dietary supplements by large numbers of the population provides an increasing market for the producers and many of the latter are keen to know what chemical species are most efficacious, and also to have standard materials available to provide some degree of quality control during production.

The analytical difficulties encountered in measuring species in food matrices such as yeast (which is frequently used for supplementation but for which the Se speciation varies with yeast strain and is still not clearly understood), have been systematically addressed by the teams of Lobinski and Szpunar. In their elegant studies of Se in yeast,[7,8] they show what can be achieved by carefully thought out experimental protocols and a combination of chromatographic and mass spectrometric techniques. Thus, for example, by evaluating eight solid–liquid extraction procedures for the recovery of Se species from industrially produced Se-enriched *Saccharomyces cerevisiae*, they were able to conclude that a sequential leaching procedure might be appropriate for the evaluation of Se species in yeast without the need for coupled techniques.[7] This type of approach is valuable when the cost of more sophisticated techniques is considered (see also Section 4 and Chapter 19 for further examples of the work of Lobinski *et al.*).

The more complex techniques are often however, the only way to confirm species identity and Uden's group at the University of Massachusetts, USA, have been studying the principal Se compounds in garlic and yeast since the early to mid-1990s. They have reported recent studies using ion-pair liquid chromatography with ICP-MS and electrospray ionisation mass spectrometric detection.[9] The principal Se compounds identified, and which accounted for 85% and 95% of the Se respectively in Se-enriched yeast and garlic, were selenomethionine and *Se*-adenosyl-selenohomocysteine in yeast, and γ-glutamyl-*Se*-methyl-selenocysteine with possibly γ-glutamyl-selenomethionine in garlic. It is apparent that successful confirmation of species identity is the result of combinations of expertise and technology.

2.3   Arsenic

The concept of speciation also applies to potentially toxic elements such as As, Cd, Sn, Hg, and Pb which can be naturally elevated in some foodstuffs or may occur as adventitious contaminants. Arsenic bioaccumulates in the marine food chain resulting in higher levels of this element in fish than in most other foods and is the element most frequently cited in discussions of speciation and the food chain. It is known that inorganic forms of As are more toxic than organic species with As(III) being more toxic than As(V). In fish the predominant forms are the organic compounds arsenobetaine and arsenocholine. Knowledge of the chemical form of a contaminating element might influence the way in which the problem is dealt with. This issue is very well exemplified and discussed, with reference to dietary As in the Basque country of Spain, in Chapter 16. In addition, As is used by Larsen and Berg in Chapter 17 to show how speciation is taken into account when producing a position paper for Codex.

2.4   Tin

Tin is not normally considered problematic in foods with the exception of occasional elevated levels of total Sn in some canned products, usually acidic foods. However, when considering the effect of the accumulation of an element in the food chain, the saga of tributyltin (TBT) in the marine ecosystem illustrates how one particular form of an element used for a specific purpose – in antifouling ship paints – can, over time, become an issue.

2.4.1   *Tributyltin from Industrial Activity in the Food Chain*

TBT is highly toxic to fish, crustaceans, and molluscs, causes the occurrence of male characteristics in some female gastropods (*imposex* or *intersex*), as well as other effects such as reduced growth and reproduction, and immunotoxicity.[10] Organotin compounds are also generally considered to be endocrine disrupters. In their recent publication,[10] Belfroid *et al.* report that when the limited data (from 22 countries) for organo-Sn levels in seafood was considered, in relation to tolerable average TBT residue levels for a 60 kg consumer (derived from the tolerable daily intakes discussed in 1993 by Penninks[11]), more than one-third of the countries exceeded this level in one or more of the samples.

At the moment no maximum residue limits exist for TBT in seafood products and thus it is not routinely covered by surveillance exercises conducted by national food safety organisations. The authors argued that a Maximum Residue Limit is needed which should take into account both the variation in seafood consumption internationally as well as the variation in average consumer weight.[10]

3 Requirements for Better Risk Assessment of Dietary Trace Elements

Essential elements can also be toxic if taken in excess. There is usually a wide margin between deficiency and toxicity. The margin can be narrower for some elements (Fe and Se) than for others (Mg and Ca). As alluded to in the discussion of TBT earlier, deciding what represents the average dietary requirements for a population can be difficult because of large inter- and intra-individual variations and because so much fundamental information is missing. Good measures of nutritional status are lacking for many elements. Whilst there has been a considerable increase in the understanding of the biochemical effects of metals and some biomarkers of toxicity can now be recognised at levels which do not produce obvious clinical effects,[12] there is still a need to ascertain the exposure levels at which biological requirements are exceeded and those at which toxicity may occur, if risk assessments are to be realistic. Therefore, a better understanding of the predominant chemical forms of trace elements in foods and their subsequent behaviour in the gut would assist those who have to make decisions concerning dietary requirements and related legislation.

4 Analytical Requirements

The analytical requirements for trace element speciation and food legislation should, in an ideal world, stem from both the regulatory requirements and the most up-to-date analytical capabilities. However, there is a need to be pragmatic and to make use of the best *widely available* methods if regulations are to be monitored. For speciation measurement in foodstuffs, the procedures for the determination of individual trace element species are still being developed and validated. Many of the analytical capabilities rely on mass spectrometric detection preceded by a variety of chromatographic separation systems, all of which can be prohibitively expensive for some surveillance and industry laboratories. Whilst these more expensive facilities are very slowly becoming more widely available, there is a need for more methods to be tested by different laboratories covering a range of analytical capabilities.

The legislative and contractual requirements with respect to analytical data quality that laboratories undertaking food regulatory work are required to observe are outlined in Chapter 5. These requirements apply to all analyses *including the determination of total trace element concentrations and their chemical species*. In the future, legislation will favour a method performance based approach where specific criteria must be met for the method to be considered valid, and this will be applicable to the speciation area. Thus the systematic approach exemplified by some workers for example[7,8] is of considerable benefit when developing methods.

5 Conclusions

The study of human nutrition and health requires more information about trace element species in food. Proper dissemination of accurate information gives consumers informed choice, it allows industry (see Chapters 14 and 15 for food industry viewpoints), to make the best use of foods, supplements and processing (which might alter chemical species), and it provides governments with the basis for good advice and legislation. Therefore, speciation studies in foods must be multi-disciplinary and not carried out in isolation. The policies which support the need for research, the subsequent project planning and analysis, and the interpretation and dissemination of information all require expertise in different subjects. As the techniques available become more sophisticated and almost certainly more expensive, further developments will depend not only upon collaboration between researchers and regulators but between countries too and projects such as the EC Speciation 21 Network have facilitated this process.

6 References

1. H.M. Crews, *Spectrochimica Acta Part B*, 1998, **53B**, 213.
2. S.J. Fairweather-Tait and R.F. Hurrell, *Nutr. Res. Rev.*, 1996, **9**, 295.
3. L. Hallberg and E. Bjorn-Rasmussen, *Scand. J. Haemotol.*, 1972, **9**, 193.
4. J.D. Cook, S.A. Dassenko and S.R. Lynch, *Am. J. Clin. Nutr.*, 1991, **54**, 717.
5. L. Hulten, E. Gramatouski, A. Gleerup and L. Hallberg, *Eur. J. Clin. Nutr.*, 1995, **49**, 794.
6. British Nutrition Foundation, *Iron: Nutritional and Physiological Significance*, Chapman and Hall, London, 1995.
7. C. Casiot, J. Szpunar, R. Lobinski and M. Potin-Gautier, *J. Anal. At. Spectrom.*, 1999, **14**, 645.
8. C. Casiot, V. Vacchina, H. Chassaigne, J. Szpunar, M. Potin-Gautier and R. Lobinski, *Anal. Commun.*, 1999, **36**, 77.
9. M. Kotrebai, M. Birringer, J.F. Tyson, E. Block and P.C. Uden, *Anal. Commun.*, 1999, **36**, 249.
10. A.C. Belfroid, M. Purperhart and F. Ariese, *Organotin Levels in Seafood in Relation to the Tolerable Daily Intake (TDI) for Humans*, Report No. E-99/12, Institute for Environmental Studies, The Netherlands, 1999.
11. A.H. Penninks, *Food Add. Contam.*, 1993, **10**, 351.
12. M.G. Cherian and R.A. Goyer, in R.A. Goyer and M.G. Cherian, eds., *Toxicology of Metals: Biochemical Aspects*, Springer-Verlag, Germany, 1995.

CHAPTER 14

Trace Metal Speciation – A View from Inside the Food Industry

SIMON BRANCH

1 Introduction

All industries conduct research and development for similar reasons, a non-exhaustive list including:

- to ensure products comply with legislation
- to gain competitive advantage
- to maintain or improve product quality
- to develop new products
- to improve production processes.

At present trace metal speciation does not obviously confer any of the above benefits, although there are exceptions, *e.g.* there are some examples of speciation legislation and determination of species may be relevant to ensuring product safety. However, the absence of the stated benefits means that most people working within the manufacturing and retailing components of the food industry will have little awareness of trace metal speciation. Indeed, for most technical personnel the term 'speciation' will be presumed to refer to the biological genus of a raw material, *i.e.* is it beef or lamb?, mandarin or tangerine? Of course this is not true for the largest multinationals who have sufficient resources to employ specialists, nor for legislators and regulatory bodies within the industry, many European representatives of which have participated in the Speciation 21 project.

A further consideration is that most small and medium sized companies will not have the analytical tools to conduct speciation determinations at meaningful levels. The United Kingdom Arsenic in Food Regulations[1] set a general limit of $1\,mg\,kg^{-1}$ arsenic in foodstuffs. Assuming that ingredients and finished products are at or below these levels, an inductively coupled plasma-optical emission spectrometer (ICP) or an inductively coupled plasma-mass spectrometer (ICP-MS) will be required to determine the species present. Most quality control laboratories are unlikely to be equipped with anything beyond FAAS or possibly ETAAS, the latter of which does not lend itself to speciation. In fact surveillance by the UK Ministry of Agriculture, Fisheries and Food[2] has shown that most samples contain less than $0.02\,\text{mg kg}^{-1}$, yielding an average daily intake of less than 81 μg arsenic, well inside guideline intakes. Since these samples comply with the existing legislation, there is no compulsive reason for industry to conduct speciation measurements.

A review of speciation publications is instructive. *The Analyst*[3] carried papers from the 3rd International Symposium on Speciation in Toxicology, Environment and Biological sciences. A total of 28 papers were published, of which 18 originated from academia, five from national institutes and five from, what might loosely be called, government bodies and regulators. Not a single paper came from within any industry, let alone the food industry. Faced with this apparent indifference, no clear commercial benefit and no legal necessity, one must ask if industry has any role or any need to pay the subject of speciation any heed. The view of this author is unequivocally yes, for the following reasons.

2 Reasons for Speciation in Industry

2.1 Legislation Based on Speciation

Firstly, speciation legislation is inevitable, although probably not for a number of years. The Essential Packaging Requirements (1998)[4] are the first such example, for which it is necessary to determine chromium(VI), lead, cadmium and mercury in packaging intended for food. This legislation requires that packaging should not contain a cumulative total in excess of $100\,\text{mg kg}^{-1}$ of the four elements. Such high levels allow a simple approach to determining the Cr^{VI}, namely to determine Cr, since if the total of this element plus Cd, Pb and Hg is less than $100\,\text{mg kg}^{-1}$ there is no necessity to determine Cr^{VI}. This approach is justified since the guidance notes accompanying the legislation fail to identify an appropriate methodology. It is ironic that with the wealth of papers on speciation, no method to allow compliance with the first Europe-wide legislation has been published and subsequently validated. The UK guidance notes on implementation suggest methodology based on US EPA methods 3060 and 7196 or 7199 – methods intended for sewage sludge and similar waste!

Although there are no EU regulations governing trace metal species in food itself, the EU has issued draft maximum residue limits for lead and cadmium in food. Subsequent speciation legislation is inevitable, though, unless there is a sudden change in knowledge on toxicity or benefit, it may be many years until it appears.

2.2 Dietary Supplementation and Fortification of Foods

Secondly, there are moves in the UK to follow the lead of countries such as Finland and increase the range of fortified foods available. Specifically, there has been concern that the UK diet has become selenium deficient.[5] The calculated dietary intake of selenium was 34 μg day^{-1} compared with a reference nutrient intake of 75 and 60 μg day^{-1} for men and women, respectively. The authors concluded that the move from Canadian to European wheat for breadmaking purposes had led to a decline in UK intakes. By comparing their results with previous work they found that levels in UK bread flour had dropped significantly in the period 1974–1993. These findings were supported by the 1995 UK total diet study which found estimated daily intakes in the range 29–39 μg.[6] These findings were reviewed by the UK Committee on Medical Aspects of Food who concluded that there was no evidence of adverse health effects due to selenium deficiency, but that more research was required on whether current intakes are adequate.

The apparent decline in selenium intakes has led to suggestions that selenium in the diet should be supplemented, not just in the UK but in most of Europe.[7] Since speciation plays a key role in bioavailability,[8] the food industry needs to start evaluating appropriate carriers and species in the event of supplementation becoming mandatory.

2.3   Existing Speciation Measurements

Thirdly, there is already some evidence of speciation measurements, depending on how loosely speciation is defined:

- The bread and flour regulations[9] require that most wheat flours contain 1.65 mg 100 g^{-1} of iron, the iron to be added in the form of ferric ammonium citrate, green ferric ammonium citrate, ferrous sulfate or iron powder. Compliance with the regulations demands knowledge of the form of the iron.
- To determine the efficacy of sulfur dioxide as a preservative it is necessary to measure both free and bound SO_2. Suitable methods have been described in detail.[10]
- It is now widely recognised that nitrates and nitrites have the potential to form nitrosamines. Regulations to control the levels of sodium nitrate and nitrite in cured meats and cheeses were introduced in recognition of this[7] and there has been considerable method development to ensure compliance.
- Sodium performs many functions in foods, depending on the species present. Sodium glutamate is a flavour enhancer, sodium nitrate, nitrite and chloride are preservatives and sodium bicarbonate and acid sodium pyrophosphate are raising agents. Determination of sodium alone tells the food scientist nothing about the form or functionality of the sodium-containing molecule present. It is impractical to hope that methods for simultaneous determination of these species can be developed, but suitable low-cost specific methods should be feasible.

3   Conclusion

In conclusion, speciation measurements exist within the food industry, normally being specific to one analyte and usually for commercial or functional purposes.

The analytical methods have evolved over many years and if the industry is to be ready to meet legislative requirements then governments, regulators and the industry have to work in conjunction now. Critically, methods have to be developed to suit the instrumentation found in quality control laboratories, not state-of-the-art equipment found only in the largest laboratories. Without these developments, speciation will remain only within the domain of the largest multinationals and university research departments.

4 References

1. *The Arsenic in Food (Amendment) Regulations*, 1973, Statutory Instrument No. 1052, HMSO, London.
2. *Survey of Arsenic in Food*, Ministry of Agriculture, Fisheries and Food, HMSO, 1982.
3. *Analyst*, 1998, **123**, 5.
4. *The Essential Packaging Requirements* (1998) *Regulations*.
5. M.N.I. Barclay, A. MacPherson and J. Dixon, *J. Food Comp. Anal.*, 1995, **8**, 307.
6. *Food Safety Information Bulletin No* 89, Ministry of Agriculture, Fisheries and Food, 1997.
7. M.P. Rayman, *Br. Med. J.*, 1997, **314**, 387.
8. J.A. Vinson *et al.*, *Relative Bioavailability of Trace Elements and Vitamins Found in Commercial Supplements*, in *Nutrient Availability: Chemical and Biological Aspects*, D.A.T. Southgate, I.T. Johnson and G.R. Fenwick, eds., Royal Society of Chemistry, Cambridge, 1989.
9. *The Bread and Flour Regulations*, 1984, SI No. 1304.
10. *Pearson's Composition and Analysis of Foods*, R.S. Kirk and R. Sawyer, eds., Longman Scientific and Technical, 1991, pp. 70–75.
11. *Preservative in Food Regulations*, 1989.

CHAPTER 15

Trace Element Speciation in Food: A Tool to Assure Food Safety and Nutritional Quality

PETER VAN DAEL

1 Introduction

The safety and nutritional quality of food are determined by both the total level and the speciation, *i.e.* the chemical form(s), of trace elements in foods.[1] The utilisation by living organisms of essential trace elements, for example Se and Zn, and potentially toxic trace elements, such as As and Hg, is determined by their total levels and speciation in food.[1] Dietary requirements and upper safe intake levels for essential trace elements, and tolerable dietary intake levels for non-essential, potentially toxic, trace elements, have been issued by food authorities.[2-4] Since trace element bioavailability is highly variable between chemical forms, information on trace element speciation, as well as total levels in food, is essential to ensure food safety and to determine nutritional quality. Trace element speciation can provide a better understanding on how the absorption and bioavailability of elements can be reduced for toxic elements or improved for essential nutrients. As a consequence, the validating of present risk assessment procedures may be undertaken. Therefore, the food industry and regulatory authorities have a mutual interest in speciation in foods and its implications. To achieve these objectives, speciation in foods requires accurate, reproducible and precise methodologies, including well-documented procedures for food sampling and sample pre-treatment which maintain the initial species intact. Although speciation research has boomed during the last decade, speciation methodology and its applications are still in a developmental stage, which is reflected in the limited number of reference materials available for method validation and the virtual lack of standard procedures.

2 Speciation and Food Safety

The speciation of trace elements, in particular non-essential trace elements, to ensure food safety, may be considered at present to be the main topic of interest in speciation methodology for the food industry and regulatory authorities. The total level of trace elements in foods is no guarantee of food safety, since chemical forms often differ considerably in their adverse effects, which can range from essential or non-toxic to highly toxic. As a consequence, speciation is becoming an increasingly important tool in defining and assuring food safety. The speciation of essential and non-essential, potentially toxic, trace elements, and its applications to food, will be outlined using the examples of As, Cr and Hg.

2.1 Arsenic

Arsenic is a non-essential, potentially toxic, trace element to man. The toxicity of high levels of As compounds is well known, and a wide variety of analytical procedures have been reported to determine inorganic and organic As species in water and in biological matrices, together with methods for total As determination. Inorganic As compounds are more toxic than organic As species, and the following descending order has been established for the toxicity of As compounds: elemental As, As^{3+}, As^{5+}, monomethylarsonic acid (MMA), dimethylarsinic acid (DMA) and trimethylarsine oxide (TMAO).[5] In general the toxicity decreases as the degree of methylation increases. Arsenic is concentrated in the marine food chain and total As levels are very high in marine organisms such as fish and crustaceans.[6] However, several of the organic As species found in seafood, such as arsenobetaine and arsenocholine, are considered to be non-toxic.[5] Arsenobetaine has been identified as the major As species in a variety of seafood products, including several species of clams[7] and many species of fish.[8-10] Arsenobetaine is thought to be the final metabolite of As in food chains,[11] although it is not present in all fish species,[12] and the transformations that As undergoes in marine food chains are still being studied.[13-14] Therefore food safety assessment requires data on total As levels, as well as on the As species present in a specific food.

During the past decade, a number of methods have been developed for As speciation in water and biological matrices, many of which involved coupling a gas- or liquid-chromatographic technique with an element specific-detector.[11-20] Initially AAS and ICP atomic emission spectrometry (AES) were the most popular element specific detectors.[15-17] In dealing with volatile As species and other species that can be derivatised into corresponding volatile arsines, cold trapping of arsines is used. A significant comparative study of various coupled chromatography–atomic spectrometry methods for As has been published.[23] Both GC and HPLC were used as methods for separation, and the detectors were based on FAAS, flame atomic fluorescence spectrometry (FAFS) and ICP-AES. The conclusions were: (1) that hydride generation with cryogenic trapping with GC–FAAS was the most sensitive of the methods studied and (2) that HPLC separation, followed by hydride generation (HG) and FAAS, was the simplest method for higher levels of As compounds. A critical evaluation of the HG–cold trapping FAAS (with a silica furnace) method for As species, was reported by Van Cleuvenbergen *et al.*[24] The parameters of the method were optimised for best sensitivity and comparisons were made with other methods.

In recent years, ICP-MS has become a more popular detection technique for As speciation, coupled to either HPLC or capillary electrophoresis.[17–22] The simultaneous analysis of major As species, namely As^{3+}, As^{5+}, MMA, DMA, arsenobetaine, trimethylarsine oxide, arsenocholine and the tetramethylarsonium ion, has been performed by coupled anion- and cation-exchange HPLC–ICP-MS.[22]

Certified reference materials (CRMs) for As species are necessary for validation of the entire speciation procedure, *i.e.* from sample pre-treatment to analysis, in order to enable accurate safety assessments of As species in food. Recently As species present in two dogfish reference materials of the National Research Council of Canada DORM 1 and DORM 2 have been determined.[22]

Hyphenated speciation techniques offer adequate selectivity and sensitivity for As speciation in food and drinking water. The further development of As speciation, together with increasing knowledge of the toxicity of different As species, will definitely lead to more stringent safety norms for As levels in food, including norms for levels of As species. Access to commercial certified reference materials, to validate As speciation data and to standardise speciation methodology, is indispensable if regulatory guidelines for tolerable and safe levels of As, and its species in food, are to be established and implemented.

2.2   Chromium

Chromium is an essential trace element for man and animal, but in view of the toxic properties of Cr^{6+}, as compared to the less toxic Cr^{3+}, Cr speciation has attracted scientific and regulatory interest.[25] As a consequence, differentiation between Cr^{3+} and Cr^{6+} is important for assessing food safety. Most of the studies have reported Cr speciation in water samples.[26] Total Cr levels have been determined, mostly by AAS.[25] Initial speciation work was performed using ion-exchange, solvent extraction and electrochemical methods, with a two-stage determination.[27–30] Simultaneous determination of Cr^{3+} and Cr^{6+} has also been achieved, using reversed-phase ion-pair HPLC with UV detection, after complexation of Cr^{3+} with EDTA.[25] Hyphenated techniques, such as HPLC coupled to AAS, ICP-AES or ICP-MS have been used more recently for Cr speciation.[31–36]

Current hyphenated speciation techniques offer sufficient selectivity and sensitivity for Cr speciation in drinking water. Further development of Cr speciation will definitely result in safety norms for Cr species in drinking water and in foods. At present commercial certified reference materials and validated Cr speciation methods are lacking. These are necessary in order to establish and implement regulatory guidelines for tolerable levels of Cr species in drinking water and food.

2.3 Mercury

Mercury is a non-essential, potentially toxic, trace element for living organisms. The difference in toxicity between elemental, organic and inorganic Hg species has resulted in a continuous interest in Hg levels and species in environmental and biological materials. Methylmercury (MeHg), one of the most dangerous environmental pollutants, acquired notoriety through the Minamata disaster in Japan, which, in an unfortunate way, demonstrated the high risk related to the consumption by man of seafood containing toxic levels of MeHg. Biomethylation of Hg by marine organisms and its consequent accumulation in seafood[37] can generate a serious food safety problem, as fish have the capacity to concentrate Hg by a factor of 10^5–10^7. This can lead to dangerous levels in seafood, even in areas with tolerable aquatic Hg^{2+} levels.[38] Guidelines for the presence of Hg derived from MeHg in seafood have been established; the US Food and Drug Administration set a guideline for MeHg in seafood at $1\,\mu g\,g^{-1}$ (on edible portion or wet mass).[39] Consequently, suitable analytical methodology for routine Hg and MeHg analysis by control laboratories has been developed.

The AOAC method[40] and species-selective analysis based on hydride generation[41-43] (or ethylation),[44-46] followed by purge-and-trap thermal desorption cold vapour (CV) AAS,[41,44] CVAFS,[45] FT-IR,[42] MIP–AES,[43] or furnace atomisation plasma emission spectrometry,[46] are the most widely used methods for the determination of MeHg and Hg^{2+} species. Detection limits of 20 and 80 ng g^{-1} dry mass for MeHg and Hg^{2+} respectively have been reported, and are sufficiently sensitive to comply with food safety assessments.[40-46]

Reference materials are available [*e.g.* CRM 464 (tuna fish), DORM-1 (dogfish muscle)] to validate the analytical methodology for MeHg speciation in seafood. Both hyphenated techniques and more classical speciation methods offer sufficient selectivity and sensitivity for Hg speciation in food in order to comply with quality control regulations for seafood.

3 Speciation and Nutritional Food Quality

The total level of essential trace elements present in a particular food is an important determinant of its nutritional value. In many countries, this information is given on food labels as part of the communication of nutritional information to the consumer of the food. However, in addition to total nutrient levels, the nutritional value depends largely on the bioavailability of the nutrient and, in the case of trace elements, on their chemical form(s) present in the food.[47] Speciation of essential trace elements in food is an important parameter in assessing and improving the nutritional quality of food but, unlike non-essential, toxic trace elements, not yet associated with food regulation. It is likely that food regulators will issue guidelines in the near future, to achieve desirable bioavailability, and these will probably rely upon speciation data to assess the nutritional quality of food. However, these recommended guidelines for nutritional quality are less stringent than the values established for safety evaluation for potential toxic elements. In the area of food supplements, such as trace element-enriched yeast or bacteria, trace element speciation may be required in order to assess and certify the nutritional value, identified by the required trace element level and species. The nutritional implications and applications of trace element speciation in food will be discussed for Se.

3.1 Selenium

Selenium is an essential trace element for man and animals.[4] It is an integral part of the seleno-enzymes glutathione peroxidase and iodothyronine deiodinase and has also been demonstrated in several other so-called seleno-proteins, for which the biological functions are still to be elucidated.[48] Dietary requirements for Se have been set by food authorities.[2-4] At high Se intakes, toxicity has been reported and an upper safe intake level has been established at 400 μg day^{-1} for adults.[4] The Se status is determined by dietary Se intake and the bioavailability of dietary Se.[49] The bioavailability of Se, defined as the amount of dietary Se utilised in biochemically active compounds, and the toxicity of Se depends on the Se species present in the diet.[49] Selenium found in food may be inorganic, *e.g.* selenite (Se^{4+}) and selenate (Se^{6+}), or organic compounds, such as sele-nomethionine (Se-Met) and selenocysteine (Se-Cys).[49] Selenium absorption has been shown to be higher from organic Se species compared with inorganic Se compounds.[49-51] Se retention has been reported to be higher for organic Se, which is a direct consequence of differences in the metabolism of inorganic and organic Se species.[49-51] Therefore Se speciation and distribution in foods is important to improve our understanding of Se utilisation by man and animals.

The analytical techniques applied to determine total Se levels have been reviewed.[52] Fluorimetric, AAS and ICP-MS methods are the more widely used for total Se analysis. Research on Se speciation analysis has been rapidly developing during the last decade. Two recent reviews document methods, distinguishing between (1) non-volatile and volatile Se species,[53] mainly applied to environmental and geological samples, and (2) inorganic and organic Se species,[54] related to speciation research in biological matrices and food. In view of the objective of this chapter, only the latter methods will be discussed.

Speciation of Se^{4+} and Se^{6+} has been achieved by HPLC coupled to conductimetry,[55,56] UV spectrometry,[57,58] HG–AAS,[59] ICP-AES[60] or ICP-MS.[61,62] More recently, the coupling of CE to ICP-MS showed very promising results.[63] Determination of Se^{4+} and Se^{6+} levels by differential analysis of total Se and Se^{4+}, the more reactive inorganic Se compound, has also been utilised for the Se speciation of water. Based on the reactivity of Se^{4+} and using specific reduction reactions, both inorganic species have been differentiated by GC–ECD,[64,65] fluorimetry,[66,67] HG–AAS,[68-70] HG–AES,[71] HG–ICP-AES,[72] ICP-MS[73] and HG–ICP-MS.[74] Modern techniques, especially those involving ICP-MS, are able to determine Se^{4+} and Se^{6+} at picogram levels in water.[62,63,73]

Speciation of organic Se compounds has been rapidly growing in recent years due to the availability of hyphenated speciation techniques.[53] Speciation of organic species such as Se-Met has been reported, using HPLC coupled to

AFS,[75] ICP-MS,[76,77] or HG–ICP-MS,[78] and CE coupled to ICP-MS. Suzuki *et al.*[77] reported that HPLC–ICP-MS can be used as a specific and sensitive technique for the speciation of Se-containing biological constituents.

Speciation of Se and its distribution in foods has been reported for selected food items.[80,81] Selenium has been identified as Se-Met in soyabean lectin[80] and associated with soyabean proteins and globulin polypeptides.[81] Olson *et al.* reported that Se-Met was a major fraction ($\sim$ 50%) in the gluten hydrolysate of high Se wheat.[82] Five Se species, namely Se^{4+}, Se^{6+}, Se-Met, Se-Cys and Met-Se-Cys, as well as several unknown Se compounds, were isolated from Se-enriched garlic, onion and broccoli by HPLC–ICP-MS.[83] Se-Met was identified as a major Se compound in Se-rich yeast, which is often used as a Se supplement.[84,85] Se-Met was also reported to be responsible for the high Se utilisation from Se-rich yeast. However, the amount of total yeast Se as Se-Met represented only 30%[84] to 50%,[85] and it has been suggested that the levels may vary between different Se-yeast products or production batches, consequently influencing the nutritional utilisation of yeast Se.[86] In order to assure the nutritional value of Se-rich yeast, data on the percentage of total Se present as Se-Met may be necessary. Calomme *et al.* characterised the bacterial seleno-compound of *L. delbrueckii* subsp. *bulgaricus* as Se-Cys.[87] In milk, Se was shown to be mainly associated (> 95%) with milk proteins for human,[88,91] cow's,[88,89,91] goat's[88,90] and sheep's milk,[91] and only a small fraction of total milk Se was found in the lipid fraction.[88-91] The Se distribution in protein fractions was different between human and cow's milk, and this has been linked to the better utilisation of Se in human milk.[88,91] Se bioavailability from fish was reported to be low and therefore Se distribution studies in different fish have been conducted, in order to evaluate the association between Se species and Se bioavailability.[92] Crews *et al.*[92] investigated the Se speciation in cooked cod by HPLC–ICP-MS and found that approximately 12% of the total Se was present as Se^{4+}, whilst the remaining Se species, although unidentified, were considered to be organic, in view of their elution characteristics.

Current speciation techniques for Se species offer good selectivity and sensitivity for Se speciation and Se distribution in drinking water and food. The further development of Se speciation will definitely result in quality control norms for Se species levels in drinking water and for the forms or distribution in food. Food Se distribution data can assist in the development of specific functional ingredients and optimisation of their nutritional value. Furthermore speciation of Se in foods will allow us to improve our understanding of Se utilisation and metabolism. At present, no specific certified reference materials are available to support and validate Se speciation. The progress of Se speciation work would definitely benefit from the availability of such materials.

4   Conclusion

During the last decade, speciation research has been growing rapidly. The application and importance of trace element speciation to evaluate and assure safety and nutritional quality of food is now well recognised. Speciation methodology for trace element speciation in general, and in food in particular, has been developed or can be adapted, in order to evaluate food safety and quality. However, validated speciation methods and standard reference materials are required in order to implement trace element speciation as a tool to assure food safety.

5 References

1. H.M. Crews, *Spectrochim. Acta Part B*, 1998, **53**, 213.
2. National Research Council, *Dietary Recommendations*, 10th Edn., National Research Council, Washington, DC, 1989.
3. European Commission, *Scientific Report on Human Nutrition – 31st Series*, Luxembourg, 1994.
4. World Health Organisation, *Trace Elements in Human Nutrition and Health*, Geneva, 1996.
5. T. Kaise and S. Fukui, *Appl. Organomet. Chem.*, 1992, **6**, 155.
6. K.A. Francesconi and J.S. Edmonds, *Oceanogr. Mar. Biol. Rev.*, 1993, **31**, 111.
7. W.R. Cullen, and M. Dodd, *Appl. Organomet. Chem.*, 1989, **3**, 79.
8. E.H. Larsen, G. Pritzl and S.H. Hansen, *J. Anal. At. Spectrom.*, 1993, **8**, 557.
9. D. Beauchemin, M.E. Bednas, S.S. Berman, J.W. McLaren, K.W.M. Siu and R.E. Sturgeon, *Anal. Chem.*, 1988, **60**, 2209.
10. E.H. Larsen, G.A. Pedersen and J.W. McLaren, *J. Anal. At. Spectrom.*, 1997, **12**, 963.
11. W. Goessler, W. Maher, K.J. Irgolic, D. Kuehnelt, C. Schlagenhaufen and T. Kaise, *Fresenius' J. Anal. Chem.*, 1997, **359**, 434.
12. J.S. Edmonds, Y. Shibata, K.A. Francesconi, R.J. Rippingale and M. Morita, *Appl. Organomet. Chem.*, 1997, **11**, 281.
13. K. Hanaoka, T. Motoya, S. Tagasawa and T. Kaise, *Appl. Organomet. Chem.*, 1991, **5**, 427.
14. W.R. Cullen, L.G. Harrison, H. Li and G. Hewitt, *Appl. Organomet. Chem.*, 1994, **8**, 313.
15. L. Ebdon, S.J. Hill and R.W. Ward, *Analyst*, 1986, **111**, 1113.
16. L. Ebdon, S.J. Hill and R.W. Ward, *Analyst*, 1987, **112**, 1.
17. O. Muñoz, D. Vélez, M.L. Cervera and R. Montoro, *J. Anal. At. Spectrom.*, 1999, **14**, 1607.
18. S.C.K. Shum, R. Neddersen and R.S. Houk, *Analyst*, 1992, **117**, 577.
19. G.E.M. Hall, J.C. Pelchat and G. Gauthier, *J. Anal. At. Spectrom.*, 1999, **14**, 205.
20. K.L. Ackley, C. B'Hymer, K.L. Sutton and J.A. Caruso, *J. Anal. At. Spectrom.*, 1999, **14**, 845.
21. M.L. Magnuson, J.T. Creed and C.A. Brockhoff, *Analyst*, **122**, 1057.
22. W. Goessler, D. Kuehnelt, C. Schagenhaufen, Z. Sljkovec and K.J. Irgolic, *J. Anal. At. Spectrom.*, 1998, **13**, 183.
23. L. Ebdon, S.J. Hill, A.P. Walton and R.W. Ward, *Analyst*, 1988, **113**, 1159.
24. R. Van Cleuvenbergen, W.E. Van Mol and F.J. Adams, *J. Anal. At. Spectrom.*, 1988, **3**, 169.
25. J.C. Van Loon and R.R. Barefoot, *Analyst*, 1992, **117**, 563.
26. J.F. Jen, G.L. Ou-Yang, C.-S. Shen and S.-M. Yang, *Analyst*, 1993, **118**, 1281.
27. J.F. Pankow, D.P. Leta, J.W. Lin, S.E. Out, W.P. Shum and G.E. Januuer, *Sci. Total Environ.*, 1977, **7**, 17.

28. G.E. de Jong and U.A.Th. Brinkman, *Anal. Chim. Acta*, 1978, **98**, 243.
29. M.H. Hushmi, R.A. Ahmed, A.A. Ayaz and F. Azam, *Anal. Chem.*, 1965, **37**, 1027.
30. R.M. Issa, B.A. Abdel-Nabey, and H. Dadek, *Electrochim. Acta*, 1968, **13**, 1827.
31. I.S. Krull, D. Bushee, R.N. Savage, R.G. Schleicher and B. Smith, Jr., *Anal. Lett.*, 1982, **15**, 267.
32. I.S. Krull, K.W. Panara and L.L. Gershman, *J. Chromatogr. Sci.*, 1983, **21**, 460.
33. A.G. Cox, I.G. Cook and C.W. McLeod, *Analyst*, 1985, **110**, 331.
34. S. Ahmad, R.C. Murthy and S.V. Chandra, *Analyst*, 1990, **115**, 331.
35. W.-L. Fong and J.-C. Wu, *Spectrosc. Lett.*, 1991, **24**, 931.
36. K.E. Collins, C.H. Collins, M.E. Dequeiroz, P.S. Bonato and C. Archundia, *Chromatographia*, 1988, **26**, 160.
37. Förstner and G.T.W. Wittmann, *Metal Pollution in the Aquatic Environment*, Springer, Heidelberg, 1983, p. 18.
38. K. Reisinger, M. Stöppler, and H.W. Nürnberg, *Fresenius' Z. Anal. Chem.*, 1983, **316**, 612.
39. W. Holak, *J. Assoc. Off. Anal. Chem.*, 1989, **72**, 926.
40. AOAC, *Official Methods of Analysis of the Association of Official Analytical Chemists*, 1995.
41. C.M. Tseng, A. De Diego, F.M. Martin, D. Amouroux and O.F.X. Donard, *J. Anal. At. Spectrom.*, 1997, **12**, 743.
42. M. Filippeli, F. Baldi, F.E. Brinckman and G.J. Olson, *Environ. Sci. Technol.*, 1992, **26**, 1457.
43. C. Gerbersmann, M. Heisterkamp, F.C. Adams and J.A.C. Broekaert, *Anal. Chim. Acta*, 1997, **350**, 273.
44. R. Fisher, S. Rapsomanikis and M.O. Andrea, *Anal. Chem.*, 1993, **65**, 763.
45. L. Liang, M. Horvat, E. Cernichiari, B. Gelein and S. Balogh, *Talanta*, 1996, **43**, 1883.
46. M.S. Jimenez and R.E. Sturgeon, *J. Anal. At. Spectrom.*, 1997, **12**, 597.
47. M.J. Jackson, S.J. Fairweather-Tait, H. van den Berg and W. Cohn, *Assessment of Bioavailability of Micronutrients – Proceedings of an ILSI Europe workshop, Eur. J. Clin. Nutr.*, 1997, **51**, Supplement 1.
48. B.A. Zachara, *J. Trace Elem. Health Dis.*, 1993, **6**, 137.
49. S.J. Fairweather-Tait, *Eur. J. Clin. Nutr.*, 1997, **51**, S20.
50. S.C. Vendeland, J.A. Butler and P.D. Whanger, *J. Nutr. Biochem.*, 1992, **3**, 359.
51. A.R. Mangels, P.B. Moser-Veillon, K.Y. Patterson and C. Veillon, *Am. J. Clin. Nutr.*, **52**, 621.
52. R.M. Olivas, O.F.X. Donard, C. Camara and Ph. Quevauviller, *Anal. Chim. Acta*, 1994, **286**, 357.
53. K. Pyrzynska, *Analyst*, 1996, **121**, 77R.
54. X. Dauchy, M. Potin-Gautier, A. Astruc and M. Astruc, *Fresenius' J. Anal. Chem.*, 1994, **348**, 792.
55. K.F. Nieto and W.T. Frankenberger, Jr., *Soil Sci., Soc. Am. J.*, 1985, **49**, 592.
56. U. Karlson and W.T. Frankenberger, Jr., *Anal. Chem.*, 1986, **58**, 2704.
57. R.G. Gerritse and J.A. Adeney, *J. Chromatogr.*, 1985, **347**, 419.
58. S. Goyal, A. Hafez and D.W. Rains, *J. Chromatogr.*, 1991, **537**, 269.
59. D.R. Roden and D.E. Tallman, *Anal. Chem.*, 1982, **54**, 307.
60. F. Nakata, H. Sunahara, H. Matsuo and T. Kumamaru, *Bunseki Kagaku*, 1986, **5**, 439.
61. J.J. Thomson and R.S. Houk, *Anal. Chem.*, 1986, **58**, 2541.
62. S.C.K. Shum and R.S. Houk, *Anal. Chem.*, 1993, **65**, 2972.
63. M.L. Magnuson, J.T. Creed and C.A. Brockhoff, *Analyst*, 1997, **122**, 1057.
64. J.K. Irgolic, R.A. Stockton, D. Chakraborti and W. Beyer, *Spectrochim. Acta Part B*,

1983, **38**, 437.
65. C.I. Measures and J.D. Burton, *Anal. Chim. Acta*, 1980, **120**, 177.
66. Y. Nakaguchi, K. Hirachi, Y. Tamari, Y. Fukunaga, Y. Nihikawa and T. Shigematsu, *Anal. Sci.*, 1985, **1**, 247.
67. Y. Tamari, R. Hirai, H. Tsuji and Y. Kusaka, *Anal. Sci.*, 1987, **3**, 313.
68. M.Q. Yu, G.Q. Liu and Q. Jin, *Talanta*, 1983, **30**, 265.
69. J.L. Fio and R. Fujii, *Soil Sci. Soc, Am. J.*, 1990, **54**, 363.
70. D. Mayer, S. Haubenwaallner and W. Kosmus, *Anal. Chim. Acta*, 1992, **268**, 315.
71. Y. He, J. Moreda-Piñeiro, M.L. Cervera and M. de la Guardia, *J. Anal. At. Spectrom.*, 1998, **13**, 289.
72. L.R. Parker, N.H. Thioh and R.M. Barnes, *Appl. Spectrosc.*, 1985, **30**, 45.
73. N. Jakubowski, C. Thomas, D. Stuewer, I. Dettlaff and J. Schram, *J. Anal. At. Spectrom.*, 1996, **11**, 1023.
74. W.T. Buckley, J.J. Budac and D.V. Godfrey, *Anal. Chem.*, 1992, **64**, 724.
75. E. Puskel, Z. Mester and P. Fodor, *J. Anal. At. Spectrom.*, 1999, **14**, 973.
76. S.M. Bird, H. Ge, P.C. Uden, J.F. Tyson, E. Block and E. Denoyer, *J. Chromatogr. A*, 1997, **789**, 349.
77. K.T. Suzuki, M. Itoh and M. Ohmichi, *J. Chromatogr. B*, 1995, **666**, 13.
78. J.M. González Lafuente, M. Dlaska, M.L. Fernández Sánchez and A. Sanz-Medel, *J. Anal. At. Spectrom.*, 1998, **13**, 423.
79. B. Michalke and P. Schramel, *Electrophoresis*, 1998, **19**, 270.
80. S.K. Sathe, A.C. Mason, R. Rodibaygh and M. Weaver, *J. Agric. Food Chem.*, 1992, **40**, 2084.
81. Z. Wang, S. Xie and A. Peng, *J. Agric. Food Chem.*, 1996, **44**, 2754.
82. O.E. Olson, E.J. Novacek, E.I. Whitehead and I.S. Palmer, *Phytochemistry*, 1970, 1181.
83. H. Ge, X.-J. Cai, J.F. Tyson, P.C. Uden, E.R. Denoyer and E. Block, *Anal. Comm.*, 1996, **33**, 279.
84. M. Korhola, A. Vainio and K. Edelmann, *Ann. Clin. Res.*, 1986, **18**, 65.
85. Z. Ouyang and J. Wu, *Biomed. Chromatogr.*, 1988, **2**, 258.
86. S. Moesgaard and R. Morrill, *The Need for Speciation to Realise the Potential of Selenium in Disease Prevention*, Chapter 18 in this book.
87. M.R. Calomme, K. Van den Branden and D.A. Vanden Berghe, *J. Appl. Bacteriol.*, 1995, **79**, 331.
88. B. Debski, M.F. Picciano and J.A. Milner, *J. Nutr.*, 1987, **117**, 1091.
89. P. Van Dael, G. Vlaemynck, R. Van Renterghem and H. Deelstra, *Z. Lebensm. Unters. Forsch.*, 1991, **192**, 422.
90. P. Van Dael, G. Vlaemynck, R. Van Renterghem and H. Deelstra, *Z. Lebensm. Unters. Forsch.*, 1992, **195**, 3.
91. P. Van Dael, *Comparative Study on the Distribution of Se in Cow's, Goat, Sheep and Human Milk*, PhD dissertation, University of Antwerp, Belgium 1992.
92. H.M. Crews, P.A. Clarke, D.J. Lewis, L.A. Owen, P.R. Strutt and A. Izquierdo, *J. Anal. Atom. Spectr.*, 1996, **11**, 1177.

CHAPTER 16

Arsenic Intake in the Basque Country (Spain): A Real Need for Speciation

I. URIETA, M. JALÓN AND M.L. MACHO

1 Introduction

The Basque Country is a small region in the north of Spain, 7261 km^2 in area with a population of 2.2 million. In 1990, the Basque Government initiated a Total Diet Study (TDS) (market basket approach) as an important part of its monitoring programme for chemical contaminants in the food supply[1] and this investigation pioneered this type of study in Spain. The primary purpose of the TDS is to obtain estimates of the average intake of constituents of concern. This provides a picture of the 'norm' and useful reference data are obtained. It also shows trends in intakes and, occasionally, gives information about unexpected sources of food contamination, checking the effectiveness of regulations and initiatives relative to levels of chemicals in foods. In addition, the results serve to guide other monitoring programmes. Apart from the estimation of the dietary intake of chemical substances, which is a basic aspect of the food chemical surveillance, 'selective controls' to determine residues of a particular contaminant in a particular food item are also conducted on a regular basis.

The design of the Basque Total Diet Study has been described in detail previously.[2] The types and quantities of foods that make up the average Basque 'Total Diet' are based on the results of a Food Survey carried out between 1988 and 1990.[3] This survey, based on a representative sample of the adult population (25–60 years), employed a '24 h recall' interview and an individual food frequency questionnaire ($n = 2348$). Information was collected on meals consumed inside and outside the home, and on alcoholic and non-alcoholic beverages.

The main features of the study (Figure 1) can be summarised as follows: using the information provided by the Food Survey, the average diet of the population was established. Then the food list was prepared and the 91 food items included were purchased at monthly intervals at different locations in the Basque

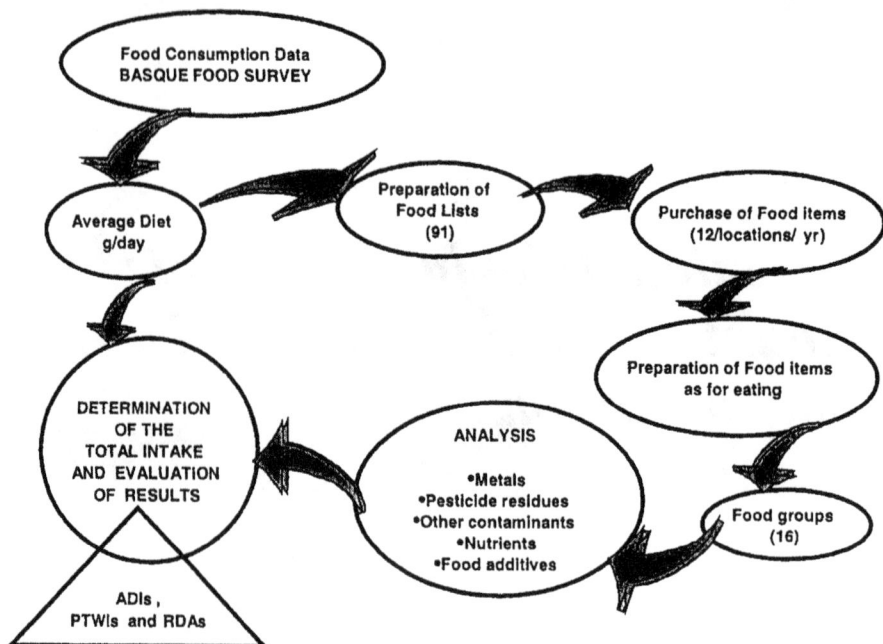

Figure 1   *Scheme of the Total Diet Study in the Basque Country*

Table 1   *Total diet study food groups*

Food group	Average intake weight (g/person/day)	% of total
Eggs	41	2
Meat	118	6
Meat products	45	2
Fish	89	4
Milk	294	15
Dairy products	58	3
Bread	122	6
Cereals	62	3
Pulses and nuts	27	1
Potatoes	90	5
Vegetables	159	8
Fruits	377	19
Sugar and preserves	34	2
Fats and oils	45	2
Non-alcoholic beverages	198	10
Alcoholic beverages	243	12
Total	2002	100

Energy content (kcal/person/day) 2485.

Country. After preparation and cooking, the foods were combined in groups (16 food groups) (Table 1) and analysed for the substances of interest. Finally, the intakes were calculated by a combination of these data with those of consumption, and compared with appropriate reference values and also with data from other countries.

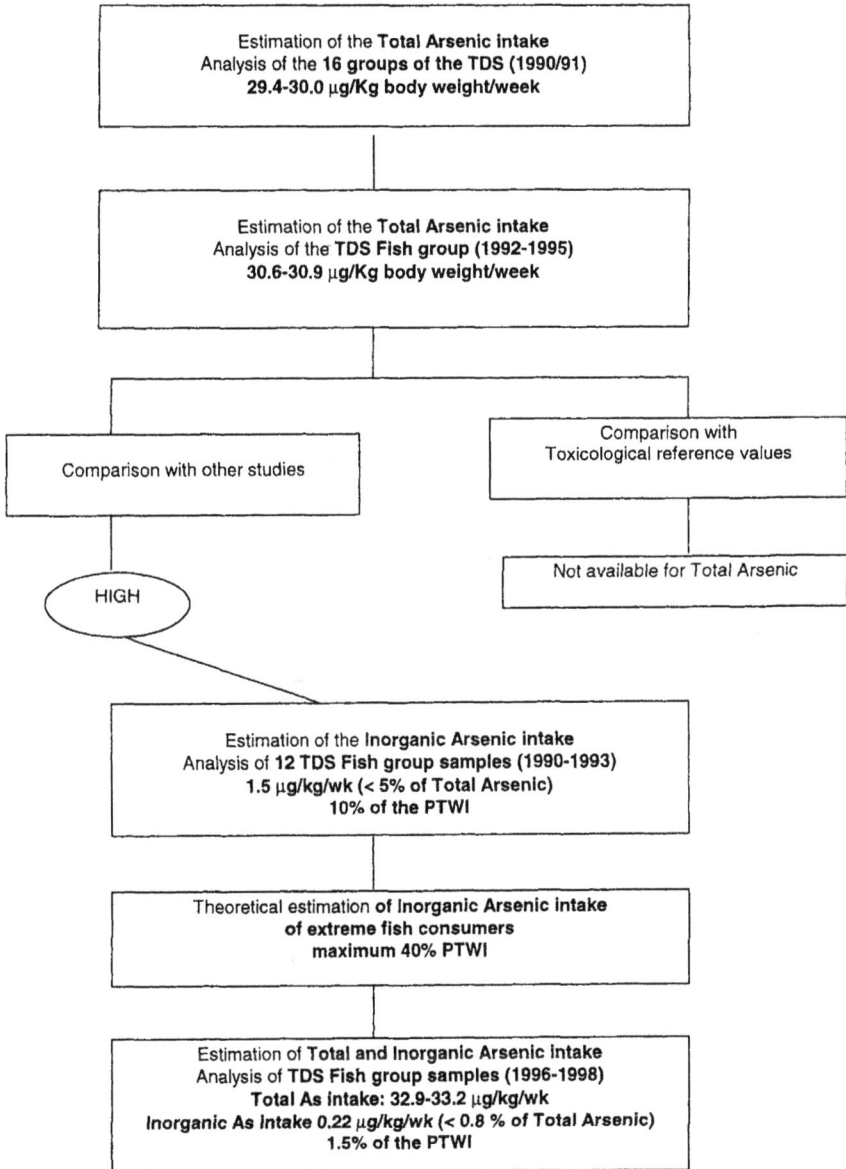

Figure 2 *Surveillance of the dietary intake of arsenic in the Basque Country, 1990–1998*

2 Dietary Intake of Total Arsenic

Total arsenic has been determined in TDS samples since 1990. Figure 2 shows the process followed to assess the health risk associated with arsenic exposure through the diet. Initially, the 16 food groups into which the total diet was divided were analysed. Values above the limit of quantification (LoQ) were only obtained in the 'fish' and 'alcoholic beverages' groups, but mean concentration levels in the latter group were very low $(4 \mu g \, kg^{-1})^4$ in comparison to those obtained for the fish group.

According to the recommendations of GEMS-Food,[5] as the proportion of results less than the LoQ exceeds 80%, two estimates have been produced using zero and the LoQ for all the results less than the LoQ. The fact that fish was the only group where significant concentrations above the LoQ were determined, together with the low LoQ established in the remaining food groups which gave a small difference between both intake estimates, led to the decision to determine arsenic only in this food group of the TDS in 1992. Since then, total arsenic has been determined continuously (12 samples per year) only in the TDS 'fish' group, the composition of which is shown in Table 2. Total arsenic intake has been estimated by adding $0.3 \mu g \, kg^{-1}$ body weight/week to that obtained by analysis of the fish group. This value was obtained by combining the appropriate LoQ in each food group with its consumption data when the analyses were carried out (1990–1991). The LoQ in the various food groups was lowered in 1991 and the appropriate value for each food group has been used since then.

The average total arsenic intake in the period 1990–1995, as shown in Table 3, lies in the range 30.2–30.5. All intakes were reported in $\mu g \, kg^{-1}$ body weight/week using a body weight of 68 kg, that being the average weight of the participants in the Food Survey.

Table 2 *Composition of the fish TDS group*

Fish species	Amount consumed (g/person/day)
Hake	7.6
Whiting	22.1
John Dory	7.2
Other white fish	11.8
Anchovies	6.9
Horse mackerel	3.2
Sardines	3.5
Other blue fish	5.3
Salted cod	7.0
Canned tuna/sardines	2.9
Squid	2.8
Mussels/Clams	5.6
Crustaceans	3.4
Total	89.3

Table 3 *Total arsenic in TDS samples 1990–1995*

	Concentration (fish group) ($\mu g\,kg^{-1}$)		Intake (fish group) ($\mu g\,kg^{-1}$ bw/week)	Total intake (< LoQ = 0 – < LoQ = LoQ) ($\mu g\,kg^{-1}$ bw/week)
	Mean	Range		
1990 n = 4	2800	2100–3300	25.6	(25.6–26.3)
1991 n = 12	3633	2200–5500	33.3	(33.3–33.6)
1992 n = 12	3008	2300–3400	27.6	(27.6–27.9)*
1993 n = 12	3133	2400–5000	28.7	(28.7–29.0)*
1994 n = 12	3842	2800–4700	35.2	(35.2–35.5)*
1995 n = 12	3375	2400–5000	30.9	(30.9–31.2)*
1990–1995 n = 64	3299	2100–5500	30.2	(30.2–30.5)*

*Since 1992 total arsenic was only determined in the fish group.

Total arsenic intakes were very high in comparison to other countries. Average intake in Japan[6] (280 μg/person/day), where the consumption of fish is also very high, was lower than the estimate for the Basque Country (293–296 μg/person/day). In most countries total arsenic intakes are usually very low, and when compared directly with reference values for inorganic arsenic it is concluded that there is no risk associated with the intake of this element, even in the hypothetical and unlikely case that all the arsenic present was inorganic. This argument thus avoids the necessity of the determination of inorganic arsenic, which is more tedious. However, in the Basque Country, this approximation could not be made because total arsenic intakes were very high and, if all the arsenic present were inorganic, it would exceed the tolerable intake (PTWI for inorganic arsenic is 0.015 mg kg^{-1} body weight/week).[7]

3 Inorganic Arsenic Intake of Average Consumers

In order to conduct a proper risk assessment, inorganic arsenic was analysed in 12 selected TDS fish group samples (1990–1993). Results indicated that 84–100% of the arsenic was present as arsenobetaine and the maximum amount of reducible forms of arsenic (*i.e.* toxic forms) detected was 5% of total arsenic. Assuming that all reducible forms of arsenic were inorganic arsenic, the maximum inorganic arsenic intake in the period 1990–1995 would be 1.5 $\mu g\,kg^{-1}$

body weight/week, which accounts for 10% of the PTWI (analyses were conducted by the Department of Environmental Sciences, University of Plymouth, UK[8]).

The estimated intake of inorganic arsenic appeared to pose no health risk for the average consumer. However, in order to estimate the possible health risk of extreme fish consumers, the theoretical total and inorganic arsenic intakes were estimated.

4 Theoretical Intake of Inorganic Arsenic of Extreme Fish Consumers

Table 4 shows the total and inorganic arsenic intakes for twice and three times the average fish consumption. As a rough rule of thumb, extreme consumers of food are, in general, unlikely to have an intake of food that exceeds twice the average consumption of the population as a whole.[9] This factor should be applied whenever an assessment of the exposure to a contaminant of such consumers is required and only average consumption data are available. If a contaminant is mostly confined to an individual foodstuff within the diet, then 'extreme' consumers of that foodstuff within the population are unlikely to consume more than three times the amount consumed on average by the population. This factor should be applied to the average data to determine the likely range of intake of the population.

According to these assumptions, and considering the maximum inorganic arsenic percentage detected of 5%, the theoretical inorganic arsenic intake for extreme fish consumers would be around 30% of the PTWI. Another estimation was done by considering the 95th percentile of fish consumption of the Basque Food Survey. In that case, inorganic arsenic intake would be below 40% of the PTWI. Exposure to arsenic through the diet appeared not to pose appreciable risk to the health of extreme fish consumers of the Basque population, even with this worst case assumption.

Table 4 *Theoretical estimation of total and inorganic arsenic intake by extreme consumers of fish*

Fish consumption	Total As intake from the fish group ($\mu g \, kg^{-1}$ bw/week)	Theoretical intake of inorganic arsenic ($\mu g \, kg^{-1}$ bw/week) % of the PTWI
Average, 89 g day^{-1}	30.2	1.5 (10%)
Twice, 178 g day^{-1}	60.4	3.0 (20%)
Three times, 267 g day^{-1}	90.7	4.5 (30%)
95 percentile (Basque food survey) 355 g day^{-1}	120.5	6.0 (40%)

Average concentration (1990–95) of arsenic in the fish group: 3299 $\mu g \, kg^{-1}$.

5 Dietary Intake of Total and Inorganic Arsenic: 1996–1998

Since 1996, both total and inorganic arsenic have been determined regularly in the 'fish' group of the TDS. Table 5 shows the dietary intake of total and inorganic arsenic for the period 1996–1998. It must be considered that water is not included in the TDS as a specific group and distilled water is used for preparation and cooking. Arsenic levels in drinking water are regularly monitored and the levels found are always below the LoQ of $1 \mu g l^{-1}$.

The average estimated intake of inorganic arsenic in that period was $0.22 \mu g kg^{-1}$ body weight/week and accounts for 1.5% of the PTWI. The maximum estimated intake accounted for 2.5% of the PTWI. The USA EPA RDA for inorganic arsenic is $0.0003 mg kg^{-1}$ body weight/day.[10] Inorganic arsenic intake in the Basque Country accounts for 10.5% of the US EPA standard.

As opposed to the method for the earlier data (which were based on 12 TDS fish group samples only, as mentioned in Section 3), the method used since 1996 gives a good estimation of the concentration of inorganic arsenic compounds (*i.e.* As^{III} and As^V) in the sample.[11] With this method of sample preparation (chloroform extraction of As^{III} after reducing As^V to As^{III}), virtually no organically bound arsenic is retained (analysis were conducted in the *Instituto de Agroquímica y Tecnología de Alimentos* (*CSIC*), Valencia, Spain[11]).

Figure 3 shows the intake of inorganic arsenic in the period 1996–1998, together with the distribution of the individual set of results for each year. No extreme values were found and there appears to be no trend in intakes, the values being at very low levels.

The Codex Committee on Food Additives and Contaminants, at its thirty-first session in March 1999, discussed a position paper on arsenic.[12] In that document, speciation analysis for arsenic was strongly recommended. It was also stated that further research was needed to clarify the fate of organic arsenic, including arseno-sugars and arsenobetaine in humans, particularly with the aim of elucidating whether any arsenic species of toxicological concern are formed during human metabolism of the compounds ingested *via* seafood. Although we do not know what the current margin of safety for dietary exposure of the general population to arsenic is, the erosion of this margin remains an on-going phenomenon. Contrary to what some people may believe, the so-called maximum tolerable exposure dose has no relevance whatsoever in protecting the general population from the effects of long-term exposure to high levels of arsenic in the environment. As far as we know, there are few, if any, early warning or biochemical indicators of subclinical arseniasis[13] and the maximum permissible dose pertains to observed symptoms of arseniasis. In the Basque Country, this is particularly relevant since fish consumption is very high in comparison to other countries. The dietary intake of organic arsenic compounds is very high and the analysis of organic species of arsenic in TDS samples of the fish group is currently under way.

Table 5 Total and inorganic arsenic in the fish group of the TDS 1996–1998

| | Total arsenic | | | | | Inorganic arsenic | | | | |
| | Concentration (fish group) ($\mu g\,kg^{-1}$) | | | Intake ($\mu g\,kg^{-1}$ bw/week) | | Concentration (fish group) ($\mu g\,kg^{-1}$) | | | Intake ($\mu g\,kg^{-1}$ bw/week) |
Year	Mean	Median	Range	Fish group	Range	Mean	Median	Range	Fish group
1996 n = 12	3493	3190	2350–4790	32.2	32.2–32.5	22.5	21.1	16–37	0.21
1997 n = 12	4076	3937	1860–8820	37.4	37.4–37.7	25.5	26.4	17–36	0.23
1998 n = 12	3173	2729	1547–5719	29.2	29.2–29.5	24.0	25.1	18–29	0.22
1996–1998 n = 36	3580	3248	1547–8820	32.9	32.9–33.2	24.0	24.1	16–37	0.22

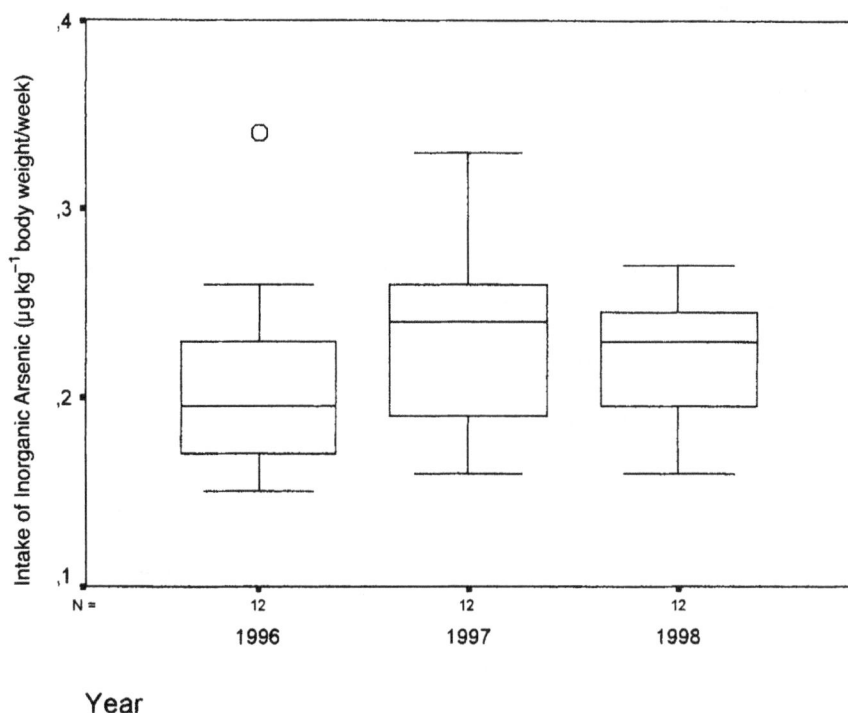

Figure 3 *Inorganic arsenic in TDS samples 1996–1998*

6 Conclusions

1 Dietary intake of inorganic arsenic estimated through the TDS in the period 1996–1998 was low (0.22 μg kg^{-1} body weight/wk) and accounted for 1.5% of the appropriate PTWI. Notwithstanding the very high fish consumption in the Basque Country, the estimated intake does not pose a health risk for the Basque population.

2 The proportion of inorganic arsenic to total arsenic in the TDS fish group was very low and was in the range 0.3–1.8% (mean 0.8%).

3 Inorganic arsenic species should be the species to measure (and, if required, limited by legislation) and therefore analytical methods should be standardised for As^{III} and As^{V}, and other toxic species, such as MMA and DMA, if appropriate.

7 References

1. M. Jalón, I. Urieta, M.L. Macho and M. Azpiri, *Vigilancia de la Contaminación Química de los Alimentos en la Comunidad Autónoma del País Vasco, 1990–1995* (*Food Chemical Surveillance in the Basque Country*, 1990–1995) book written in Spanish

with an English summary of 30 pp. and Figures and Tables in Spanish/English, Servicio Central de Publicaciones del Gobierno Vasco, 1997.

2. I. Urieta, M. Jalón, J. Garcia and L. Gonzalez de Galdeano, *Food Addit. Contam.*, 1991, **3**, 371.

3. Departamento de Sanidad, *Encuesta de Nutrición de la Comunidad Autónoma del País Vasco*, Servicio Central de Publicaciones del Gobierno Vasco, Vitoria-Gasteiz, 1994.

4. I. Urieta, M. Jalón and I. Eguileor, *Food Addit. Contam.*, 1996, **13**, 29.

5. GEMS/Food, Proceedings of the GEMS/Food-Euro workshop, 1998.

6. T. Tsuda, T. Inoue, M. Kojima and S. Aoki, *J. AOAC Int.*, 1995, 1363.

7. WHO, *Evaluation of Certain Food Additives and Contaminants*, Thirty-third report of the Joint FAO/WHO Expert Committee on Food Additives. WHO Technical Report Series, No 776, Geneva, 1989.

8. S. Branch, L. Ebdon and P.J. O'Neill, *J. Anal. At. Spectrom.*, 1994, **9**, 33.

9. FAO/WHO, *Guidelines for the Study of Dietary Intakes of Chemical Contaminants*. WHO offset publication No 87, Geneva, 1985.

10. US EPA, IRIS (Integrated Risk Information System), Washington DC, 1999.

11. O. Muñoz, D. Vélez and R. Montoro, *Analyst*, 1999, **124**, 601.

12. Codex Committee on Food Additives and Contaminants, *Position Paper on Arsenic*, CX/FAC 99/22, 1998.

13. J.O. Nriagu and J.M. Azcue, in *Food Contamination from Environmental Sources*, eds. J.O. Nriagu and M.S. Simmons, John Wiley & Sons, Inc, New York, 1990, Vol. 3, p. 121.

CHAPTER 17

Trace Element Speciation and International Food Legislation – A Codex Alimentarius Position Paper on Arsenic as a Contaminant

ERIK H. LARSEN AND TORSTEN BERG

1 Introduction – What is a Codex Position Paper?

In international legislation concerning trace elements in food, in the environment or in occupational health, most regulations are based on the total element contents, and are frequently given as maximum limits or guideline levels. Only a few regulations pay attention to the molecular species in which the elements are bound. The generation of species-specific analytical and toxicological data is necessary before it will become possible to lay down species-specific regulations in all the cases where it is considered reasonable from a scientific point of view.[1]

International food legislation is laid down under the auspices of the Codex Alimentarius Commission (CAC), a United Nations organisation established jointly by the FAO and the WHO. The statutes of the CAC describe the purpose of the Food Standards Programme to be *inter alia*: 'Protection of the consumers and ensuring fair practices in the food trade'.[2] The CAC has, since 1962, developed specific standards for individual foods or group of foods, called vertical or commodity standards. The trend, however, over the last decade has been to develop horizontal or general standards, such as the General Standard for Contaminants and Toxins in Food (GSCTF) by the Codex Committee for Food Additives and Contaminants (CCFAC).[3,4] These standards are of an ever increasing importance as, under the Sanitary and Phytosanitary (SPS) Agreement of the World Trade Organisation (WTO), the Codex Alimentarius Standards form the basis of any decisions, by a WTO panel, in disputes concerning international trade in food. The Codex GSCTF hence provides the framework for future international legislation on trace elements as contaminants in food.

The scarcity of species-specific regulations in legislation does not necessarily imply that speciation was not considered in the scientific basis for the legislation. Background papers, such as those presented by the International Agency for the Research on Cancer and the FAO/WHO Joint Expert Committee on Food Additives and Contaminants (JECFA), contain much relevant information concerning the chemical nature and the fate of the various species ingested.[5,6]

The Codex GSCTF is, at present, a legislative framework to be filled with maximum limits for relevant contaminants in specific foods and food groups. The basis for these maximum limits and for other provisions is termed a Position Paper, with information including relevant general chemistry, toxicology and levels found in different foods, as well as speciation considerations, intake in different parts of the world and fair trade considerations. A Position Paper on Arsenic[7] for use by the CCFAC in connection with the Codex GSCTF has been developed by the authors, as a part of the work under the auspices of the EU-funded Thematic Network Speciation 21.[8] International consensus regarding the Position Paper on Arsenic was reached during the 31st Session of the CCFAC in 1999, and therefore forms the basis for the future legislative work on arsenic in the framework of the CCFAC.

2   Position Paper on Arsenic

2.1   Occurrence of Arsenic in Food

Arsenic ranks 20th in abundance among the elements in the Earth's crust. Arsenic is introduced into the environment from natural sources (*e.g.* volcanic activity and weathering of minerals) and from anthropogenic activity (*e.g.* ore smelting, burning of coal, pesticide use) and the ratio between the two types of sources has been estimated at 60:40.[9] As a result of natural metabolic processes in the biosphere arsenic occurs as a large number of organic or inorganic chemical forms (species) in food.

The different chemical and toxicological characteristics of the various molecular species and oxidation states occurring in food make it necessary to distinguish between them, in order to present a full picture of the content of arsenic in food and what impact the intake of arsenic in food has on the consumer's health.

In the marine environment, the total arsenic concentrations, which typically range from 0.5 to 50 mg kg^{-1} (wet weight), are found in animals and plants, including seaweed, fish, shellfish and crustaceans.[9,10] The high concentration of arsenic in seafood has been known since the beginning of the 20th century.[11] The study of the metabolic routes of arsenic in the marine environment has led to the understanding of some of the conversions of inorganic arsenic, found in ocean water, into the significantly higher concentration of organic forms of arsenic present in seafood.[10] In fish from freshwaters, arsenic is present at much lower concentrations in comparison with their oceanic counterparts, typically at less than 10 μg kg^{-1}.[12]

In the terrestrial environment arsenic is generally found at low concentrations in crop plants which typically range from 0 to $20\,\mu g\,kg^{-1}$, with the exception of rice, at $150-250\,\mu g\,kg^{-1}$, and certain edible mushroom species, which contain arsenic at several $mg\,kg^{-1}$ taken up from the soil.[13] Information is generally scarce regarding the molecular species of arsenic found in crop plants. In livestock the arsenic concentration is similar to that in plants. A notable exception is arsenic in poultry which ranges from 0 to $100\,\mu g\,kg^{-1}$. The arsenic originates from arsenic-containing fishmeal used in the poultry fodder or possibly arsenic-containing growth promoters, which are used in some countries. Increased arsenic concentrations in plants (tobacco) have also been observed where dimethylarsinate was used as a pesticide. Levels and speciation of arsenic in drinking water (including natural mineral and other bottled water) is a matter of concern in many countries as arsenic levels exceeding $200\,\mu g\,l^{-1}$ have been reported.[14] Several reports in the literature of similar or even higher arsenic concentrations in well and in ground water reflect this problem.

Atmospheric fall-out of arsenic has contaminated crop plants cultivated near industrial point sources. The arsenic concentration found in such crops depends on a number of factors including the chemical form and bioavailability of arsenic in the soil and on the atmospheric deposition rate, and can therefore not be generalised. Finally, increased arsenic concentrations have been observed in crop plants when cultivated in soil with a naturally high arsenic content, or in soil contaminated by spills after chromium, copper, arsenic (CCA) wood impregnation.[15] No direct carry-over, however, has been observed, for example, in potatoes stored in bins made from CCA impregnated wood.[16]

2.2 Toxicological Evaluations

The most toxic forms of arsenic found in food and water are the inorganic arsenic-(III) and -(V),[17] and the International Agency for Research on Cancer (IARC) has classified inorganic arsenic as a human carcinogen.[18] The inorganic arsenic trioxide has a well-known history as a poisonous compound, often used in cases of homicide. The methylated forms, however, *e.g.* dimethylarsinate, are low in acute toxicity[19] while arsenobetaine, the principal arsenic species found in fish and crustaceans, is considered non-toxic.[17] In shellfish, molluscs, and seaweed, the dimethylarsinylriboside derivatives, also known as 'arseno-sugars', are the quantitatively dominating arsenic species. Their possible toxicity to humans is not known in any detail. Organic arsenicals generally dominate in food and only a few per cent of the total arsenic in fish is present as inorganic forms.[20,21]

A provisional tolerable daily intake of inorganic arsenic *via* food and water was established by WHO at $2\,\mu g\,kg^{-1}$ body weight in 1983[22] and later changed to the equivalent provisional tolerable weekly intake (PTWI) at $15\,\mu g\,kg^{-1}$ body weight in 1988.[23] The epidemiological data used for this risk assessment refer to inorganic arsenic in drinking water. However, WHO could not establish a similar recommendation for organic arsenic species in food due to lack of appropriate toxicological data. This recommendation has been repeated and substantiated by the US Agency for Toxic Substances and Disease Registry.[24]

The recommended WHO guideline level for inorganic arsenic in drinking water is $10 \, \mu g \, l^{-1}$,[25] and according to these guidelines 20% of the PTWI is allocated to drinking water.

In contrast to its toxicity, the possible essential role of arsenic[26] is a matter of controversy.[27] The underlying animal experiments, however, do not allow for any conclusions regarding a possible essential function of arsenic in humans.

Inorganic arsenic-(III) and -(V) are well absorbed from the gastro-intestinal tract. The absorption rates of methylated arsenic, and that of arsenobetaine, are also high but tissue retention of the latter is much lower as studied by excretion of radio-labelled arsenobetaine. Inorganic arsenicals are transported to the liver where methylation may lead to the formation of monomethylarsonate and dimethylarsinate. However, a fraction of the inorganic arsenic may accumulate in hair, nails and skin.[28] No transformation of arsenobetaine has been reported to take place in mammals but arsenocholine ingested *via* seafood can be oxidised to arsenobetaine. Elimination of inorganic and organic forms of arsenic mainly takes place *via* the urine. Only a few studies exist regarding the fate of arsenobetaine in the human body. The question remains, however, whether arsenobetaine is stable in the presence of anaerobic bacteria in the gastro-intestinal tract. In the marine environment such bacteria have been shown to metabolise arsenobetaine to lower molecular weight compounds. Obviously, this possible catabolic route in man needs investigation.

Epidemiological studies have shown human health effects, after long-term oral exposure to inorganic arsenic species in drinking water from wells in areas where the soil is geochemically rich in arsenic. A daily intake of inorganic arsenic at $10-50 \, \mu g \, kg^{-1}$ body weight contributed to vascular problems, which may ultimately lead to necrosis and gangrene of hands and feet ('black foot disease').[23] Inorganic arsenic may also cause skin lesions and skin cancer. Besides the risk of developing skin cancer, internal types of cancer have been reported for patients who already had developed arsenic-induced skin cancer.

The United States Environmental Protection Agency (US EPA) has estimated the excess skin cancer risk from life-time exposure to arsenic *via* water[29] containing $1 \, \mu g \, l^{-1}$ of inorganic arsenic at 7×10^{-5}. Therefore drinking water concentrations of arsenic exceeding these levels is a matter of concern. At the $10 \, \mu g \, l^{-1}$ WHO guideline level of arsenic in drinking water the estimated lifetime risk for arsenic-induced skin cancer has been estimated[24] at 6×10^{-4}.

2.3 Analytical Data

The analysis of total arsenic in food has, to date, suffered from difficulties with respect to accuracy and precision. However, the situation has improved somewhat in recent years with the advent of modern analytical techniques such as atomic absorption spectrometry (AAS), with hydride generation or graphite furnace atomisation, and inductively coupled plasma mass spectrometry (ICP-MS).

However, it is generally advisable to disregard data older than approximately

10 years and also reported data, the quality of which have not been justified by, for example, the use of certified reference materials. Data for arsenic in food at low concentration levels may be seriously biased because of possible interferences, or because of other disturbances such as laboratory contamination or loss of analyte during sample work-up. Consequently, international intercomparison exercises are needed to test the proficiency of analytical laboratories, particularly for the analysis of arsenic in seafood.

The analytical difficulties encountered for arsenic are non-specific absorbance from matrix constituents in AAS, or insufficient sample mineralisation if hydride generation AAS has been used. In ICP-MS polyatomic interferences and matrix effects may hamper the accuracy of the analyses. For speciation work, however, ICP-MS is an excellent detector for coupling with separation techniques such as HPLC.[30]

There is a particular need for speciation data for arsenic because of the large difference in toxicity to humans of various chemical forms of arsenic. At least 25 different chemical forms of arsenic, which are polar or ionic in molecular structure, have been detected, particularly in seafood samples.[10] Furthermore a number of non-polar (fat-soluble) arsenolipids constitute 1–18% of the total arsenic content in seafood.[31] Due to the different toxicology of these species a value for the total arsenic content in a food sample is of little value. First of all procedures that provide information on the content of the toxic inorganic arsenic species are needed. Secondly, data on the methylated species are of interest because they are the immediate conversion products of ingested inorganic arsenic. Finally, more sophisticated speciation procedures based on, for example, the HPLC–ICP-MS combination are needed in toxicological studies such as those suggested above for the arseno-sugars and arsenobetaine.

2.4 Intake Data

Humans are mainly exposed to arsenic *via* the diet and drinking water, whilst a small proportion of the population is additionally occupationally exposed. The arsenic level in these individuals must be specifically monitored in order to prevent a possible health risk. The highest average arsenic concentration is found in seafood products followed by poultry and cereal products. Combined with data on food consumption, the average adult intake has been estimated in Denmark at $118 \mu g \, day^{-1}$ with a 90th percentile intake of $233 \mu g \, day^{-1}$.[13] Seafood contributes to the average arsenic intake by $86 \mu g \, day^{-1}$ (72% of the total intake) in spite of the fact that fish only accounts for a small fraction of the daily food consumption. In the USA, the mean intake of total arsenic by adults was found to be $38.6 \mu g \, day^{-1}$, of which seafood contributed $34.1 \mu g \, day^{-1}$, or 88% in 1986–91, and in 1991–97 the figures were $56.6 \mu g \, day^{-1}$, of which $51.9 \mu g \, day^{-1}$, or 92%, were from seafood.[32,33] The mean adult dietary intake for UK consumers is $63 \mu g \, day^{-1}$ and fish constitute $56 \mu g \, day^{-1}$ or 89% of this value.[34] In Canada the mean daily intake was $49 \mu g \, day^{-1}$ and the contribution from fish constituted $32 \mu g \, day^{-1}$ or 64% of the total.[35] In Australia, the mean adult dietary intake of arsenic has been estimated at 63 μg day^{-1} with seafood contributing 63% of the total, while the 95th percentile intake of total arsenic was 119 μg day^{-1}.[36] An estimate of the mean adult dietary intake of inorganic arsenic in Australia was 0.77 μg day^{-1} or 0.5% of the PTWI value. The equivalent 95th percentile value was 1.25 μg day^{-1} for inorganic arsenic. The reported data reflect that the total arsenic intake in a given population is largely determined by the amount of seafood consumed. In Japan, where the arsenic-rich seafood, seaweed and rice traditionally constitutes a large part of the diet, the daily arsenic intake *via* the diet has been estimated at 985 μg day^{-1}.[37]

In order to arrive at a meaningful estimate of the intake, which can be evaluated using the existing toxicological standards, arsenic species information is strongly needed. However, such results for market foods are rarely available in the literature.

2.5 Potential Health Risks

If it were (erroneously) assumed that all arsenic in the food is inorganic the average arsenic intake for the Danes would amount to 79% of the PTWI value of 1050 μg for inorganic arsenic for a person of 70 kg body weight. This high value is, however, not of immediate relevance because it is well known that arsenic does not occur as inorganic arsenic in the diet, except in low amounts. A more relevant estimate is therefore based on the assumption that a maximum of 5% of the arsenic ingested *via* seafood is inorganic.[21] The Australian estimate of the inorganic arsenic intake at 1% of the total arsenic intake confirms the generally low content of inorganic arsenic in food including seafood.

Assuming that arsenic in non-marine food items and in water is present as inorganic species, the weekly intake of inorganic arsenic amounts to 250 μg based on data from the Danish investigation.[13] This is equivalent to 24% of the PTWI value. Results of similar magnitude would be obtained based upon the UK and Canadian intake data, whereas the situation in Japan would merit a closer investigation of the intake of the arsenic species *via* the traditional diet.

This estimated intake of inorganic arsenic implies that a large majority of consumers are not in any danger of exceeding the PTWI value, except in cases of high concentrations of inorganic arsenic in the drinking water. Furthermore, individuals who have been exposed to arsenic-containing particles or arsines in the air may be at risk. In such cases measures must be taken to reduce the risk by *e.g.* improved processing of the water or if possible by reduction or removal of the source of the emission of arsenic.

However, if the diet contains an appreciable amount of seaweed, which is customary in some countries, the intake of inorganic arsenic from this source may increase strongly, and the risk of exceeding the PTWI value for inorganic arsenic must be considered. In such cases, a risk assessment aimed at the relevant population groups must be carried out.

With the aim to investigate whether any arsenic species of toxicological concern are formed following consumption of seafood, research is needed to clarify the fate of, particularly, arseno-sugars and arsenobetaine in humans. Furthermore, such studies should also show whether these arsenic species are accumulated in any given tissue.

Finally, more research is needed to clarify whether the natural arsenic species in seafood undergo conversions into more toxic compounds during industrial processing (canning or freezing) and during storage.

2.6 Fair Trade Considerations

Several countries have established maximum levels (MLs) for arsenic in food commodities. There are a number of Codex MLs for arsenic which, however, do not cover the whole range of national MLs. The foods for which Codex MLs have been established are fats, oil, juice, nectar and sugar, cocoa and chocolate.

In general, trade problems are not prevalent. In some markets, however, the national MLs for total arsenic in seafood products are set at relatively low values in comparison with the naturally occurring concentration levels of arsenic. This is prevalent for a commodity such as cold-water shrimp. It is likely that such national MLs are erroneously based on the assumption that a substantial amount of the arsenic in the seafood product is present as the toxic inorganic arsenic.

Until maximum limits for specific toxic arsenic compounds in food are laid down by the Codex Alimentarius Commission it is suggested that foods that are at risk of rejection are analysed and assessed on a case by case basis in the light of the information presented in this Position Paper.

2.7 Conclusions and Recommendations

The industrial uses of arsenic-containing preparations such as those used for wood preservation should be phased out and they should be replaced by other less toxic agents. Similarly, the arsenic emissions from burning fossil fuels should also be reduced by improving the efficiency of smoke cleaning.

There is no apparent risk of exceeding the PTWI value for inorganic arsenic from food and water, except in regions with high arsenic levels in drinking water and/or extreme seafood and seaweed consumption. The levels of inorganic arsenic reported in some drinking and in some bottled waters, however, may cause concern.

Organo-arsenic compounds dominate the total intake of arsenic from food (with the exception of water). This can not be changed because of the natural biogeochemical cycles of this element in nature, particularly in the marine environment. Until further knowledge regarding the content of the naturally occurring arsenic species in a wide range of foods and the toxicity of these species has been established, and until methodologies have been developed for their control, there seems not to be sufficient basis to decide whether Codex MLs are needed for these species. The information available, however, indicates that such MLs should not be necessary.

It is, however, recommended that further study on the analytical methodology with special emphasis on the arsenic species of known, or suspected, toxicological concern is undertaken. To make analytical methods readily available to laboratories other than expert laboratories, such analytical development should ideally be based on inexpensive instrumentation. Furthermore, the absorption rate, bioavailability and possible toxic effects of arsenic species need further study. This applies particularly to the arseno-sugars.

Water is a special case that may cause concern because of possible high levels of inorganic arsenic that may contribute to the risk of skin cancer. For packaged water moving in international trade, including natural mineral waters, it is recommended that the Codex Alimentarius considers replacing the current Codex ML of $50 \, \mu g \, l^{-1}$ for total arsenic with the WHO guideline level of $10 \, \mu g \, l^{-1}$ for inorganic arsenic in drinking water.

The current Codex Alimentarius MLs in some foods are based on total arsenic and do not take into consideration which chemical forms of arsenic are present. Consequently, they do not cover the situation adequately, and it is suggested that they should be suspended, or amended, as above, for natural mineral water. Foods moving in international trade should instead be analysed and assessed based upon the relevant scientific information available as, for example, presented in this paper.

The PTWI value was established by WHO saying that sufficient knowledge was unavailable to set a similar tolerable intake value for organic forms of arsenic. Arsenobetaine and the group of arseno-sugars which originate from various marine food products represent the quantitatively important organic arsenic species and their fate and possible toxicological effects to man must be elucidated. We now have laboratory methods and instrumentation to hand that allow such investigations to be carried out. Ultimately, the calculation of the dietary intake of arsenic species may constitute the basis for human health risk evaluation.

Future legislative limits for arsenic must be based on those arsenic species, which are of toxicological concern, such as As(III) and As(V). As a result of the suggested research topics, the need for more detailed legislation, focusing on those commodities that have potential to cause health concern regarding arsenic species, will emerge.

3   References

1. T. Berg and E.H. Larsen, *Fresenius' J. Anal. Chem.*, 1999, **363**, 431.
2. *Codex Alimentarius Commission Procedural Manual*, 10th Edn., FAO and WHO, Rome, 1997.
3. *Codex Alimentarius General Standard for Contaminants and Toxins in Food*, Food Safety Information Centre RIKILT-DLO, Wageningen, The Netherlands, 1998.
4. T. Berg, in *International Control of Food Safety: WTO, Codex and Regional Standards*, eds. N. Rees and D. Watson, Aspen Publishers, Inc., Gaithersburg, Maryland (in press).
5. *Some Metals and Metallic Compounds*, IARC Monographs on the Evaluation of the

Carcinogenic Risk of Chemicals to Humans, International Agency for Research on Cancer, Lyon, 1980.

6. WHO Technical Report Series, 46th Report of the Joint FAO/WHO Expert Committee on Food Additives, WHO, Geneva, 1997.

7. Codex Alimentarius Commission, *Position Paper on Arsenic*, CX/FAC 99/22, December 1998, The Hague, The Netherlands, including comments received in CX/FAC 99/22 – Add. 1.

8. R. Cornelis, C. Camara, L. Ebdon, L. Pitts, B. Welz, R. Morabito, O. Donard, H. Crews, E.H. Larsen, B. Neidhart, F. Ariese, E. Rosenberg, D. Mathé, G.M. Morrison, G. Cordier, F. Adams, P. Van Doren, J. Marshall, B. Stojanik, A. Ekvall and P. Quevauviller, *Fresenius' J. Anal. Chem.*, 1999, **363**, 435.

9. W.R. Cullen and K.J. Reimer, *Chem. Rev.*, 1989, **89**, 713.

10. K.A. Francesconi and J.S. Edmonds, *Oceanogr. Mar. Biol. Annu. Rev.*, 1993, **31**, 111.

11. C. Chapman and H. Linden, *Analyst*, 1926, **51**, 563.

12. Danish Veterinary and Food Administration, *Danish Fresh Water Fish. Contents of Trace Elements, PCB and Chlorinated Pesticides*, Publication No. 138, Søborg, 1986.

13. National Food Agency of Denmark, *Food Monitoring in Denmark. Nutrients and Contaminants 1983–1987*, Publication No. 195, Søborg, 1990.

14. G. Farmer and L.R. Johnson, *Env. Geochem. Health*, 1985, **7**, 124.

15. E.H. Larsen, L. Moseholm and M.M. Møller, *Sci. Tot. Environ.*, 1992, **126**, 263.

16. Jorhem and K. Nilsson, *Storage of Potatoes in Impregnated Wood Barrels*, National Food Administration, SLV-rapport nr. 1, Uppsala, 1992.

17. World Health Organisation, *Arsenic*, Environmental Health Criteria 18, Geneva, 1981.

18. ARC Monographs, Suppl. 7, International Agency for the Research on Cancer, Lyon, 1980.

19. International Programme on Chemical Safety, *Dimethylarsinic Acid, Methanearsonic Acid, and Salts*. Health and Safety Guide No. 69, WHO, Geneva, 1992.

20. P. Buchet, R. Lauwerys and H. Roels, *Int. Arch. Occup. Environ. Health*, 1981, **48**, 71.

21. S. Edmonds and K.A. Francesconi, *Mar. Pollut. Bull.*, 1993, **26**, 665.

22. Food and Agriculture Organisation, World Health Organisation, WHO Food Addit. Ser., No. 18, Geneva, 1983.

23. Food and Agriculture Organisation, World Health Organisation, WHO Food Addit. Ser., No. 24, Geneva, 1989.

24. *Toxicological Profile for Arsenic*, US Department of Health and Human Services, Agency for Toxic Substances and Disease Registry, Atlanta, 1991.

25. World Health Organisation, *Guidelines for Drinking Water Quality*, Geneva, 1993.

26. O. Uthus, *Environ. Geochem. Health*, 1992, **14**, 55.

27. Department of Health, *Report on Health and Social Subjects*, **41**, London, 1991.

28. M. Vahter and E. Marafante, in *The Biological Alkylation of Heavy Elements*, eds. P.J. Craig and F. Glocking, Special Publication No. 88, Royal Society of Chemistry, London, 1988, p. 105.

29. United States Environmental Protection Agency, *Special Report on Ingested Inorganic Arsenic. Skin Cancer; Nutritional Essentiality*, EPA-625/3-87-013, Washington DC, 1988.

30. E.H. Larsen, *Spectrochim. Acta Part B*, 1998, **53**, 253.

31. E.H. Larsen, G. Pritzl and S.H. Hansen, *J. Anal. At. Spectrom.*, 1993, **8**, 1075.

32. E.L. Gunderson, *J. AOAC Int.*, 1995, **78**, 1353.

33. S.-H. Tao and P.M. Bolger, *Dietary Intakes of Arsenic in the United States*, Third International Conference on Arsenic Exposure and Health Effects, 1998.

34. Ministry of Agriculture, Fisheries and Food, *1994 Total Diet Study: Metals and other Elements*, Food Surveillance Information Sheet No. 131, London, 1997.
35. R.W. Dabeka, A.D. McKenzie, G.M.A. Lacroix, C. Cleroux, S. Bowe, R.A. Graham, H.B.S. Conacher and P. Verder, *J. AOAC Int.*, 1993, **76**, 14.
36. Australian Quarantine and Inspection Service, *Diet Survey*, Canberra, 1994.
37. S. Shinichiro, A. Hayase, M. Murakami, L. Hatai, K. Higashigawa, C.S. Moon, Z.-W. Zhang, T. Watanabe, H. Igushi and M. Ikeda, *Fd. Addit. Contamin.*, 1996, **13**, 775.
38. *Gezondheitsraad, Arsenic*, Advisory Report Issued by the Committee on Risk Assessment of Substances of Health Council of the Netherlands, Report No. 1993/02, The Hague, 1993.

CHAPTER 18

The Need for Speciation to Realise the Potential of Selenium in Disease Prevention

SVEN MOESGAARD AND RICHARD MORRILL

1 Introduction

Speciation in its original meaning refers to the process of evolution by which diverse animal or plant species evolve from common ancestors. In the context of selenium and cancer research, speciation refers to the process of determining and identifying the various chemical forms of selenium in foodstuffs and dietary supplements, as well as in relevant body fluids and tissues.

The current status of selenium research reflects a curious instance of clinical research results leading, indeed outpacing, knowledge gained from basic research. It points to the need for additional investment in basic research at the present time. Recently, the Linxian Study,[1] the Clark Study,[2,3] and, by inference, the Bonelli Study[4-6] have presented evidence from carefully designed and controlled clinical trials, showing statistically significant reductions in cancer incidence and mortality in humans treated with selenium supplements, as compared with placebo. Clinicians and researchers alike are now asking the following questions: which chemical form(s) of selenium can these results be attributed to, and what is the mechanism by which specific chemical forms of selenium have had an effect on cancer incidence and mortality?

These questions illustrate the need for basic research on the speciation of the various chemical forms of selenium in foodstuffs and in dietary supplements, as well as in the organism following ingestion. The current understanding of the composition and action of the components of the selenium ingested in the aforementioned intervention trials is far too limited. Present knowledge of the distribution of the specific chemical forms of selenium in geographical areas of relatively high and relatively low dietary intakes, as well as how this relates to cancer incidence and mortality, is equally tenuous.

We assert the need for knowledge of the speciation of selenium in order to realise the full potential of selenium in disease prevention. By this, we mean not only the speciation of selenium in the foodstuffs and dietary supplements that are ingested, but also the determination of the chemical forms of selenium in seleno-proteins and selenium-containing proteins in the organism following ingestion.

Of particular interest is the question of whether the chemical form L-sele-nomethionine, the form that is generally most abundant in human foodstuffs and that gives the highest rate of absorption, the highest rate of retention in tissues and the highest level of incorporation in the selenium-dependent GSH-Px en-zymes, is as effective at preventing cancer as is the combination of organic selenium compounds that are found in high-selenium yeast supplements.[7,8]

2   Historical Background of Selenium

Selenium was discovered in Sweden in 1817 by Berzelius. Selenium, an element with the atomic number 34, is found in Group VI of the periodic table, just below sulfur and just above tellurium, with which it shares a similar chemistry. Sele-nium combines with both metals and non-metals and can form both organic and inorganic compounds. It can function both as an oxidant and as a reductant. It can become toxic in higher doses. Its distribution in soil and its availability to vegetation, and by extension to animals and humans who eat plants, varies considerably.[9] Generally speaking, selenium is found in high concentrations in the more alkaline soils of the more arid regions of the Earth.

Today, selenium is recognized as an essential trace element in animal and human diets,[10] and it has been seen to be an integral constituent of several enzymes including the glutathione peroxidase (GSH-Px) enzyme system, *i.e.* the enzymes that remove hydrogen peroxides, lipid and phospholipid hydroperox-ides.[11]

2.1   Recognition as an Essential Trace Element

Selenium in the diet has not always been viewed so favourably as it is today. In the 1930s, selenium was widely regarded as highly toxic when consumed by humans, and in the 1940s, it was even thought to be a carcinogenic substance. As late as 1957, Schwartz and Foltz were able to demonstrate that trace amounts of selenium could be used to prevent liver necrosis in vitamin E deficient rats, and, shortly thereafter, widespread use of selenium fortification and supplementation was initiated in the field of animal husbandry for the purpose of preventing various conditions in livestock and poultry.[12]

Beginning in 1966 and continuing through the 1970s and into the 1980s, Shamberger, Clark, Salonen and others investigated and reported on the rela-tionship of selenium to cancer.[13–15] The methodology of these investigations involved the correlation of selenium levels in crops intended for human con-sumption with the corresponding cancer rates for specific geographical areas and the correlation of serum selenium levels in humans with cancer risk.[13–15] Much evidence was accumulated that pointed to an inverse relationship between selenium intakes/levels and risk of cancer at several different sites, sufficient to warrant the funding of intervention trials. It is interesting to note that the investigations of this period focused on selenium as an entity, and were not concerned with the speciation of the various chemical forms of the element.

The first solid evidence of deficiency conditions in humans came in 1979, when Chinese researchers established a negative association between selenium status in humans and Keshan disease, a disease that primarily affects children and women in their childbearing years. An intervention trial involving many thousands of Chinese children was then undertaken which confirmed that selenium could be used to prevent the disease. Chinese researchers later established a negative association between selenium status in children and the incidence of Kashin–Beck disease, a form of osteoarthritis.[12]

Thus, selenium today is characterized as an essential trace element in humans as well as in animals, with low intakes associated with deficiency symptoms* and high intakes associated with toxicity symptoms. The margin between effective level and toxic level is relatively narrow. In between the two extremes of deficiency and toxicity lies a range of selenium intakes that, generally, serves to reduce damage from oxygen free radicals that would otherwise cause the destruction of polyunsaturated fatty acids in cell membranes.[16]

2.2   Anti-carcinogenic Agent

The history of selenium as an anti-carcinogenic agent can be traced back to 1912 when a report appeared in the French periodical *La Province Medicale* on the intravenous injection of 'colloidal selenium A' to treat two cancer patients successfully.[17] In more modern times, the investigation of selenium as an anti-carcinogenic agent begins in 1966 with Shamberger's epidemiological study. Experimental investigations of selenium's role in the prevention of cancer began in laboratory animals and were quickly followed by epidemiological studies of human populations, primarily in the 1970s and 1980s.

2.2.1   Animal Studies

Animal studies have typically focused on the effects of selenium supplementation on induced cancer incidence and mortality, rather than on the effects of selenium depletion on the development of cancer in test animals. Most of the animal studies indicate that intervention with selenium is effective in inhibiting the development of chemically induced, spontaneous, and transplanted tumours in laboratory animals. These studies have been summarized by Ip;[18] most, it should be noted, involved the use of what might be characterized as relatively high doses of selenium. In a typical experiment from this period, Overvad *et al.* showed that an intervention with drinking water containing high levels of selenium inhibited

---

* It should be noted that pure selenium deficiencies, in the traditional sense, are difficult to induce and difficult to diagnose.

the incidence of skin cancer induced in hairless mice exposed to ultraviolet light.[19] Ip investigated the interaction of vitamin E and selenium in laboratory rats that had been given tumour-inducing 7,12-dimethylbenz[a]anthracene (DMBA). High levels of vitamin E were seen to improve the protective effect of selenium while low vitamin E levels lessened the anti-carcinogenic effects of selenium.[20] Many of the experiments with laboratory animals called into question the assumption that selenium achieved anti-carcinogenic effects solely by virtue of its function in the antioxidant activity of the GSH-Px enzymes, as the anti-carcinogenic effect was seen to increase at levels beyond saturation of these enzymes.

2.2.2   Human Studies

The early work on selenium and cancer in humans was of an epidemiological nature. In the 1970s, Schrauzer *et al.*[21–24] surveyed selenium intakes in more than 20 countries and found that an inverse association existed between the estimated level of selenium intakes and the incidence of various forms of cancer. Other correlation studies conducted by Shamberger and colleagues in the same period, and by Clark in the 1980s, also indicated that selenium levels were inversely related to cancer mortality rates.[21–25] Similarly, comparisons of blood or plasma selenium levels in cancer patients with selenium levels in healthy controls showed a generally consistent negative association between blood/plasma concentrations and cancer risk.[26,27] Similar relationships have been established between selenium levels and risk of cardiovascular disease.[28] While these epidemiological studies provided useful information, they could not be used to establish a cause–effect relationship, since other factors may have had an influence on the result.

3   Clinical Intervention Trials

3.1   Linxian Study

In 1993, the results from a large nutritional intervention trial conducted in Linxian County in rural China were published in the *Journal of the National Cancer Institute*.[1] 29 584 individuals aged 40–69 years had been recruited in 1985, and their cancer incidence and mortality rates had been followed until 1993. The arm of the trial in which participants had received a combined antioxidant supplement containing 50 μg of selenium (*as selenium yeast*), 30 mg of vitamin E and 15 mg of β-carotene daily showed a 9% statistically significant ($P = 0.03$) lower total mortality. Cancer mortality in this arm was reduced by 13% ($RR = 0.87$; 95% $CI = 0.75–1.00$) (Figure 1).

3.2   Clark Study

In 1996, Clark and colleagues reported the results of the National Prevention of

Participants: (29,584)   Age: 40 - 69   Duration: 5¼ years

Supplements	Endpoints	% Reduction
Selenium (50 µg)	Total mortality	9*
Beta-carotene (15 mg)	Cancer mortality	13*
Vitamin E (30 mg)	Gastric cancer mortality	21*
	Mortality from other cancers	20*

*Statistically significant results.

Blot, WJ et al, J Natl Cancer Inst,1993, 85, 1483-1492

Figure 1   *Nutrition intervention trials in Linxian, China: general population study results*

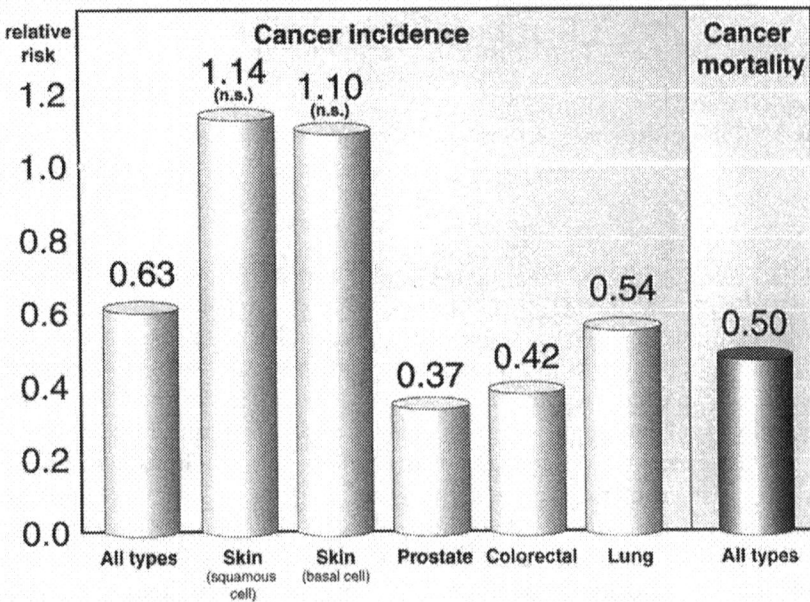

Larry C. Clark et al, JAMA, 1996, 276, 1957-1963

Figure 2   *Effects of selenium supplementation for cancer prevention*

Cancer (NPC) Study in the *Journal of the American Medical Association.*[3] 1312 skin cancer patients had been enrolled in a multi-centre, randomized, placebo-controlled, double-blind trial in 1983, administered 200 µg of selenium (*as selenium yeast*) or placebo daily, and followed until 1993 for a total of 8271 person-years. There was no effect of the selenium intervention on the trial's primary endpoints, the incidence of basal and squamous cell carcinoma, but there were statistically significant reductions in several secondary endpoints as follows: 37% in total cancer incidence, 63% in prostate cancer incidence, 58% in colorectal cancer incidence, 46% in lung cancer incidence, and 50% in total cancer mortality (Figures 2 and 3).

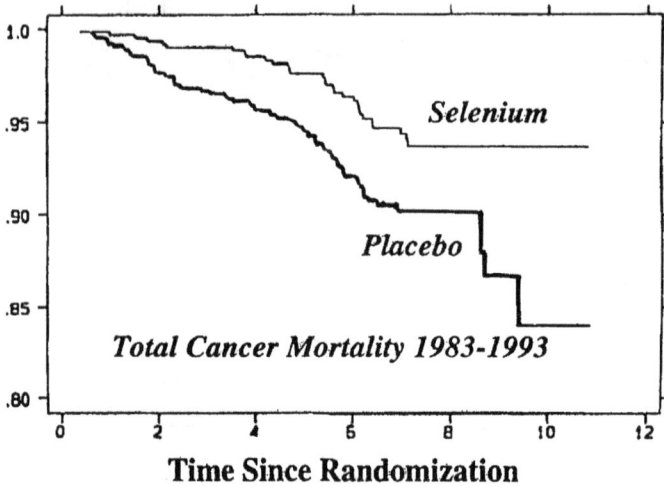

Clark LC et al., JAMA: 276: 1957-1963 (1996)

Figure 3   *Kaplan–Meier survival estimates, by treatment group*

Clark *et al.* reported the NPC prostate cancer results separately in the *British Journal of Urology*.[2] The most interesting aspect of the prostate cancer results was the relationship between the baseline plasma selenium level in the study participants and the treatment effect. Those patients who started the trial with plasma selenium concentrations in the lowest tertile and in the middle tertile had statistically significant reductions in incidence of prostate cancer (see Figure 4). The prostate cancer results elicited positive comment from Giovannucci in *The Lancet*, to the effect that 'even if the other potential benefits of improved selenium status were disregarded, the evidence available for prostate cancer seems to justify the further assessment of increasing selenium intake in the population as a priority for public health'.[29] Taylor and Albanes wrote in an editorial in the *Journal of the National Cancer Institute* that '. . . we have arrived at a new plateau of promise with . . . a strong desire to get to the "final chapter in the story" regarding selenium, vitamin E, and prostate cancer. The chapter may be as close as one randomized trial away'.[30]

3.3   Bonelli Study

At the 8th annual research conference of the American Institute for Cancer Research in Washington DC in 1998, Bonelli *et al.* reported a statistically significant reduction ($RR = 0.51$, $CI = 0.27$–0.95) in the recurrence of meta-chronous adenomas of the large bowel, through intervention with 200 μg of selenium (*as* L-*selenomethionine*), 30 mg of vitamin E and other antioxidants daily for a period of five years in a placebo-controlled, double-blind, multi-centre trial

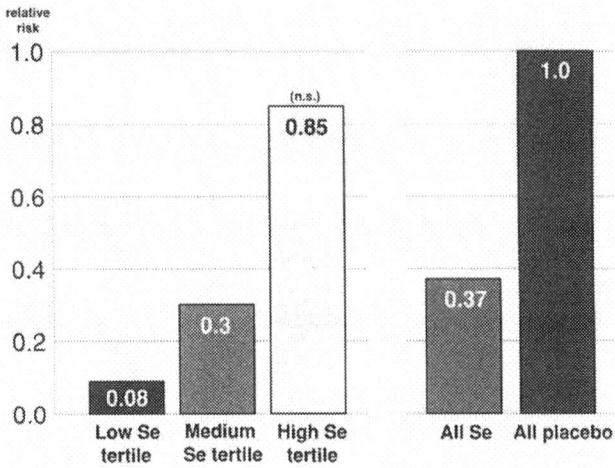

Clark LC, Dalkin B, Krongrad A, Combs GF, Turnbull BW, *et al.* **British Journal of Uroglogy**, 81, 730-734.

Figure 4 *Prostate cancer incidence: treatment effect according to baseline Se level*

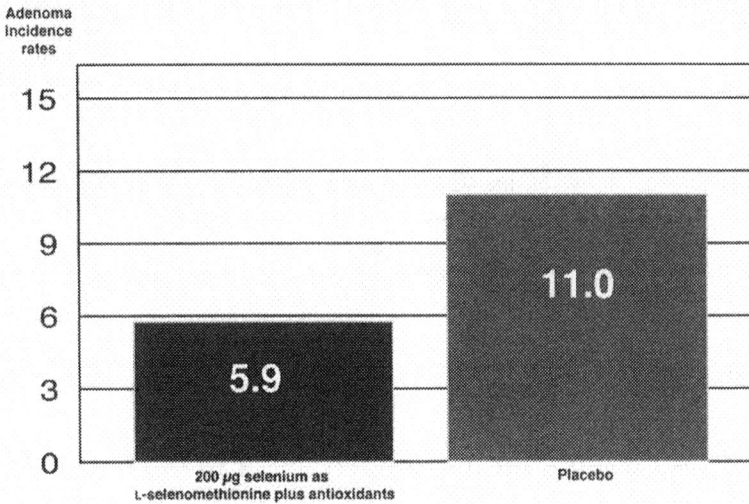

Bonelli L, Camoriano A, Ravelli P, Aste H, et al. Proceedings of the 8th Annual Research
Conference, American Institute for Cancer Research, Washington D.C., 1998, 5.

Figure 5 *Chemoprevention of metachronous adenomas of the large bowel by means of
selenomethionine and antioxidant vitamins: a randomized, double-blind,
placebo-controlled trial*

in northern Italy (Figure 5).[6,*] These results are interesting because it is known
that most cases of colorectal cancer arise from adenomatous polyps, and because

* The dosage was two *Bio-Selenium + Zinc*[R] tablets per day. Each tablet contained 100 μg selenium
as L-selenomethionine, 15 mg zinc, 15 mg vitamin E, 90 mg vitamin C, 2.2 mg vitamin B6, and 1 mg
vitamin A.

colorectal cancer is the second leading cause of cancer morbidity and mortality in Italy. They are also interesting in that they involve the use of selenium not as selenium yeast but in the form of L-selenomethionine.

3.4 Summary

Of the primary and secondary cancer prevention trials, only one is conclusive with regard to selenium as a single substance (Clark *et al.*),[2,3] as the other two studies (Linxian[1] and Bonelli[4-6]) also used other antioxidants. All three studies did, however, use organic forms of selenium (Table 1).

4   The PRECISE Trial

At the time of writing (2001), attempts are being made to find additional funding for a definitive study to clarify whether selenium supplements, in the form of high selenium yeast tablets (*SelenoPrecise* tablets®, Pharma Nord, Vejle, Denmark, can reduce the risk of cancer. The PRECISE (Prevention of Cancer through Intervention with Selenium) Study, a randomized, placebo-controlled, double-blind study enrolling 32 000 healthy participants in the age group 60–74 years, equally distributed between men and women, will be conducted at centres in Denmark, Great Britain, Sweden, and the USA. Trial participants will receive daily supplements of 100, 200 or 300 μg selenium or placebo for at least five years. Enrolment of participants began at Odense University Hospital in Denmark in November 1998 and at centres in Great Britain in the autumn of 1999, with centres in Sweden scheduled to come on line as soon as funding has been secured. The PRECISE trial will have 90% power to detect a 15% reduction in total cancer incidence and a 32% reduction in prostate cancer incidence. If the results of the PRECISE Trial confirm the results of the Clark Study and the Linxian Study, this will have significant implications for public health policy-making in the 21st century.

5   Species of Selenium in Foodstuffs and Dietary Supplements

5.1 Selenium in Foodstuffs

There is wide variation in the content of selenium in foodstuffs, but it should be noted that the most frequently consumed foodstuffs generally have a low content of selenium. Nuts, in particular brazil-nuts, offal and seafood provide the highest yields. There is also considerable variation in selenium content in cereals and grains from region to region, depending upon the extent to which selenium is available in the soil. Rayman has reported that daily selenium intakes in Great Britain fell from an average of 60 μg in 1975 to 34 μg in 1994 and attributes this substantial fall, in part, to the shift from imported selenium-rich North American wheat for flour to relatively selenium-deficient European and British varieties.[31]

Table 1 *Intervention trials with selenium*

Trail	Enrollees	Form of selenium	Duration	Cancer sites
Linxian study	(29 584)	50 µg as selenium yeast	5 years	Total cancer mortality reduced
Clark study	1312	200 µg as selenium yeast	10 years	Prostate, lung and colorectal cancer incidence reduced; total cancer mortality reduced
Bonelli study	304	200 µg as 1-selenomethionine	5 years	Colorectal adenomas reduced

Generally speaking, little information exists on the proportional levels of the various organic and inorganic forms of selenium in foodstuffs. The proportion of L-selenomethionine in wheat and yeast seems to be high.[32] Schrauzer has stated that 'selenium in cereals is found mainly in the form of L-selenomethionine'.[33] It is difficult to assess which foodstuffs may have a beneficial effect against cancer, since so little is known about the quantities and relationships of different chemical forms in products made from wheat, meat or fish.

The chemical forms of selenium most frequently ingested by humans include, but are not necessarily limited to, the following organic and inorganic compounds:

- Selenocysteine (mostly from meat products), $HSeCH_2CH(NH_2)COOH$
- Selenocystine (oxidized form of selenocysteine), $HOOCCH_2(NH_2)SeSeCH(NH_2)COOH$
- Selenomethionine (mostly from plant foods), $CH_3SeCH_2CH_2CH(NH_2)COOH$
- Selenate, $SeO_4{}^{2-}$
- Selenite, $SeO_3{}^{2-}$
- Selenium yeast, structure unknown, not specified
- Elemental selenium, Se (not biologically available)
- Aromatic selenium compounds, *e.g.* triphenylselenonium ion, diphenyl selenide, methyl phenyl selenide

5.2 Selenium in Dietary Supplements

Nève reviewed known selenium supplementation studies in 1995 and concluded that the chemical form of selenium in supplements 'varied widely, ranging from simple inorganic (sodium selenite and selenate) and organic (selenomethionine) forms, to more complex derivatives such as Se-enriched yeast or Se-rich food (wheat or meat)'.[34] Most studies of selenium supplementation with isolated selenium species have involved the inorganic form sodium selenite or the organic form L-selenomethionine, which are the two most commonly used standardized, single-form selenium supplements. It is typical of selenium supplementation studies, that there is greater variation in the chemical forms of the supplement than there is in the dosages used, generally 100–300 µg per day. The inorganic forms (sodium selenite and selenate) and the organic L-selenomethionine form used in selenium supplements are single chemical entities with known compositions.*

This is not true of the selenium-yeast supplements which are produced by culturing brewers'/bakers' yeast in a selenium-rich medium of molasses and then sterilizing and harvesting the micro-organisms that have incorporated and metabolized the selenium. Nève contrasts Finnish-produced selenium yeast tablets allegedly containing approximately 50% selenomethionine and unspecified amounts of selenoglutathione, selenodiglutathione, selenocysteine, sodium selenite and other unidentified compounds with selenium yeast tablets produced in the USA and thought to contain approximately 85% selenomethionine and 15% other unspecified compounds.[35] Ip, in a later review article, indicates that more sophisticated methods to determine the content of the different forms of selenium in selenium yeast reveal an actual L-selenomethionine content of no more than 20%; the remainder of the selenium yeast consists of selenocystine, *Se*-methylselenocysteine, selenoethionine (also approximately 20%), and other unspecified forms that, combined, represent nearly half of the total content of selenium in selenium yeast.[36] It is entirely conceivable that one or other composition of selenium yeast will yield a better anti-carcinogenic effect than other formulations, since the chemical form of the selenium has been seen to have definite consequences for the absorption, transport, retention and elimination of the substance following ingestion.

5.3   Selenium in Body Fluids and Tissues

The other side of the coin is, of course, the speciation of the various chemical forms of selenium following ingestion. Thomson has pointed out that 'animal and human studies have established that the bioavailability of the selenium depends upon the chemical form, which also influences the distribution of selenium in the body'.[37] This is not to say that chemical form alone is decisive; other factors, including a person's selenium status, dietary habits, physiological condition *etc.* also influence the bioavailability and distribution of ingested selenium.

The chemical form of the selenium has an important effect on the absorption and distribution of selenium in plasma, serum and whole blood, as well as on platelet GSH-Px activity. The distribution of selenium in human body fluids has also been seen to differ from the distribution of selenium in animal body fluids, due at least in part to differences in the form of selenium ingested.[38] In tissue, most of the retained selenium is bound to proteins; for this reason, speciation of

---

* It should be noted that selenocysteine is too unstable for use in supplemental tablets. L-seleno-methionine can be used in supplements, but, due to its characteristic odour, it is difficult to disguise in double-blinded trials. As opposed to the above substances, selenium yeast is sometimes seen to give rise to reports of allergic reactions.

selenium in tissue is primarily a matter of determination of selenocysteine and selenomethionine in the various selenoproteins and selenium-containing proteins.[39]

Increasingly, there is a need for research into the ways in which the various forms of inorganic and organic selenium are incorporated into proteins, which compounds in body fluids and tissues the various selenium species are associated with, what functions (especially protection against tumourigenesis) the various forms of selenium fulfil, and by what mechanisms they work. Particularly of interest in this regard is the question of how specific species of selenium change or remain stable following absorption.

6 Selenium Metabolism in Humans and Animals

As mentioned above, much of the selenium in the body is bound to proteins. It is useful to distinguish between selenoproteins, of which GSH-Px enzymes and selenoprotein P are examples, and selenium-containing proteins. Animals are capable of synthesizing selenocysteine from inorganic selenium forms but cannot synthesize selenomethionine, which they must obtain from exogenous sources. Selenoproteins are metabolically active proteins that contain endogenously synthesized selenocysteine. Proteins, *e.g.* in skeletal muscle, that incorporate exogenously obtained selenomethionine or selenocysteine, are called selenium-containing proteins. Exogenously acquired selenocysteine can be incorporated into the selenium-containing proteins but not directly into the selenoproteins.[40]

6.1 Absorption

Selenium is seemingly ingested in many forms. The most important forms seem to be the selenomethionine form (from plants) and the selenocysteine form (from animal proteins). Two things can be said about the absorption of selenium forms: little is known about the manner in which they are absorbed in the intestinal tract, and all forms seem to be absorbed easily under normal conditions as follows: the organic form L-selenomethionine > the inorganic form selenate > the inorganic form selenite.[7] Research into the absorption process as it relates to chemical form and into other factors affecting absorption (for instance, positive or negative interactions) is needed.

6.2 Transport

Metabolic pathways for selenium are not well characterized. Selenium is transported bound to proteins, but which proteins bind and transport which forms is not well defined. Selenomethionine presumably has a metabolic fate similar to that of methionine, and it is known to be easily substituted for methionine in the selenium-containing proteins, particularly in skeletal muscle.[7]

6.3 Storage

Selenium seems to be present in all human tissue. The content varies from approximately 3 to 6 mg of whole body selenium in selenium-poor New Zealand, to 13 to 20 mg in areas of more adequate selenium quantities. Concentrations are highest in the liver and the kidneys, but these organs only account for 8% and 4%, respectively, of the total body content.[7] Research results indicate that the metabolism and storage of selenium varies according to the chemical form ingested, and that the storage of the organic form selenomethionine in skeletal muscle protein accounts for a significant percentage (approaching 50%) of the total body selenium pool. Interestingly enough, cardiac tissue has a higher concentration level than skeletal muscle tissue.[7]

The body's store of selenium appears to be divided between two primary compartments or pools: the one containing the incorporated selenomethionine in various selenium-containing proteins, and the other consisting of selenocysteine in the various selenoproteins. Levander and Burk[40] suggest that decline in selenium concentration in the body will lead to a depletion of the stores of selenomethionine, in order to maintain the more tightly regulated stores of selenocysteine, which is the form known to be biologically active. Thus, selenomethionine ingested in plant foodstuffs will eventually make its way to both tissue compartments: first by incorporation into selenium-containing proteins in which it substitutes for methionine and, secondly, as selenocysteine in selenoproteins once it has been catabolized. Selenium ingested in the form of selenocysteine or in inorganic forms will only be found in the second of these two compartments, *i.e.* in selenoproteins.[40]

In summary, it seems probable that organic selenomethionine is readily stored in skeletal muscle whereas inorganic selenite and selenate forms are not as easily stored but remain in the second body pool for immediate use in the synthesis of selenoproteins.[7] The extent to which selenomethionine stored in tissue can be released and utilized in the biologically active selenoproteins is of much interest.

6.4 Elimination

Selenium is excreted primarily in the urine. Selenium excreted in the faeces is usually selenium from unabsorbed dietary selenium. Dermal and respiratory excretion are not significant except in the case of toxic levels of intake. Urinary elimination correlates strongly with level of intake. Intakes of organic selenomethionine results in lower urinary elimination rates than do intakes of inorganic selenite and selenate.[7]

7 Selenium Intake Levels

Estimated total dietary selenium intakes vary considerably from region to region depending upon the amount of selenium in soil available for uptake by plants. A survey undertaken by the UK Ministry of Agriculture, Fisheries and Foods in

1994 found that the intake of selenium in Britain was approximately 34 μg per day.[2,31] This level of daily selenium intake would fall within the range of the lowest tertile of patients in the National Prevention of Cancer (NPC) Trial conducted in the south-eastern USA where the average dietary selenium intake is estimated to be approximately 90 μg.[2,31] By way of contrast, the Eskimo population in northern Greenland is reported to have a calculated selenium intake of 1284 μg per day.[41]

7.1 Safe and Adequate Levels

An indication of adequate levels of intake can be derived from the data in the NPC intervention trial, in which an average daily dietary intake of 90 μg, combined with a dietary supplement of 200 μg of high-selenium yeast in tablet form daily, could be seen to be a sufficient and safe amount for the statistically significant reduction of cancer incidence and mortality rates.[2] Whether it was necessary to raise the selenium intakes of grown men and women by 200 μg per day in order to obtain the same results remains unclear; it is to be hoped that the PRECISE Trial in Europe and North America, with its more differentiated dosages (100, 200, 300 μg per day) and equal distribution between the sexes will provide an answer to this question.

7.2 Recommended Levels

The Recommended Dietary Allowances established in the USA set intakes for selenium at 70 μg for men and 55 μg for women.[42] Current dietary reference values for the UK are set at 75 μg per day for men and 60 μg per day for women.[43] The 1996 Nordic Nutritional Recommendations prescribe 50 μg per day for men and 40 μg per day for women.[44] The recommendation for selenium for the Nordic countries seems to have been set for practical reasons to equal actual dietary intake: 'A higher intake than 60 μg per day cannot be obtained from Swedish food nor from Danish food, without enrichment or unrealistic increases in fish consumption'.[45] [Author's translation of the Swedish text] In all cases, the recommendations can be seen to fall well below the level that proved to be effective and safe in the NPC Trial. Furthermore, there is no specification as to the form of selenium ingested in any of the recommendations.

7.3 Upper Safety Levels

The intake level of approximately 290 μg per day on average that was achieved in the NPC Trial resulted in no signs of toxicity[3] and remained well under the 'maximum safe intake' level of 450 μg per day set by the European Union.[46]

8 Uptake Variability of Selenium Species

Absorption can be defined as the rate and extent to which the active substance or therapeutic moiety is taken up and assimilated by the body, rather than passed through and eliminated in an unchanged form. Bioavailability, on the other hand, means 'the rate and extent to which the active substance or therapeutic moiety is absorbed from a pharmaceutical form and becomes available at the sites of action'.[47] A review of the literature on selenium reveals that some work has been done on the absorption of selenium, although relatively little is known about the mechanism of absorption in the intestine; considerably less work has been done on the bioavailability of selenium in the sense of the above definition. The difficulty obviously lies in a finding a consensus identification of 'the sites of action' for selenium.

Two trends can be seen in the available research literature on the topic of the absorption and bioavailability of selenium:

- the use of different forms of selenium results in differing responses
- the greatest variability is to be found in the response to selenium yeasts

We will illustrate these two points by way of reference to the following studies.

In 1989, Clausen *et al.* published the results of a comparison of ten selenium supplementation products, selected to represent all of the major types commercially available at the time.[48] Clausen's study enrolled 135 healthy individuals aged 20–62 years who were randomly assigned to ten experimental groups. Three months of supplementation with 200 μg of selenium were followed by four months of withdrawal. Whole blood selenium content and GSH-Px activity (whole blood and erythrocyte) were measured at onset, after one and three months of supplementation and at the end of the withdrawal period. The study was double-blinded with regard to which preparations were given to which participants.

Clausen's results showed an effect on whole blood selenium concentrations as follows:

organic L-selenomethionine > selenium yeast > inorganic selenate > inorganic selenite,

i.e. preparations containing organic forms of selenium gave a better response than did preparations containing inorganic forms; furthermore, the values for the organic forms continued to rise throughout the three-month period of supplementation while values for the inorganic forms stagnated after one month of supplementation. After four months of withdrawal, the whole blood levels in participants administered organic forms were still higher than they had been at baseline; this was not true for participants who had received inorganic forms (Figure 6).

As regards GSH-Px activity, Clausen noted that the difference in GSH-Px activity caused by preparations based on L-selenomethionine and those based on inorganic forms was statistically significant in favour of the L-selenomethionine preparations.[48] The yeast forms gave varied responses.

Clausen summarized his results thus: 'The present data clearly indicate the significant differences between the response from organic and inorganic Se,

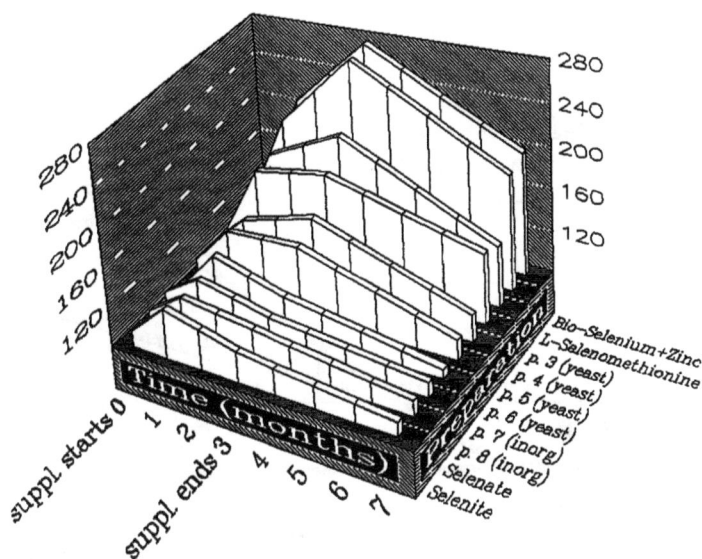

Figure 6 *Changes in whole blood selenium content (µg l⁻¹) from the intake of different species of 200 µg selenium*

respectively. L-Se-methionine had a stronger and more stable effect on Se status than Se-yeast and inorganic Se. Se-yeast tended to produce a mixed response – probably reflecting an actual composition of these preparations as a combination of organic and inorganic Se compounds'.[48]

Other workers have since pointed out that while L-selenomethionine is superior in raising and maintaining blood and GSH-Px concentrations, its use as a supplement in studies with laboratory animals does not necessarily confer as high a degree of cancer inhibition as do other forms of selenium.[49,50] It seems as though the metabolism of the differing chemical forms may be responsible for differing anti-carcinogenic actions.

In a Norwegian study conducted by Hommeren,[51] three groups of healthy men aged 20–49 years were given a commercial preparation containing selenium in the form of L-selenomethionine, a commercial preparation containing 'pea selenium in which the selenium is bound in the form of selenocysteine' (*sic*), or placebo. Supplementation with L-selenomethione increased whole blood selenium levels significantly more than did supplementation with pea selenium. In fact, the effect of supplementation with the pea selenium did not differ at all from the effect of the placebo. The Hommeren study illustrates that the use of an 'organic selenium' is no guarantee for good absorption or bioavailability and suggests that there is excessive variability in the quality of selenium preparations on the market. It also makes clear the need, on the part of producers and regulatory agencies, for research on selenium speciation.

A study conducted by Teherani and associates in Austria in 1991, for registration files for the preparation *Bio-Selenium + Zinc®*, determined the selenium concentration in several body compartments: whole blood, urine, lymphocyte

micrograms

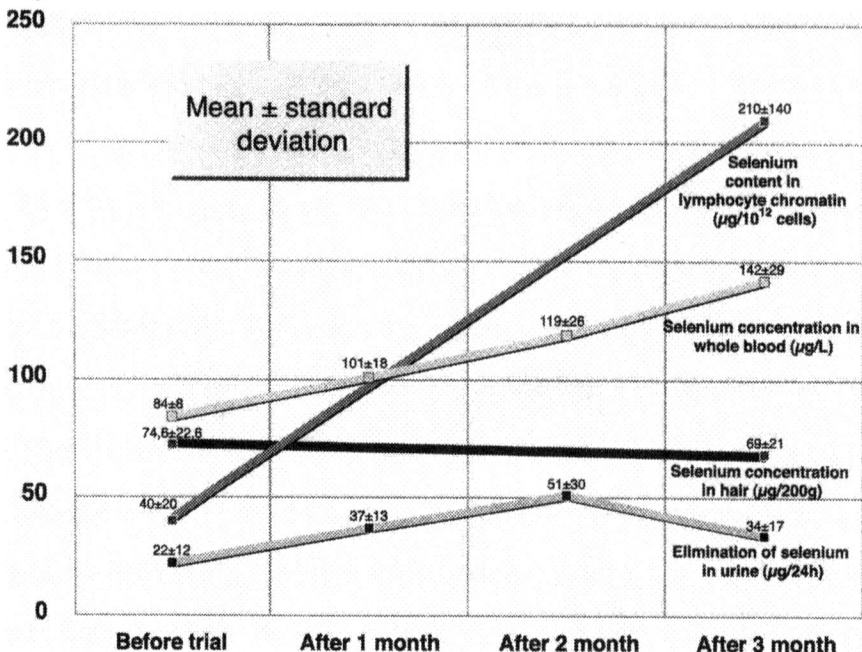

D. Teherani, Unpublished Registration data, Bio-Selenium+Zinc, Seibersdorf, Austria, 1991.

Figure 7   *Se concentrations in various compartments during 3 month supplementation with 100 μg Se as L-selenomethionine*

chromatin and hair in adult human subjects supplemented with 100 μg of selenium as L-selenomethionine per day for a period of three months.[52] Selenium concentrations in whole blood, urine, and lymphocyte chromatin increased throughout the entire period of supplementation. In contrast, hair concentrations did not change. Figure 7 shows the average rise in selenium content in each of these three compartments.

Generally speaking then, selenium is well absorbed. The organic form L-selenomethionine gives the best absorption, and the absorption of selenium yeast varies from brand to brand. The mechanisms of absorption for the inorganic forms differ, of course, from the mechanism for L-selenomethionine, which seems to share the same mechanism as methionine. Selenite seems to be absorbed through passive diffusion in the small intestine while selenate is thought possibly to share the same pathway as sulfur. Under optimal conditions, the rates of absorption are approximately 95–98% for L-selenomethionine, 90–95% for selenate and about 60% for selenite.[7] Different forms of selenium yeast vary greatly in absorption and bioavailability.

The lack of identification of active sites for selenium in cancer prevention, and the lack of consensus as to which species of selenium are needed in order to influence active sites, makes it necessary to focus, in clinical trials, on the forms and species of selenium that have actually been proven to be clinically active in human studies. Standardization of methods for production of selenium yeast which yield the same proportions between the species over time, and identification of these species, is a matter of priority for progress at this stage. The counting of selenium atoms as a standard for the declaration of selenium is clearly not sufficient as we can see in the examples above.

9 Speciation of Selenium

For a discussion of the various species of selenium, see Section 5. As regards speciation of selenium, various methods for the determination of selenium concentration in body fluids and tissues have been employed with varying degrees of success: fluorimetry, neutron activation analysis, atomic absorption, mass spectrometry, HPLC *etc.*[40] In latter years, researchers have recognized, however, that the problem with these methods lies as much in the interpretation of the results as it does in the methods. Thomson[37] points out that total selenium composition, for example, 'is less important than levels of individual seleno-proteins' and can be misleading. What is needed, she says, are new methods of assessment that determine the availability of various forms of selenium for specific biochemical functions, in particular, the levels and distribution patterns of the various hierarchical selenoproteins and how they respond to 'deficient, adequate, and toxic levels' of selenium intake.[37]

In this context, it should also be noted that significant differences exist between animal species and also within the same animal species with regard to the metabolic fate of differing chemical forms of selenium. For example, Thomson cites the case of high levels of selenoprotein W in sheep heart and nearly non-existent levels in rat heart,[37] and Whanger *et al.*[38] note that, in rats, selenite supplementation leads to increases in GSH-Px in red blood cells, while supplementation with selenomethionine leads primarily to increases in the selenium content associated with haemoglobin in the red blood cells. For obvious reasons, it is not acceptable to extrapolate effects from studies of other species to humans.

9.1 Selenium in Foodstuffs

Selenium becomes a part of the human food chain whenever it is incorporated into compounds in plants that otherwise would contain sulfur. It is present in plants primarily in the form of selenomethionine but also, to a much lesser degree, in the form of selenocysteine and selenium analogues of sulfur amino acids. It is present in the form of selenocysteine in animal proteins. The extent to which inorganic forms of selenium occur in foodstuffs is uncertain. One of the problems with studies examining food selenium intakes and cancer risk has been the failure to specify which selenium compounds were ingested in which proportions.

9.2 Selenium in Dietary Supplements

Failure to specify the form of selenium has not been a problem in studies in which supplements straightforwardly containing sodium selenite or L-selenomethione were used; it has given rise to many questions, however, concerning studies in which selenium yeast has been employed. Put bluntly, the extent to which the various selenium yeast preparations contain L-selenomethionine, as well as other organic and inorganic forms of selenium, is too uncertain to permit the drawing of any conclusions as to which forms, or combinations of forms, may have an effect on the element's metabolic fate in the body, its bioavailability for various biochemical functions including a cancer inhibiting role and, presumably, even its toxicity.

The validity of the clinical trial results is thus limited to the actual preparation put to the test, and to the extent to which it is possible to reproduce, exactly, the same preparation. Consequently, precision in raw material production, batch consistency, and stability – also with regard to speciation – is of crucial value in the use of these results. A major task for the PRECISE trial was the manufacture of a pharmaceutical-grade species-constant yeast (SelenoPrecise) using pure pharmacopoeia raw materials.

9.3   Selenium in Body Fluids and Tissues

Speciation of selenium in the organism itself is, seemingly, more important than speciation in the ingested materials, although the two would seem to be related to a considerable degree. A variety of methods has been developed to determine selenium concentrations in the body's fluids and tissues; techniques have also been developed for the speciation of the chemical forms of the element in the body.[38] Of special interest is the stable isotope method in which selenium yeast is grown with the rare stable isotope [74]Se by which it can be detected in various body compartments by the use of isotope ratio measurements. Accurate speciation of selenium forms would seem to require a good understanding of the varying metabolic fates of the various chemical forms, just as selenium speciation should be regarded as a tool for the charting of the distribution of selenium in various proteins in the body.

9.4   Methods of Speciation

Whanger *et al.*[38] summarize the major findings of speciation studies of body fluids of adult humans in the USA (Oregon), New Zealand and China in a 1994 review article and discuss the methods used to assess selenium content in blood fractions. Thomson,[37] in a 1998 article, discussed the implications of the recent work in selenium speciation.

Behne *et al.*[39] outlined in a 1998 article the strategies and procedures used 'in the identification, characterization, and determination of the selenium species present in the mammalian organism'.

Michalke and Schramel reported the use of capillary zone electrophoresis–inductively coupled plasma mass spectrometry and capillary isoelectric focusing–inductively coupled plasma mass spectrometry for the speciation of both organic and inorganic forms of selenium in standard mixtures and in body fluids.[53] Marchante-Gayon *et al.*[54] reported on the use of electrothermal atomic absorption spectrometry for the speciation of selenium in human colon cancer tissue. Suzuki *et al.*[55] report on the speciation of selenium in biological samples using HPLC–ICP-MS, as do Shibata *et al.* (Japan).[56]

9.5 Pharmaceutical Quality Control

A simple method, routinely used in the Pharma Nord Quality Control Department in Denmark for the speciation of inorganic selenium compounds in selenium yeast preparations, is based on the knowledge that selenite (SeO_3^{2-}) can be reduced to elemental selenium by ascorbic acid and that both selenite and selenate (SeO_4^{2-}) can be reduced to elemental selenium by stannous chloride $(SnCl_2)$. The elemental selenium appears as an orange/reddish precipitate. Since the organic selenium compounds cannot be reduced to elemental selenium by either ascorbic acid or stannous chloride, this method provides useful information concerning the extent and proportion of the inorganic selenium compounds in selenium yeast preparations. This method was derived from the work of G. Gissel-Nielsen.[57]

During the 20 years that we have been working with selenium preparations, we have seen great variation in the quality of the raw material. Particularly in the 1980s, it was common to find yeasts 'salted' with selenite or selenate.[58] Fraudulent raw material labelled L-selenomethionine has been found to consist of a mix of methionine, valine, and selenite, or to be selenomethionine in the DL-form.[59] Careful quality control has been, and continues to be, a crucial part of production, but it should be extended to include speciation of selenium.

10 Public Health Implications of a Positive Population Study

If the PRECISE Study outlined above should yield statistically significant, positive results concerning the role of selenium in cancer inhibition at some time in the middle of the next decade, this would seem to have far-reaching implications for public health policy. As soon as steps are taken, however, to devise a method of increasing selenium intakes in the general population in geographical regions with low naturally occurring selenium levels, problems will present themselves. It will be logical to ask, for example, whether dosages that have proven effective in persons aged 60–74 years also are appropriate for younger persons and whether there have been gender-specific variations in dosage responses.

Questions about which organic forms in the *SelenoPrecise*® formulation have been absorbed and have had which effects in the body will assume even greater importance if the PRECISE Study yields statistically significant, positive results, as will questions about which forms of selenium are found at various stages of the food chain and questions about which forms should and may be used in functional foods.

There is an additional, foreseeable problem of a legalistic nature: in accordance with current European Union regulations (EEC-65/65), substances that have a preventive effect against recognized diseases may only be sold as a registered drug. Enforcement of this regulation could lead, in an extreme scenario, to the banning of the specific dietary supplement formulation that has been shown to be effective, while leaving other unproven formulations, more or less unregulated, on the market. At some point, it will be necessary to decide whether clinically tested selenium preparations are to be handled as prescription drugs, over-the-counter drugs or dietary supplements, and to decide what to do about selenium preparations that have not been tested in clinical settings. These questions are important, because the present state of public health legislation and regulation in most countries of the European Union only permits the making of health claims for drugs and not for supplements. This means, in practice, that many Europeans are being 'protected from' access to information about the effects of substances contained in dietary supplements.

When the makers of public health policy sit down to consider their options *vis-à-vis* selenium and cancer prevention, they will be confronted with, essentially, three possibilities: fortification of foodstuffs and feedstuffs, enrichment of crop fertilizers, or selenium supplementation. Regardless of the method chosen, many questions concerning the form of selenium to be employed will have to be answered, and the answers will require a broader and deeper base of knowledge – including speciation – than we have today.

10.1   Fortification of Foodstuffs and Feedstuffs

Fortification of commonly ingested foodstuffs, *e.g.* bread or milk, with substances such as calcium or vitamin D is sometimes suggested as a way of ensuring good compliance. Fortification is seen as a way to deal with some of the problems with supplements; for example, people tend to forget to purchase them or to take them once they have purchased them.

Fortification is not, however, a panacea. It would be folly to assume that each and every individual will eat a quantity of the fortified foodstuff that precisely delivers the required dosage. In the case of selenium, this can be a significant issue: the margin between selenium's potent dosage and its toxic dosage is relatively narrow. If bread is the chosen medium for fortification, teenagers who eat large amounts of bread may risk ingesting, involuntarily, daily dosages that are excessively high for their age and stage of development. Elderly persons who eat very little bread will get too little of the fortified substance. Pregnant and lactating women and children represent other obvious problem groups.

The main problem, however, is to determine the forms of selenium that could be added to fortified foodstuffs. Addition of inorganic salts of selenium can no longer be regarded as an ethical approach if the only forms of selenium that have been shown to be effective are specific organic forms.

10.2   Enrichment of Fertilizers

In Finland, the enrichment of multi-mineral crop fertilizers with selenium in the form of inorganic sodium selenate (Na_2SeO_4) has seemingly proven to be an effective, safe, and controlled way of bringing the selenium intake of the whole population up to target levels. The interesting question now is whether increased ingestion of selenium-fertilized cereal foods and of food products from animals that have been fed these same cereal grains can be seen to result in reduced rates of cancer incidence and mortality.

Most of the increased selenium intake in Finland seems to derive from foods of animal origin, so direct enrichment of animal feeds with *organic* selenium compounds, as opposed to the production of animal feeds from crops fertilized with a selenium component, might also be a way to increase the selenium concentration in foods and the selenium intake of targeted populations.[60] It is not a method that would help vegetarians and people who primarily eat fish for the meat course.

Enrichment of fertilizers as a strategy suffers from the same problem as food fortification: it is a shotgun approach that makes differentiation of intakes based on individual differences such as age, gender, eating habits, condition of health *etc.* impossible. Further details on the chemical forms of selenium attained through fertilization needs to be compared with the forms that research shows to be effective in reducing cancer risk.

10.3 Selenium Tablets (Dietary Supplement or Drug)

Supplementation of the individual diet with selenium tablets would seem to provide the possibility of the most varied response to the problem of increasing selenium intakes, assuming that public health publicity campaigns would ensure good compliance. The chemical forms and the dosage levels of the selenium in the tablets, and the extent to which the tablets should be available over-the-counter, are matters for public health policy discussion. The supplementation solution would have the advantage that it can be made with the species of selenium shown to be effective, and could be given to the appropriate user groups based on gender, age or other criteria as indicated by research results.

11 Need for Research and Legislation/Regulation

Two types of research are needed at present. On the one hand, there is a clear need for clinical research that attempts to replicate and further explore the results of Clark's NPC Study. This is what the PRECISE Trial is intended to do. The importance of the PRECISE Trial cannot be overestimated. Ip has pointed out that the NPC Study is, by far, the most successful of all of the human cancer intervention studies that have been completed to date.[61]

At another level, there is a very acute need for basic research into the relationships that link selenium speciation, selenium biochemistry, and the mechanisms of selenium's anti-carcinogenic activity. We need to know more about the metabolic fate of various forms of selenium, and about the response of neoplasms at various stages of development to differing forms of selenium. Stewart *et al.*,[8] for example, have begun to examine the extent to which prooxidative effects of different forms of selenium can induce apoptosis in cancer cells.[62] Davis *et al.*[49] have also begun to investigate the differences in chemical form as they relate to differences in metabolic fate and to cancer prevention. In the future, we may well see governmental regulations specifying the activity of different selenium species, analogous to the expression of the biological activity of various forms of vitamin E in international units.

At the present time, the testing of literally hundreds of substances for possible cancer-inhibiting effects is being funded. The potential of selenium for the prevention of cancer has consistently been demonstrated in epidemiological and clinical studies. It may well be that the time has come for a concerted two-pronged investigation (basic research and clinical trials) of the potential of selenium. This would seem to require the appropriation of research funds earmarked for selenium and cancer prevention by national legislative bodies. Especially in view of the non-patentability of selenium preparations, in contrast to many other cancer-inhibiting substances, it would seem that public funding of research into the role of selenium in cancer prevention is needed.

12   Conclusion

The absorption, metabolic fate and anti-carcinogenic activity of selenium can be seen to be closely bound to the chemical form of the trace element in the diet and in supplements. At present, we do not know whether it is one or more of the specific selenium forms found in selenium yeast that has the anti-carcinogenic effect, or whether it is the case that various selenium forms found in selenium yeast can be converted in the body to give a more effective, anti-carcinogenic response. For this reason, researchers and clinicians need to use the broad-based, multi-spectrum approach to cancer prevention that a standardized, speciated selenium yeast makes possible. There might be a future stage in which isolated, specific forms of selenium, with newly discovered mechanisms of action, and with the ability to directly influence the active sites of prevention and cure, can be identified. Early and adequate funding of selenium speciation research is therefore an urgent prerequisite if we are to avoid running into a dead-end in the future. The situation today, in which only the elemental form is taken into consideration, is no longer tenable.

13   References

1. W.J. Blot, J.Y. Li, P.R. Taylor, W.D. Guo, S. Dawsey, G.Q. Wang, C.S. Yang, S.F. Zheng, M. Gail, G.Y. Li, Y. Yu, B.Q. Liu, J. Tangrea, Y.H. Sun, F.S. Liu, J.F. Fraumeni, Y.H. Zhang and B. Li, *J. Natl. Cancer Inst.*, 1993, **85**, 1483.
2. L.C. Clark, B. Dalkin, A. Krongrad, G.F. Combs, B.W. Turnbull and E.H. Slate, *Br. J. Urol.*, 1998, **81**, 730.
3. L.C. Clark, B. Dalkin, A. Krongrad, G.F. Combs, B.W. Turnbull and E.H. Slate, *J. Am. Med. Assoc.*, 1996, **276**, 1957.
4. L. Bonelli *et al.*, *Proceedings of the 6th International STDA Symposium*, Scottsdale,

Arizona, 1998, pp. 91–94.

5. L. Bonelli *et al.*, *Proceedings of the 2nd International Conference on Natural Antioxidants*, Helsinki, Finland, 1998, p. 6.

6. L. Bonelli *et al.*, *Proceedings of the 8th Annual Research Conference, American Institute for Cancer Research*, Washington DC, 1998, p. 5.

7. L.A. Daniels, *Biol. Trace Elem. Res.*, 1996, **54**, 190.

8. M.S. Stewart, J.E. Spallholz, K.H. Neldner and B.C. Pence, *Free Radic. Biol. Med.*, 1999, **26**, 42.

9. M. Ihnat, *Occurrence and Distribution of Selenium*, CRC Press, 1987, pp. 1–347.

10. *The Merck Index*, Merck & Co, Rahway, New Jersey, 1989, p. 1337.

11. *Martindale: The Extra Pharmacopoeia*, The Pharmaceutical Press, London, UK, 1993, p. 1047.

12. O.A. Levander and R.F. Burk, in *Present Knowledge in Nutrition*, eds. E.E. Ziegler and L.J. Filer, ISLI Press, Washington DC, 1996, pp. 320, 324.

13. L.C. Clark, *Fed. Proc.*, 1985, **44**, 2584.

14. J.T. Salonen *et al.*, *Br. Med. J.*, 1985, **290**, 417.

15. R.A. Shamberger, *Nutr. Res.*, 1985, **154**, 29.

16. P. Mason, *Handbook of Dietary Supplements: Vitamins and Other Health Supplements*, Blackwell Science, Oxford, UK, 1995, p. 149.

17. M. Lancien and M. Thiroloix, *La Province Medicale*, 1912, **13**, 172.

18. C. Ip, *Fed. Proc.*, 1985, **44**, 2573.

19. K. Overvad, *Cancer Lett.*, 1985, **27**, 163.

20. C. Ip, *Adv. Exp. Med. Biol.*, 1986, **206**, 431.

21. G.N. Schrauzer *et al.*, *Bioinorg Chem.*, 1977, **7**, 23.

22. G.N. Schrauzer *et al.*, *Bioinorg Chem.*, 1977, **7**, 35.

23. R.J. Shamberger, *Crit. Rev. Clin. Lab. Sci.*, 1971, **2**, 211.

24. R.J. Shamberger, *Arch. Environ. Health*, 1976, **31**, 231.

25. L.C. Clark, *Fed. Proc.*, 1985, **44**, 2584.

26. L.C. Clark, G.F. Graham, R.G. Crounce, R. Grimson, B. Hulka and C.M. Shy, *Nutr. Cancer*, 1984, **6**, 3.

27. G.N. Schrauzer, T. Molenaar, S. Mead, K. Kuehn, H. Yamamoto and E. Araki, *Jpn. J. Cancer Res.*, 1985, **76**, 374.

28. P. Suadicani, H.O. Hein and F. Gyntelberg, *Atherosclerosis*, 1992, **96**, 33.

29. E. Giovannucci, *Lancet*, 1998, **352**, 756.

30. P.R. Taylor and D. Albanes, *J. Natl. Cancer Inst.*, 1998, **90**, 1185.

31. M.P. Rayman, *Br. Med. J.*, 1997, **314**, 387.

32. P.J. Peterson and G.W. Butler, *Austral. J. Biol. Sci.*, 1962, **15**, 126; WHO task group on Se, *Environmental Health Criteria 58: Selenium*, WHO, Geneva, Switzerland, 1987, 53f.

33. G.N. Schrauzer, *Selen: Neuere Entwicklungen aus der Biologie, Biochemie und Medizin*, Verlag für Medizin, Heidelberg, Germany, 1983, p. 20.

34. J. Nève, *J. Trace Elem. Med. Biol.*, 1995, **9**, 66.

35. J. Nève, *J. Trace Elem. Med. Biol.*, 1995, **9**, 69.

36. C. Ip, *J. Nutr.*, 1998, 1845.

37. C.D. Thomson, *Analyst*, 1998, **123**, 827.

38. P.D. Whanger, Y. Xia and C.D. Thomson, *J. Trace Elem. Electrolytes Health Dis.*, 1994, **8**, 1.

39. D. Behne, C. Hammel, H. Pfiefer, D. Rothlein, H. Gessner and A. Kyriakopoulos, *Analyst*, 1998, **123**, 871.

40. O.A. Levander and R.F. Burk, in *Present Knowledge in Nutrition*, eds. E.E. Ziegler

and L.J. Filer, ISLI Press, Washington DC, 320.

41. J.C. Hansen, Doctoral thesis, Aarhus, Denmark, 1988, p. 26.
42. National Research Council, *RDA – Recommended Dietary Allowances*, National Academy Press, Washington DC, 1989, p. 220.
43. L.H. Foster and S. Sumar, *Crit. Rev Food Sci. Nutr.*, 1997, **37**, 222.
44. B.M. Sandström *et al.*, *Nordiska näringsrekommendationer 1996*, Nordisk Forlagshus, Copenhagen, Denmark, 1996, p. 186.
45. Nordic Nutrient and Food Recommendations, II, 1988, p. 60.
46. European Community, *Reports of the Scientific Committee for Food (Thirty-first series)*, Office for Official Publications of the European Community, Luxembourg, 1993, p. 204.
47. EudraLex, *Guidelines: Medicinal Products for Human Use*, Vol. 3C in *The Rules Governing Medicinal Products in the European Union*, European Commission, Brussels, Belgium, 1998, p. 234.
48. J. Clausen and S.A. Nielsen, *Proceedings, 2nd International Congress on Trace Elements; in Medicine and Biology*, Avoriaz, France, 1988, pp. 305–314.
49. C.D. Davis, Y. Feng, D.W. Hein and J.W. Findey, *J. Nutr.*, 1999, **129**, 63.
50. C. Ip and C. Hayes, *Carcinogenesis*, 1989, **10**, 921.
51. O.J. Hommeren, *Tidsskr. Nor. Lægeforen*, 1990, **26**, 3350.
52. D. Teherani, Unpublished research data, Seibersdorf, Austria, 1991.
53. B. Michalke and P. Schramel, *J. Chromatogr. A*, 1998, **807**, 71.
54. J.M. Marchante-Gayon, J.E. Sanchez-Uria and A. Sanz-Medel, *J. Trace Elem. Med. Biol.*, 1996, **10**, 229.
55. K.T. Suzuki, M. Itoh and M. Ohnichi, *J. Chromatogr. B*, 1995, **666**, 13.
56. Y. Shibata , M. Morita and K. Fuwa, *Adv. Biophys.*, 1992, **28**, 31.
57. G. Gissel-Nielsen, *IAEA-SM*, 1977, **235**, 427.
58. S.H. Hansen and S. Bager, *Farmaci*, 1989, 144.
59. S.H. Hansen and M.N. Poulsen, *Acta Pharm. Nord.*, 1991, **3**, 95.
60. A. Aro, in *Environmental Chemistry of Selenium*, eds. W.T. Frankenberger and R.A. Engberg, 1998, Marcel Dekker, New York, pp. 81–97.
61. C. Ip, *J. Nutr.*, 1998, 1845.
62. M.S. Stewart, J.E. Spallholz, K.H. Neldner and B.C. Perie, *Free Radic. Biol. Med.*, 1999, **26**, 42.

Health

CHAPTER 19

The Importance of Speciation of Trace Elements in Health Issues

RITA CORNELIS

1 Introduction

Trace element species have their origins in the environment: air (of the indoor working place, outside, urban, rural, industrial), water, sediments, *etc.* and arise from either natural or anthropogenic sources. At this point, the concentrations may be relatively high, with a major proportion being the result of various emissions or dumping practices. For many decades, such practices have been considered totally harmless, on the premise that the contaminants would be diluted in the environment down to concentrations low enough to exhibit no adverse effects on man or the biosphere. This idea has proved to be very naive in many circumstances. It completely overlooked the specific characteristics of the various trace element species in question and the transformations they undergo in the ecosystem, which can totally alter their behaviour. From the environment, the trace element species made their way inexorably into the food chain. At that point matters became more serious since man himself became contaminated with many of the harmful trace element species.

A representative historical example concerns the episode resulting in Minamata disease, which was caused by mercury intoxication. In 1956, a most serious epidemic of methylmercury poisoning occurred amongst fishermen and their families in the town of Minamata in Japan. This was the result of the discharge into the Yatsushiro Bay of mercury and methylmercury used as a catalyst in a polymer producing plant. Furthermore, micro-organisms in the water methylated the Hg^{2+}. The resulting methylmercury then accumulated in the aquatic flora and fauna, its concentration increasing with ascending order in the food chain, a process known as bioaccumulation. As methylmercury is very soluble in fat, toxic levels were eventually reached in fish which formed part of the local human diet. Eventually the fish-eating population of Minamata was severely affected with symptoms including birth defects, mental retardation, paralysis and death.

This chapter looks at the implication of trace element species in health issues, with special emphasis on industrial hygiene. The different fields that will be considered are:

- biological monitoring of trace element species by measuring biomarkers of exposure
- the study of the kinetics in the human body of trace elements resulting from industrial exposure and the influence of the chemical form of the exposure
- identification of the chemical form of the species that will become the basis for toxicity studies

The following paragraphs will set the scene by giving examples of the topics that have already been studied and identifying those areas that need to be considered.

2 Biological Monitoring of Trace Element Species by Measuring Biomarkers of Exposure

Currently, workers in the As, Cr, Ni, Pt and other industries are monitored for exposure to one or more of the metals, or their compounds, that they may encounter in the workplace, by measuring the total concentration of these metals in their urine. These tests are not sufficiently specific to make a meaningful evaluation of the potential hazard and harm resulting from such occupational exposure. A major drawback is that biomonitoring carried out by measuring total trace element content in urine will probably be based upon spot-tests conducted at the end of a shift. This does not reveal anything about the exposure rate: did it result from a single dose, or was it spread over the working day? The dose–effect associated with these two modes of exposure will be different. In addition, the biological half-life of the element in the body may be short or long and its excretion is subject to large individual and diurnal variations. Moreover, the metal may have been inhaled or ingested by different, harmless means. Increased As concentrations in urine may have been caused by eating fish or seafood that contains elevated amounts of non-toxic, organically bound As. Besides exposure to Cr through, for example, welding dust, a metal-worker may also have taken food-supplements containing Cr which would result in elevated urinary chromium output and could lead to erroneous conclusions about his workplace exposure. Elevated Ni in urine tests may be due to absorption of highly toxic Ni-carbonyl or less toxic Ni compounds. Similarly, elevated Pt concentrations in urine may reflect exposure to toxic hexachloro-platinum, but could also be the result of the use of Pt in a dental prosthesis or tooth fillings.

The solution to these problems lies in developing a test that specifically measures a substance that is unique to the exposure for an individual to a given species of the element. This is not as difficult as it may first appear because exposure to different chemical forms of an element will probably be followed by different biological reactions in the body, leading to the production of different metabolites.

The case of arsenic has been well studied and is a good illustration. Arsenic in fish or seafood is mainly present in four different species: arsenobetaine, arsenocholine, arseno-sugars and DMA (dimethylarsonic acid). It is generally accepted that these compounds are not catabolised by humans. In the case of a biological monitoring programme for exposure to harmful inorganic arsenic compounds (arsenate, arsenite, arsine), these will show up in urine as arsenate, arsenite, MMA (monomethylarsonic acid) and DMA. It is therefore easy to separate the different As-containing species in urine to identify the cause of exposure. (DMA from organic origin will always be accompanied by arsenobetaine, which allows it to be distinguished from DMA of inorganic origin.) Separation and detection techniques to do this type of measurement have become very reliable in recent years.[1]

The problems faced when different species of the element are metabolised to the same compounds are more intractable. For instance Cr in urine as a result of inhalation or gastro-intestinal absorption of Cr(VI) cannot be distinguished from Cr(III) supplements. Inhaled Cr(VI) enters the cells, where it is reduced to Cr(III) and becomes potentially carcinogenic. Ingested Cr(VI) is reduced to Cr(III) in the mouth by saliva and, also, in the gastro-intestinal tract in the case of an excess.[2] Urinary excretion in both cases contains solely Cr(III). Similarly, exposure to various Ni- and Pt-compounds results in the same form of these elements in the urine.

When workers are exposed to harmful chemicals, it is totally unethical – if not downright criminal – to wait until they show the first clinical signs of illness before withdrawing them from the production unit and treating their symptoms. Biomonitoring to gauge the exposure to chemical species must be developed to such a point that early markers are identified and monitored on a regular basis. These will not necessarily contain the trace element. Such a marker could well be a compound that is induced, or of which the concentration is increased by exposure, such as the induction of metallothionein through exposure to Cd. It could also be the reduction or the absence of a marker compound as a result of the exposure: for example, the presence of inorganic arsenic compounds reduces the activity of a number of enzymes. The marker could also show genotoxic damage of target cells, such as nasal cells, due to exposure to Cr(VI). Extensive research is needed in the field of biomonitoring of trace element species in order to improve the measurement of potentially toxic substances, their metabolites or biochemical effects in tissues, secretions, excreta, expired air or any combination of these materials. This should allow the evaluation of occupational or environmental exposure and health risk, by comparison with appropriate reference values based on the knowledge of the probable relationship between ambient exposure and resultant adverse health effects.[3]

3 The Study of Trace Element Kinetics in the Body Due to Industrial Exposure and the Dependency on the Chemical Form of Exposure

Since the early 1960s attempts have been made to classify trace elements in man. This has resulted in the creation of three distinct categories: essential elements, non-essential elements and toxic elements.[4,5] Some trace elements have been extensively researched to explain the way in which they function in living systems, more particularly in mammals. The complexity of living systems is such, however, that a full spectrum of chemical interactions in the body has yet to be established, even for a single element. Most data were first obtained in toxicity studies that, from a very early stage, focused on toxic species of trace elements. This resulted in a diversity of doses or values: threshold limit values, absolute lethal dose, acute toxicity, chronic toxicity, acceptable daily intake, additive effect, adverse effect, *etc.* For many elements the toxic effect of the element and/or some of its compounds is known. Only in a limited number of cases has a fairly specific description of how the chemical species exert their deleterious effects been documented. The picture is even more confused when it comes to try to explain the 'essentiality' of an element, let alone of a species of that element. The third category, non-essential elements, has been even less investigated. As time progressed, some elements shifted from the category 'non-essential' to 'essential', on the basis of studies with animals that were deprived of that element for many generations.[6] Was it deprivation of just that element, and if so in which chemical form, or were there other confounding factors? Who knows?

In order to explain the reaction mechanism of trace elements and, more specifically, trace element species in the human body, much more research is needed. Knowledge about trace element metabolism and the kinetics of trace element species has only been obtained on a very limited scale.[7] These studies should describe the pathways the trace element species follows after being inhaled or orally ingested or after dermal contact. They should include possible transformation of the species, binding to various ligands, its active role played in life-sustaining biochemical processes, its storage in certain target organs and, finally, its excretion. Such studies require *in situ* measurements in cells. This concept is gaining ground, although not yet for trace element species measurements. In the meantime most research is performed utilising classic analytical separation methods and detection systems. This involves isolation of the biological sample, without disturbing the trace element species (an extremely difficult task), followed by separation of the species by chromatography, electrophoresis or analytical ultracentrifugation, and then, on the one hand, measurement of the component (*e.g.* protein) to which the trace element is linked, or on the other hand, the detection of the element. The latter may be achieved by GFAAS, ICP-MS or even by the incorporation of radiotracers or stable isotopes at an appropriate experimental stage. The former may be achieved by nephelometry for the detection of specific proteins, or by mass spectrometry in the case of the large molecules. Figure 1a–e illustrates the sequence of steps taken in the speciation of cadmium in rabbit liver cytosol.[8] The first step (Figure 1a) is a fractionation by SEC–ICP-MS that permits the discrimination of the metallo-thionein–Cd complex (MT-Cd) (fraction) from other Cd species potentially present. The second step (Figure 1b), anion exchange (AE) HPLC coupled to ICP-MS, fractionates the MT fraction into two components (MT-1 and MT-2) according to the sequence of the ligand. The third step (Figure 1c) is at the borderline between fractionation and speciation. The mechanism is reversed-

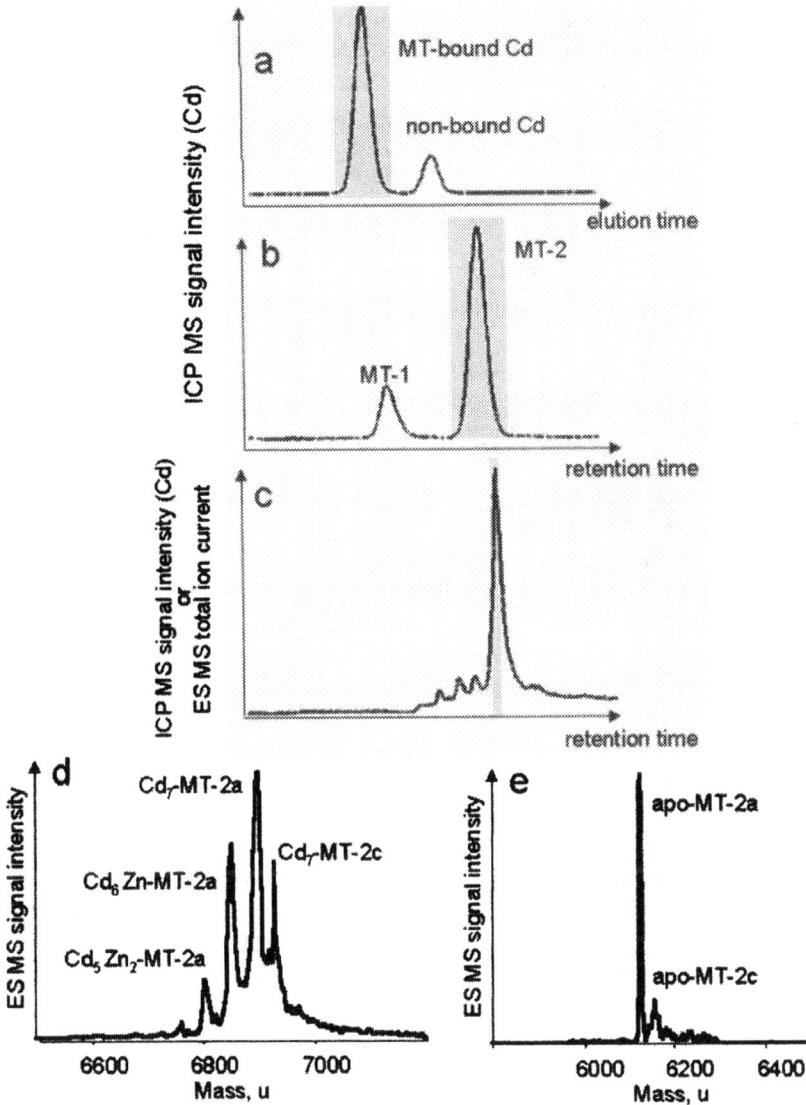

Figure 1 (a) *Fractionation by SEC–ICP-MS of liver cytosol;* (b) *Anion exchange chromatography (AE) HPLC coupled to ICP-MS: fractionation of the MT fraction into two MT-1 and MT-2 according to the sequence of the ligand;* (c) *Reversed-phase chromatography applied to the Cd-MT-2 anion-exchange fraction: each signal still contains a mixture of different species;* (d) *Mass spectrum of one of the peaks, showing a mixture of the different metal complexes with different molecular mass for each of the Cd-Zn complexes of a particular MT-2 isoform (having a unique amino acid sequence);* (e) *Mass spectrum of the complex stripped from its metal, confirming that we have different metal complexes of the same ligand* (all courtesy of R. Lobinski, University of Pau, France)

phase chromatography applied to the Cd-MT-2 anion-exchange fraction. This illustrates either speciation, provided that a peak contains just one species, or fractionation where the signal still contains a mixture of different species. A technique for the verification of the chromatographic purity is necessary. This is shown in the Figure 1d. The mass spectrum of one of the peaks identifies it as being a mixture of the different metal complexes. It is possible to attribute the molecular mass to each of the single species, each being a different mixed Cd–Zn complex of a particular MT-2 isoform (having a unique amino acid sequence). Figure 1e shows a mass spectrum of the complex stripped from its metal atom (apo-complex), confirming that we have different metal complexes of the same ligand.

A special group of trace element species comprises very stable compounds of anthropogenic origin, with which we are now contaminating water and sediments world-wide. The best examples are the trialkyl- and triphenyl-tin compounds. Their deleterious effect as an endocrine disrupter on shellfish was recognised at an early stage. What was not foreseen when these products were launched was that their degradation by UV light would be very slow. The chemical is stored in sediments and only gradually released.

Although analytical chemists have developed reliable techniques to measure tributyl and triphenyl compounds in contaminated waters and sediments, they struggle to determine such residues in humans. To the authors' knowledge, these compounds have never been documented in human body fluids or tissues of subjects not occupationally exposed. It seems highly desirable that the detection limits of the analytical methods be improved in order to measure these organotin compounds and to set up epidemiological studies to determine the possible negative effects they have on human health, including reproduction.

Fortunately, the manufacturers of chemicals have recently become conscious of the responsibility they carry not only for their workers, but also for the environment and for the fate of their products 'from cradle to grave'. Today strict rules exist that must be followed before any new pharmaceutical can be launched. It is not inconceivable that in the next decades similar rules will be imposed before a new chemical or compound can be introduced into the market and, more critically, the environment.

4 Identification of the Chemical Form of the Species That Will Become the Basis of Toxicity Studies

The chemical characterisation of environmental and industrial particulate samples is essential so that those substances that are liable to cause deleterious effects when inhaled can be described. Up to now, hazard identification has concerned stoichiometric substances. Under controlled laboratory conditions their adverse effects, target populations and conditions of exposure have been determined, taking into account toxicity data and knowledge of effects on human health, other organisms and their environment.[9] In reality, when the particles that are inhaled in hazardous circumstances (such as indoor working places) are examined using modern techniques, stoichiometric compounds of suspected elements cannot be identified. The chemical species (*i.e.* the specific form of an element with respect to molecular, complex or oxidation state condition) occur with a variety of unusual combinations (for instance chromate bonded to silicate.) This implies that most of the historical testing to evaluate the hazards of exposure should be completely re-evaluated on the basis of the chemical species of the elements as they present themselves, when inhaled by exposed workers. This was clearly shown in a recent study of particulates in a nickel refinery in Kirovsk, Russia, where many non-stoichiometric nickel compounds were identified.[10,11] The classes of nickel compounds to be considered are the soluble nickel salts, including nickel chloride and nickel sulfate, and the insoluble nickel species such as nickel oxides and sulfides. Their bioavailability depends not only on the chemical form but also on the particle size. Exposures are seldom to one single species of nickel. For example, in various refining operations, workers may encounter nickel subsulfide, nickel oxide, metallic nickel, nickel sulfate, nickel chloride and nickel carbonate. Therefore, an air filter analysed for exposure monitoring purposes may contain nickel salts deposited from aerosols as well as other inorganic forms of the metal and particulates of different sizes. While fractionation (such as separating soluble from insoluble forms) of the particle distribution provides more information than measuring total nickel only, additional useful information on species of differing carcinogenic potential can be obtained by speciation analysis. The need for speciation in occupational health, using nickel and its compounds as an example, is very well outlined in Chapter 20 in this book.

Today very specialised means of characterising particles that are relevant to speciation exist. They use a combination of methods for single characterisation in the outer layer of the particle.[12] Figure 2 gives a picture of an agglomerate of particles originating from a Ni-melt. The aerodynamic diameter is large; thus this agglomerate will deposit in the extrathoracic part of the respiratory system (mouth and nose). The individual particle is also typical for this environment (part of the extrathoracic fraction as well) with a more 'open' and larger surface. Defined phases and simple stoichiometric compounds are absent in these particles.[11] If bioavailable, these particles will contribute to the gastro-intestinal uptake of, *e.g.*, nickel. The instrumentation required for this type of analysis is complex and the cost of analysis is high and it can only be used for fundamental research. Developing cheaper routine methods of analysis is a challenge that must be addressed urgently.

In the chromium industry there is a keen interest to have better characterisation of the surface layer of metal aerosols encountered in the workplace. The metal aerosols in the chromium industry and in welding workshops need to be accurately characterised, especially for the occurrence of toxic Cr(VI) species. The chemical composition of the surface layer is of primary concern because the species liable to interact with cells, membranes and alveolar fluid of the respiratory tract are liable to be encountered there. Typical particulates encountered in ambient air in the metalworking industry consist of conglomerates or chains of up to a dozen particles. Silicates are major constituents of these aerosols. The

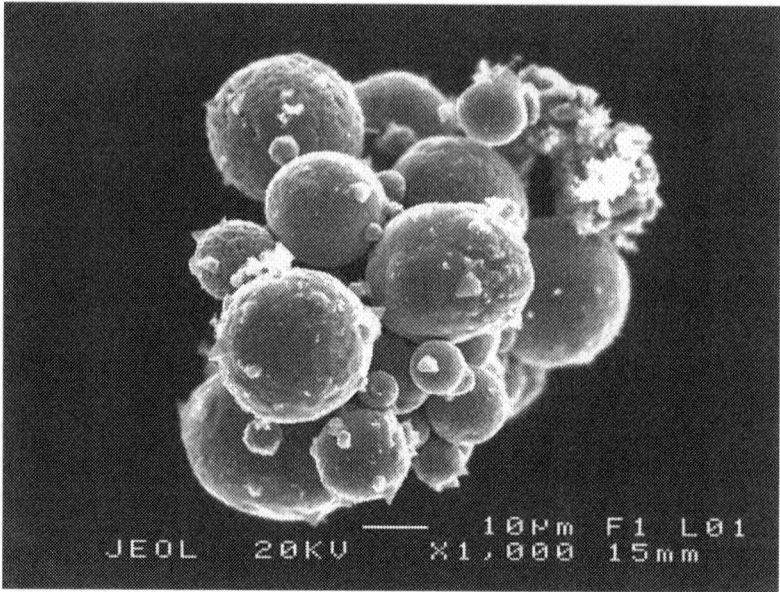

Figure 2   *An agglomerate of particles originating from a Ni-melt in the Ni anode casting department*
(unpublished micrographs, courtesy of Asbjørn Skogstad and Yngvar Thomassen, National Institute of Occupational Health, Oslo, Norway)

chemical bonding of the Cr(VI) species to matrix components, such as those silicates, may be most relevant in explaining the diverging degrees of bioavailability and toxicity of Cr(VI)-carrying particulates.

In industry, a large variety of welding processes, welding rods and steel sheet can be encountered. It is therefore necessary to study the surface layer of the metal aerosols in well-defined circumstances, so as to define the link between the chemical composition and any deleterious effects they may provoke. Toxicity caused by exposure to Cr(VI) has a latency period of 15 years and therefore it is necessary to identify the Cr(VI) species that are the most harmful. In a later stage, these species should be monitored specifically and concentrations in the air of the workplace reduced to safe levels.

Another item of primary concern is the development of new tests for exposure threshold values in metallurgy, based on the study of metal species.

The existing, so called 'organic', tests to establish threshold limit values in metallurgy need to be redefined in order to gauge better the impact of the metal aerosols in indoor air. *In vivo* and *in vitro* tests consisting of exposure to ground metals are not representative of real-life situations. The particulates emitted during metallurgical processes undergo major changes from the moment they are formed until they encounter the respiratory tract of the worker.

What is needed is a thorough study of the metamorphosis occurring as a result of interactions with gaseous substances and ambient humidity, the agglomeration of very small dust particles into larger particulates and, last but not least, the composition and speciation of the elements in the surface layer of such particles. The amounts tested must span a wide range, from very low to worst-case scenario concentrations, without, however, becoming unrealistically high. In the latter case, observations are of no practical value and only serve to add to the confusion about the possible health effects caused by metal particulates. In addition to characterisation of the relevant species for occupational health purposes, it is necessary to make a systematic study, both *in vivo* and *in vitro*, of the interaction between these species and cell membranes and the transformations that occur therein. Such studies will necessitate an orchestrated multidisciplinary approach. Well-planned co-operation between metallurgy, pharmacology, physics, toxicology, occupational health and hygiene, medicine and analytical chemistry will allow the development of new and superior test methods for threshold values of exposure in metallurgy. A major benefit of the outcome of these new developments lies in the focus on metal species of inhaled particulates.

Special attention should also be paid to the study of arsenic species in workplace air in various industrial settings. The industries processing and/or producing arsenic-containing compounds have acknowledged a need to improve the procedures for monitoring the environmental quality of the workplace. The toxicity of arsenic is entirely dependent on the chemical species in which it is present. Therefore volatile and/or particulate arsenic species created in the indoor workplace atmosphere as a consequence of an industrial process need to be well-defined, in order to allow the development of adequate methods subsequently to identify and quantify those substances and their effect on humans. The industries involved are, to name just a few, semiconductor manufacture, glass production and the production and recycling of batteries.

A systematic study of the arsenic species emanating from the production equipment (for example during maintenance operations) or at the workstations in the indoor atmosphere and their impact on man are of primary concern. This topic has to be tackled by a multidisciplinary approach: physics, toxicology and analytical chemistry. The role of the latter discipline is very important, as suitable methods have to be conceived and developed to identify and measure the relevant species.

Up to now, very few studies have dealt with gallium and indium species to which workers may be exposed in the various fields of application of such compounds. The volatile and/or particulate are not necessarily the primary components that are used. It would be interesting to identify and develop methods to quantify the concentration in the workplace air of the various gallium and/or indium derivatives and also to investigate some biomarkers of effect in exposed workers and to establish a link between the concentration in the air and any detrimental effects on their health.

Much more research is needed in the field of solid-state chemical speciation to fill the current void as there is a severe lack of sound knowledge and toxicological data. Multidisciplinary studies involving industry, researchers, regulators, toxicologists, environmentalists, public and occupational health bodies and occupational chemists are the only logical way forward. Researchers have to meet the need for rapid, simple, inexpensive methods to enable routine monitoring.

It is also vital to have a good database of background levels of trace element species in man and his environment. Biological systems (bivalves, plants, insects, *etc.*) or cell cultures may also be of value to determine exposure, examine pre-disease markers and also effect markers.

5 Conclusions

Speciation of trace elements is the only rational approach to unravel the complex mechanisms of elemental moiety interactions underlying the beneficial and harmful effects of trace elements in man. Occupational hygiene and health in general will benefit tremendously if major progress in trace element knowledge can be achieved. Only sound analytical methodology, coupled to a multidisciplinary approach, can lead to these new insights. In order to meet these challenges effectively we need to develop instrumentation, methodologies and quality assurance.

6 References

1. X. Zhang, R. Cornelis, J. De Kimpe and L. Mees, *Anal. Chim. Acta*, 1996, **319**, 177.
2. S. De Flora and K.E. Wetterhahn, *Life Sci. Rep.*, 1989, **7**, 169.
3. J.H. Duffus, *Glossary for Chemists of Terms Used in Toxicology*, *Pure Appl. Chem.*, 1993, **9**, 2003.
4. E.J. Underwood, *Trace Elements in Human and Animal Nutrition*, 4th Edn., Academic Press, New York, San Francisco, London, 1977.
5. W. Mertz, *Fed. Proc.*, 1970, **29**, 1482.
6. M. Anke, L. Angelow, M. Glei, M. Muller and H. Illing, *Fresenius' J. Anal. Chem.*, 1995, **352**, 92.
7. F. Borguet, R. Cornelis, J. Delanghe, M.-C. Lambert and N. Lameire, *Clin. Chim. Acta*, 1995, 71.
8. R. Lobinski, personal communication.
9. WHO, *Basic Terminology for Risk and Health Impact Assessment and Management*, Internal report of Working Group, 12 April 1988 (Annex 3), World Health Organisation, Geneva, 1988.
10. Y. Thomassen, E. Nieboer, D. Ellingsen, S. Hetland, T. Norseth, J.Ø. Odland, N. Romanova, S. Chernova and P. Tchachtchine, *J. Environ. Monit.*, 1999, **1**, 15.
11. S. Gunst, S. Weinbruch, M. Wentzel, H.M. Ortner, A. Skogstad, S. Hetland and Y. Thomassen, *J. Environ. Monit.*, 2000, **2**, 65.
12. H.M. Ortner, P. Hoffmann, F.J. Stadermann, S. Weinbruch and M. Wentzel, *Analyst*, 1998, **123**, 833.

CHAPTER 20

Occupational Health and Speciation Using Nickel and Nickel Compounds as an Example

SALLY PUGH WILLIAMS

1 Introduction

This chapter looks at occupational nickel exposure, nickel speciation and cancer risk. The importance of speciation of metals is discussed by illustrating that nickel species are encountered both in the general environment and in certain workplaces. Nickel species may be found in toxic and non-toxic forms. Safe levels of nickel exposure, particularly in occupational settings, can only be achieved if speciation concepts and analytical techniques are further developed.

To understand any occupational health consequences of exposure to nickel and the role of speciation, it is important to understand the sources and origins of nickel exposures in the natural environment, as well as those arising in the workplace. The way that metals, *e.g.* nickel, exert any effect in the body, is dependent on several factors:

- the way that the nickel species or form is presented to the body – this includes the route of exposure as well as physicochemical and morphological form(s), and the dose
- the body's mechanisms for transport, handling and ultimate fate of the nickel species, particularly in relation to the bioavailability of the nickel species
- natural background levels of exposure to nickel – the role of nickel as a trace element is important. Here nickel species are in a non-toxic form, and even marked elevations may lack clinical or toxicological significance.

Many metals such as nickel are ubiquitous in nature: therefore, zero exposure is not possible, desirable or achievable. Some metals are essential trace elements and have been shown to have beneficial effects at low doses for normal bodily functioning; nickel is probably an essential trace element in man.[1]

Occupational exposure settings provide exposure scenarios where normal body homeostatic mechanisms may be overwhelmed or at least challenged by additional exposure, usually mixed exposures, to a wide variety of substances. The role and contribution of nickel species to any adverse health effect in the workplace requires an ability to characterise the workplace exposures. Then exposure has to be related to any adverse health effects, taking into account both the toxic and non-toxic species.

Firstly it is important to understand what nickel is, where it comes from and what are the sources of non-occupational nickel exposure as well as those derived from the workplace.

2   Sources of Exposure to Nickel

Nickel is used extensively in a variety of commercially important industries, including stainless steel production, manufacture of nickel alloys, electroplating, battery production, manufacture of coins, jewellery and nickel catalysts. Annual world production of nickel in recent years has averaged in excess of 900 kilotonnes. 90% of all nickel is consumed in the production of different stainless steels, other nickel alloys and foundry products. 9% is used in plated products and the remaining, approximately 2%, in a number of relatively small applications including, chemicals, catalysts, batteries, coins, powders and pigments. Figures 1 and 2 show the distribution and end uses of primary nickel in 1996, indicating the wide applications and end use of nickel.[2]

Occupational exposure to nickel and nickel compounds is diverse and widespread and therefore many occupations may lead to nickel exposure. Public health exposure may occur from items made from or containing nickel, which are available to consumers, *e.g.* certain items of jewellery or kitchen ware.

It should be remembered that nickel is a naturally occurring element, occurring in soil, water and air. The average nickel content in the Earth's crust is about 0.008%. Nickel ore deposits occur in certain areas of the world such as Canada and Australia where they are found as accumulations of nickel sulfides, mostly pentlandites, and also laterites. Nickel-bearing particles in the atmosphere are mainly present as suspended particulate matter or occasionally as mist aerosols. Nickel in ambient air is primarily derived from combustion of fossil fuels, *e.g.* oil and coal burning, and to a lesser extent from metallurgical operations such as mining, processing and manufacturing nickel. Typical ambient air concentrations of nickel range from 6 to 25 ng Ni m^{-3}.[3]

In aquatic systems, degradation of rocks and soils containing nickel may occur into watercourses, as well as deposition of airborne nickel into water. Certain industrial wastewater discharges may contribute to the nickel content of water. Nickel is usually present in water as the nickel cation Ni^{2+}; the average of nickel values in water varies from 1 to 50 μg l^{-1}.[3] Nickel is also present in food; levels are generally low, < 0.5 mg kg^{-1}. Some foodstuffs have high concentrations of

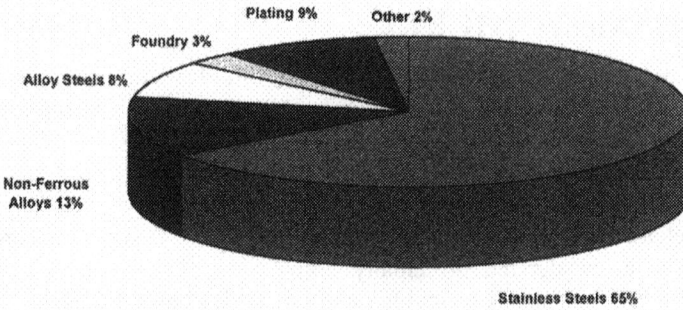

Figure 1 *Distribution of primary nickel, 1996*
(Source: NiDI, 1996)

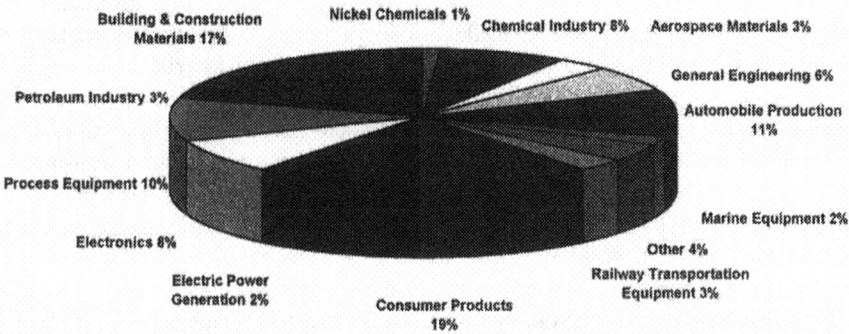

Figure 2 *End uses of primary nickel, 1996*
(Source: NiDI, 1996)

nickel, *e.g.* cocoa, soybean, dried legumes, nuts. The daily intake of nickel varies between 100 and 800 μg Ni.[3]

3 Is All Nickel 'NICKEL'?

In the natural environment as well as in the workplace, the species of nickel, whether they are in toxic or non-toxic (trace element) forms, are often not well described. Very often all exposures are described as 'NICKEL'. This generic approach for describing nickel exposures cannot describe all the different species of the element or their potential for effect, either good or bad.

Trace elements are metals that are normally present at very low concentrations in the body, derived principally from foods. Nickel occurs as a trace element, and in this form is probably an essential element in humans.[1,3] Recent helpful work has looked at setting trace element reference values in tissues from inhabitants in the European Union.[4] A knowledge of the levels of nickel as a trace element in the general population is needed to establish appropriate species-monitoring strategies for occupationally exposed groups. There has been little data available on trace element levels in human populations to date. Identifying nickel species in this respect is very important, so that the contribution of each nickel species to total nickel exposure is defined. This will allow for distribution profiles of toxic and non-toxic nickel species to be developed in terms of both exposure assessment and toxicological significance in the body.

The nickel species shown in Table 1 have been identified as important in relation to possible adverse health effects. Chemical species are often described as water soluble or insoluble; this definition may not hold as true in biological systems, where biological solubility/bioavailability may occur with a *water insoluble* substance which is of very small particle size, *e.g.* metal powders, fine particles of nickel oxides or fumes.[5] Metallic nickel may occur in a non-inhalable/insoluble form, *e.g.* pellets or brickettes; nickel metal powders are also produced which have inhalation potential and may be biologically soluble. Nickel carbonyl is a volatile liquid, acutely toxic by inhalation, with marked biological solubility but virtually insoluble in water. Physico-chemical speciation is important and is a starting point for biological speciation.

Different forms or species of nickel may display different features such as transport characteristics, bioavailability and changes in such properties as surface charge, solubility, diffusion coefficient, protein binding, surface chemistry

Table 1 *Nickel and nickel compounds – speciation/water solubility of nickel species*

Insoluble	Soluble	Sparingly soluble
Nickel oxide	Nickel sulfate	Nickel carbonyl
Nickel sulfide	Nickel chloride	
Nickel subsulfide		
Metallic nickel		

etc. Some of these changes may start to occur during exposure, *e.g.* in ambient air before entering the body, as well as the species changing once they are within the body. Species modification may also occur in the presence of concurrent exposures to other nickel species or other chemicals, due to factors such as heat or light or due to processing conditions in the workplace. Similar principles probably apply once the species enter the human body as well as inside and outside body cells. Intracellular speciation remains in its infancy.

The role of aerosol-borne particles of nickel as distinct toxic entities aside from their chemical characteristics, *i.e.* the particle effect, also needs to be better understood in relation to adverse inhalation effects (see Section 6, particle effect).

Speciation analysis is a technique which identifies and measures the quantities of one or more individual chemical species in a sample. It is very helpful to determine the distribution of defined chemical species of an element within a system, whether this is in the human body, in biological fluids or inside cells, in the environment or in food. This is key to understanding and controlling for health effects.

Some levels of exposure to certain nickel species are undoubtedly quite safe and may be essential. Factors that determine bioavailablity and biological effects of a species may render seemingly innocuous species as harmful. The converse may also be true.

It therefore is erroneous to consider that all nickel is 'NICKEL'.

4 Nickel and Health

Although nickel exposure is widespread, adverse health hazards primarily relate to inhalation exposures to certain forms and/or combinations of nickel species in certain occupational settings. Respiratory cancers, lung and nasal cancers, have been described in workers at nickel refinery operations. Many of these refinery operations are now largely obsolete and the risks of cancer appear to have abated. Detailed evaluation of the carcinogenic risks of nickel species exposure have, however, been undertaken because of the striking epidemiology data in historical nickel refinery worker cohorts, linking certain nickel exposures with cancer risk (this will be discussed in more detail later).[2,3,6-9]

Other adverse health problems encountered following nickel exposure include dermal sensitisation from nickel metal or soluble nickel exposure and, rarely, occupational asthma due to soluble nickel salts.[9-11] Nickel carbonyl is the only nickel species which is acutely toxic by inhalation and linked to a characteristic illness often described in exposed individuals.[9,12-14]

Dermal exposure to nickel may occur from everyday items, *e.g.* certain forms of nickel jewellery. However, direct and prolonged skin contact is required between a nickel-containing item and the skin so that nickel ion release occurs in a corrosion reaction between the item and sweat produced on the skin. This may lead to dermal sensitisation or nickel allergy in a small number of susceptible individuals (not all nickel containing items release nickel, *e.g.* most stainless steels are corrosion resistant). The immunological mechanism of dermal sensitisation, including the ability of nickel species to release the Ni^{2+} ion to exert its effect, has been and is the subject of much research and debate.[10,15]

Most research to date, regarding any adverse health effects due to nickel exposure, has relied predominantly on traditional methods of epidemiology and toxicology. The ability to confirm exposure scenarios with speciation analysis and link this with health effects has been limited.

Any of the adverse health effects of nickel described above could be used to understand the role of speciation and occupational nickel exposure but there is most information about nickel and cancer risks, so it is helpful to look at the respiratory cancer story in more detail.

5 Nickel and Respiratory Cancer

In 1990, the report[7] of the International Committee on Nickel Carcinogenesis in Man (ICNCM) was published; this report attempted to clarify the nature and extent of the risks of cancer that nickel and its compounds might cause. One of the deficiencies noted in the report was the inability to differentiate more precisely between nickel species. This was of particular importance as the report proved that, from the examination of ten cohorts of workers exposed to nickel, it appeared that more than one form or species of nickel gives rise to lung and nasal cancer. The authors were unable to determine a level of exposure at which nickel became hazardous.

To investigate the relationship between the workers exposure to nickel species and the development of cancer, it was necessary for the authors to estimate the airborne concentrations of nickel species in the workplaces. Nickel species in the report were divided into the following categories based on the chemistry of the industrial processes:

- metallic nickel
- oxidic nickel
- soluble nickel
- sulfidic nickel including nickel subsulfide.

There were few actual measurements of nickel in air before 1950; the authors therefore decided to perform some workplace air monitoring in current nickel refinery operations to determined concentrations of nickel and other metals in the workplace air. The results of the monitoring were used in conjunction with a knowledge of the chemistry of the industrial processes over many years to develop workplace exposure characteristics. However, whenever measurements were available, sampling was not standardised and there was inconsistency of measuring devices and analysis techniques between workplaces. This made identification of the level of exposure at which nickel species became hazardous to human health difficult.

However, the report did identify quite clearly the following;

'respiratory cancer risk seen amongst nickel refinery workers could be attributed to exposure to a mixture of oxidic and sulfidic nickel at high concentrations, exposure to large concentrations of oxidic nickel in the absence of sulfidic nickel was also associated with lung and nasal cancer risks. There was also evidence that soluble nickel exposure increased the risk of these cancers and that it may enhance risks associated with exposure to less soluble forms of nickel.

There was no evidence that metallic nickel was associated with lung and nasal cancer risks, and no substantial evidence was obtained to suggest that occupational exposure to nickel or any of its compounds was likely to produce cancer elsewhere other than the lung or nose'.[7]

This was a very important report indicating that exposure to distinct nickel species was associated with cancer, whilst the metallic nickel showed no evidence of this effect. Combinations of nickel species and certain doses of nickel species were also seen to be associated with cancer risk.

The report concluded that although it was not possible to provide dose-specific estimates of risks for individual nickel species, it was possible to comment on the cancer risks associated with the level of airborne nickel to which the general population is exposed. Respiratory cancer risks are primarily related to exposure to soluble nickel at concentrations in excess of $1 \, mg \, Ni \, m^{-3}$ and exposure to less soluble forms at concentrations greater than $10 \, mg \, Ni \, m^{-3}$. It was concluded that general population exposure to the extremely low concentrations (less than $1 \, \mu g \, Ni \, m^{-3}$) to which it is exposed in the ambient air represents a minute risk, if indeed there is any risk at all.

Since the report was published, further work has been undertaken to confirm the importance of nickel speciation and cancer risk. Nickel subsulfide was used in a two year chronic inhalation bioassay in rats and mice; the results showed clear evidence of carcinogenicity in rats but not mice.[16] Nickel oxide was tested in a similar way and showed some evidence of cancer in rats but equivocal evidence in mice.[17] Nickel sulfate tests (soluble nickel compound) gave negative results.[18]

In vitro, nickel subsulfide has been shown to effect heritable changes in the cells, whilst nickel oxide may exert its effect through inflammatory changes and/or the particle effect. Nickel sulfate may cause cytotoxicity within the lung cells, resulting in cell proliferation, and perhaps enhance carcinogenicity potential of other substances, *e.g.* insoluble nickel compounds, through a promoter effect.[2,19]

Whilst this information begins to unravel the role of nickel species and cancer, much more research is needed particularly regarding speciation at the cellular and mechanistic level as well as in exposure assessment.

6 Occupational Exposure Limits for Nickel

Many regulatory bodies, including the American Conference of Industrial Hygienists (ACGIH), have evaluated data concerning nickel and nickel compounds to set Occupational Exposure Limits (OELs), often applying additional safety factors. OELs usually group nickel species together, setting standards for soluble nickel species, insoluble species, metallic nickel and nickel carbonyl. Carcino-

Table 2   *Occupational exposure limits/threshold limit values and cancer classification for nickel and nickel compounds UK and USA 1998*

Nickel species	OEL: UK[a] mg Ni m^{-3}	TLV: USA[b] mg Ni m^{-3}	Cancer category: UK[c]	Cancer class: USA[c]
Metallic nickel	0.5	1.5	3	A5
Nickel subsulfide	as insoluble nickel	0.1	1	A1
Insoluble nickel	0.5	0.2	1	A1
Soluble nickel	0.1	0.1	3 (NiSO$_4$)	A4
Nickel carbonyl	0.24 STEL	0.12	3	

[a]OEL: UK = occupational exposure limit in the United Kingdom expressed as mg Ni m^{-3} (as Ni, 8 hour time weighted average total inhalable nickel); STEL short term 15 minute exposure limit.[20]
[b]OEL/TLV: USA = occupational exposure limit, American Congress of Governmental Hygienists, expressed as mg Ni m^{-3} (inhalable particulate fraction, 8 hour threshold limit value).[21]
[c]A1/category 1 signify known human carcinogen status; A4 signifies not classifiable as a human carcinogen; A5 signifies not suspected as a human carcinogen; category 3 signifies some animal evidence for cancer, limited human evidence, substances which cause concern in man but where evidence is inadequate for making a satisfactory assessment[22] (definitions abbreviated: for full definitions see refs 20–22).

genic classifications may also be noted based on the nickel species. Importantly, there are often differences between the OEL values, including cancer category or class (Table 2). Different measuring techniques may be employed making comparison between measurement data and exposure limits difficult (see Section 7). International standardisation is therefore being developed.

Evaluation of the epidemiology data, animal evidence, *in vitro* data, mechanistic studies and exposure estimates all contributes to scientific standards for setting OELs. In practice, such levels are often referenced to total nickel measurements, and not to the relevant nickel species and, as has been stated before, often differ between regulators and scientists. This adds to the need for verified quantitative nickel speciation data if uniform protective standards are to be achieved.

7   Air Monitoring and Biological Monitoring for Nickel

Generally available techniques for airborne workplace exposure assessments, including compliance monitoring, measure total nickel and not the nickel species.[20,21] Most workplaces have more than one species of nickel present in their processes. The contribution from each nickel species to workplace exposure may be estimated but not accurately defined during monitoring surveys of the workplace. This makes compliance with speciated OELs difficult, if not impossible, and safe levels of exposure to avoid health risk cannot be confirmed.

Airborne concentrations of particulates need to be measured as a function of particle size as well as mass concentration of substance and substance species. This is because the potential hazard of the particle often depends on the site of its deposition in the respiratory tract as well as its composition. Not all particles are inhalable in humans; larger particles (100–200 μm) may settle in the upper parts of the respiratory tract, and only half the particles with aerodynamic diameters greater than 30 μm are inhaled. A particle size of less than 5 μm has to be inhaled for deposition in the alveolar regions of pulmonary tissues. Factors such as the amount of particle deposited, particle solubility, surface characteristics and size influence the behaviour of the particles deposited in the respiratory tract.

With some aerosol particles, *the particle effect* is also of great importance; some of the toxic effects of nickel oxide exposure are thought to be due to this effect.[2] The particle effect means that possible health effects may occur due to a physical aspect of a particle, as distinct from its chemical composition. In the case of nickel, very fine particles may be toxic in their own right, regardless of or in addition to, their having nickel in their chemical composition. Retention, clearance and absorption of particles is complex; little is known about the precise pharmacokinetics of nickel in the human lung.[2]

Particle size selective OELs (inhalable particulate mass, thoracic particulate mass, respirable particulate mass) rather than total aerosol are more closely related to health hazard as well as speciation of elements. International standardisation of OELs with clearly defined reference values is urgently needed, as well as more readily available particle size selective measuring instruments and speciation analysis.

Most techniques for particle characterisation remain research tools, due to practical problems in sample collection and expense considerations. Better methods for nickel speciation, particle characterisation *etc.* are beginning to be developed. OELs will have to be reviewed accordingly with additional data of this nature with re-evaluation of health effects.

Biological monitoring and ambient air monitoring complement each other in assessing occupational exposure to nickel.[23,24] Human exposure to nickel may be measured by analysing for nickel in blood or urine, usually the latter. Biological monitoring however, has several drawbacks.[25] It measures total nickel from *all* sources of exposure whether it is oral exposure from diet or oral ingestion, from contaminated hands from workplace dusts, from inhalation exposure from the ambient air or from occupational exposure. A thorough knowledge of all exposure sources is therefore required when interpreting biological monitoring data. Even so, exposure assessments made this way are far from exact. They cannot reflect a scientific health risk assessment but rather reflect exposure status from all sources.

If it were possible to measure more specific biological markers of exposure for the relevant toxic species as distinct from the non-toxic nickel species, then this would be extremely helpful. This information would make the setting of biological exposure indices for nickel more predictive of adverse health effect, rather than relying on measuring the degree of exposure to nickel.

For both air and biological monitoring strategies, sample timing, collection, transport, storage, laboratory preparation and analysis all need to be standardised, using well-documented techniques. Certified reference materials need also to be developed.

8   Conclusion

The discussion so far has focused on occupational health and nickel, as an example for the case of speciation and speciation analysis. The issues reviewed apply to many other substances but particularly metals.

Occupational health assessments require a detailed knowledge of workplace exposures and, where possible, appropriate biological monitoring techniques. Epidemiology and toxicology studies still remain the first-line tools for studying diseases in occupationally exposed individuals, but more is needed.

Occupational health specialists need harmonised and simplified methods for measuring nickel species in workplace air and in body fluids, to complement the data obtained from traditional methods of occupational health research. Without this information, safe levels of occupational exposure to nickel and nickel compounds cannot be confirmed. Speciation analysis is therefore an important area of interest for the nickel industry so that reliable, economic, standardised techniques are developed and utilised and adverse health effects avoided.

The concept of speciation for nickel and nickel compounds in hazard identification, risk assessment and risk management cannot be overemphasised.

9   References

1. J.R. Lancaster Jr., ed., *The Bioinorganic Chemistry of Nickel*, VCH Publishers Inc, New York, 1988.
2. NiDI/NiPERA Health guide. *Safe Use of Nickel in the Workplace*, 2nd Edn., NiDi-Nickel Development Institute, 214 King Street West, Suite 510, Toronto, Ontario, Canada M5H 3S6, and NiPERA–Nickel Producers Environmental Research Association, 2605 Meridian Parkway, Suite 200, Durham, North Carolina 27713, USA.
3. World Health Organisation, IPCS *Environmental Health Criteria, 108, Nickel*, WHO, Geneva, 1991.
4. M.A. White and E. Sabbioni, *Sci. Total Env.*, 1998, **216**, 253.
5. H. Roels, R. Van de Noorde, V.M. Mata Vargas and R. Lawreys, *Occup. Med.*, 1993, **43**, 95.
6. F.W. Sunderman Sr., in *Nickel and Human Health: Current Perspectives*, eds. E. Nieboer and J.O. Nriagu, Vol. 25, Chapter 1, John Wiley and Sons, Inc., New York, 1992.
7. Report of the International Committee on Nickel Carcinogenesis in Man (ICNCM), *Scand. J. Work, Env. Health*, 1990, **16**, No. 1 (special issue).
8. World Health Organization, International Agency for Research on Cancer, *IARC Monographs on the Evaluation of Carcinogenic Risks to Humans . . . Chromium, Nickel and Welding*, Vol. 49, IARC, Lyon, 1990.
9. Health and Safety Executive, *Toxicity Review 19: The Toxicity of Nickel and its Inorganic Compounds*, HMSO, London, 1987.
10. R.H. Guy, J.J. Hostynek, R.S. Hinz and C.R. Lorence, *Metals and the Skin: Topical Effects and Systemic Absorption*, Marcel Dekker, inc., 1999.
11. J.L. Malo, A. Cartier and M. Doepner, *J. Allergy Clin. Immunol.*, 1983, **72**, 407.
12. H.W. Armit, *J. Hyg.*, 1907/1908, **7**, 525.
13. L.G. Morgan, in *Nickel and Human Health: Current Perspectives*, eds. E. Nieboer and J.O. Nriagu, Vol. 25, John Wiley and Sons, New York, 1992.

14. F.W. Sunderman, Sr., in *Nickel and Human Health: Current Perspectives*, eds. E. Nieboer and J.O. Nriagu, Vol. 25, Chapter 21, John Wiley and Sons, New York, 1992.

15. NiPERA, *Proceedings of the NiPERA Dermal Sensitisation Workshop*, Brussels, March 17–18 1997, NiPERA–Nickel Producers Environmental Research Association, 2605 Meridian Parkway, Suite 200, Durham, North Carolina 27713, USA

16. National Institutes of Health, *Report of the Toxicology and Carcinogenesis Studies of Nickel Subsulfide in F344/N Rats and B6C3F1 Mice: Inhalation Studies*, NIH publication No. 94-3369, Washington DC, 1994.

17. National Institutes of Health, *Report on the Toxicology and Carcinogenesis Studies of Green Nickel Oxide in F344/N Rats and B6C3F1 Mice: Inhalation Studies*, NIH publication No. 94-3363, Washington DC, 1994.

18. National Institutes of Health, *Report on the Toxicology and Carcinogenesis Studies of Nickel Sulfate Hexahydrate in F344/N Rats and B6C3F1 Mice*, NIH publication No. 94-3370, Washington DC, 1994.

19. A.R. Oller, M. Costa and G. Oberdorster, *Toxicol. Appl. Pharmacol.*, 1997, **143**, 152.

20. Health and Safety Executive, *EH40/99 Occupational Exposure Limits 1999*, HMSO, London, 1999.

21. ACGIH, *TLVs and BEIs: Threshold Limit Values for Chemical Substances and Physical Agents*, ACGIH Worldwide, the American Conference of Governmental Industrial Hygienists,1330 Kemper Meadow Drive, Cincinnati, Ohio 45240-1634, USA, 1998.

22. Health and Safety Commission, *Approved Guide to the Classification and Labelling of Substances and Preparations Dangerous for supply* (second edition) *Chemicals (Hazard Information and Packaging for supply) Regulations 1994 CHIP2 Guidance on Regulations*, HSE books, HMSO, London, 1994.

23. Health and Safety Executive, *Biological Monitoring in the Workplace: A Guide to its Practical Application to Chemical Exposure*, 2nd Edn., HMSO, London, 1997.

24. A. Aitio, *Biological Monitoring of Occupational Exposure to Nickel: Nickel in the Human Environment*, IARC Scientific publication No. 53, Lyon, 1984.

25. L.G. Morgan and J.G. Rouge, *Biological Monitoring in Nickel Refinery Workers: Nickel in the Human Environment*, IARC Scientific publication No. 53, Lyon, 1984.

CHAPTER 21

Surface Structure and Speciation of Metal Aerosols: A Key to the Understanding of Their Biological Effects

MARKKU HUVINEN

1 Introduction

The differences in the bioavailability of different compounds of metals are well known and documented. There are also differences in bioavailability of a given element between different production facilities. This has been reported for example in the production of chromium, nickel, lead and copper. Different concentrations of metals have been measured in biological fluids, although the exposure levels to metals have been similar. This has been shown also in animal experiments. The chemical solubility of metal particles present in workplace air may vary substantially at different stages of production, which indicates that the metals are chemically bound in different ways. The particle structure of metallurgical fumes and dusts will depend largely on temperature, process conditions like oxidation and reduction, slag compounds, other metals present and mechanical handling. It is also known that fume particles are inhomogeneous. Because metallurgical plants have many different process phases, they provide unique opportunities to study the properties of different types of metal aerosols and their potential biological effects. Chromium-containing aerosols as encountered in the production of ferrochromium and stainless steel will be described as an example of the links between the characteristics and surface structure of the particles and their biological effects.

2 Ferrochromium Production

The main use of ferrochromium is in the production of stainless steel. Ferrochromium is also used in the electrolytic production of chromium metal.

2.1 Production Process

High-carbon ferrochromium is produced by electrothermal reduction of chromite ore in an electric arc furnace. Carbon in the form of either coke, coal or charcoal is used as the reducing agent. At the high temperatures of the process, carbon reduces the oxides of chromium and iron present in the ore to an iron–chromium alloy.

Low-carbon and medium-carbon ferrochromium are produced by silicothermic reduction of chromite ore. An alternative method for producing medium-carbon ferrochromium is the decarbonisation of high-carbon ferrochromium in an oxygen blown converter.

2.2 Exposure

Chromium in chromite ore exists entirely as Cr^{3+} and is reduced to Cr^0 during the smelting process. The conditions inside a closed-top arc furnace are reducing. Chromium may be partly oxidised to Cr^{6+} only in the fume particles when released into the air during tapping.

The exposure levels to total dust at a modern ferrochromium plant in Finland have been reported to be $1.5 \, mg \, m^{-3}$ on average.[1] The dust contained an average of 5–10% of chromium. The proportion of Cr^{6+} in the total chromium was 0.1–0.3%. The highest concentrations were detected during tapping in the vicinity of the tap hole, where the proportion of Cr^{6+} was 10 times (1–3%) the level in other areas.

Biological monitoring has indicated that the urinary chromium levels of workers in the same ferrochromium smelter are low but slightly elevated.[1] The average concentration of chromium in the urine was $0.03 \, \mu mol \, l^{-1}$ and the maximum $0.14 \, \mu mol \, l^{-1}$. The average blood chromium concentration was $< 0.01 \, \mu mol \, l^{-1}$.

A study combining several non-destructive techniques[2] determined the concentration, species and bioavailability of chromium in the work environment of a ferrochromium smelter. About 50% of the total chromium in the smelter dust was reported to be bioavailable, 40% of which was Cr^{6+}. The remaining Cr was insoluble.

Dust samples from the ferrochromium plant mentioned above[1] were analysed by a field emission scanning electron microscope. The aerosols observed contained agglomerates of particles with a diameter less than 1 μm (Figure 1, Table 1). Chromium seemed to be dissolved in the silica matrix. Most particles were covered with a thin layer of zinc oxide. Zinc originates from the ore. Zinc seems to be the last condensate on the surface of the particles, due to having the lowest melting point.

3 Production of Stainless Steel

Stainless steels are defined according to the European Standard EN 10088 as

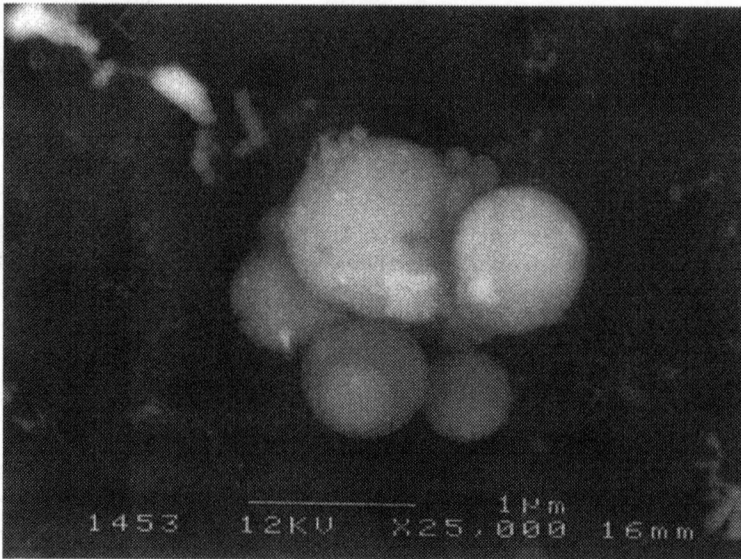

Figure 1 *Typical dust agglomerate from the ferrochromium smelter*

Table 1 *Composition of the largest particle in Figure 1 (%)*

Mg	2.7	*Mn*	0.8
Al	1.2	*Fe*	11.6
Si	13.3	*Ni*	0.2
K	2.0	*Zn*	18.6
Ca	0.4	*O*	35.0
Cr	14.3		

iron-based alloys containing at least 10.5% chromium and a maximum of 1.2% carbon. Stainless steels may contain nickel as another major alloying element, with a content up to 38%. Other alloying elements may be present including molybdenum (0–8%), manganese (0–11%), silicon and copper (0–3% each). Titanium and niobium may be used as stabilisers.[3]

3.1 Production Process

Stainless steels are produced by melting a charge containing high-carbon ferro-chromium, stainless steel scrap, steel scrap, nickel and sometimes molybdenum in an electric arc furnace. The molten metal is then refined using argon–oxygen decarbonisation. The molten metal undergoes various metallurgical operations in the ladle, followed by continuous casting and hot rolling. The hot process stages are followed by annealing, pickling and cold rolling and finally finishing.

3.2 Exposure

The production of stainless steels gives rise to airborne dusts and fumes. The components of the airborne metal aerosols vary according to the type of production process, the stage of the process and also according to the grade of stainless steel under production. Airborne particles may contain the metal constituents used to produce stainless steel, such as chromium, nickel, iron, molybdenum and manganese. Chromium is encountered predominantly as trivalent chromium, although hexavalent chromium compounds do occur.[1,4]

Other elements present in the airborne particles include silicon, calcium, aluminium, magnesium and sodium.[4] Among workers in the steel melting shop the median Cr^{6+} exposure was reported to be 0.5 μg m^{-3} and the maximum 6.6 μg m^{-3}.[1]

In a study conducted in 1999 at a Finnish stainless steel plant the dust samples were analysed by a field emission scanning electron microscope with energy dispersive elemental analyser. The particles in the aerosols encountered at the work place were predominantly metal alloys (Figure 2, Table 2). No pure chromium or nickel particles were observed. The particles had an iron oxide or iron core surrounded by chromium and nickel as alloys, silicates and oxides. These complexes are chemically tightly bound and insoluble in the biological environment. Some of the particles were covered by a thin layer of slag or zinc condensed or crystallised on the surface. The individual particles formed agglomerates having some half a dozen particles in a chain formation.

Welding fumes differ from fumes in metallurgical melting processes. Welding fumes typically contain agglomerates and long chains consisting of particles of same size (Figure 3).

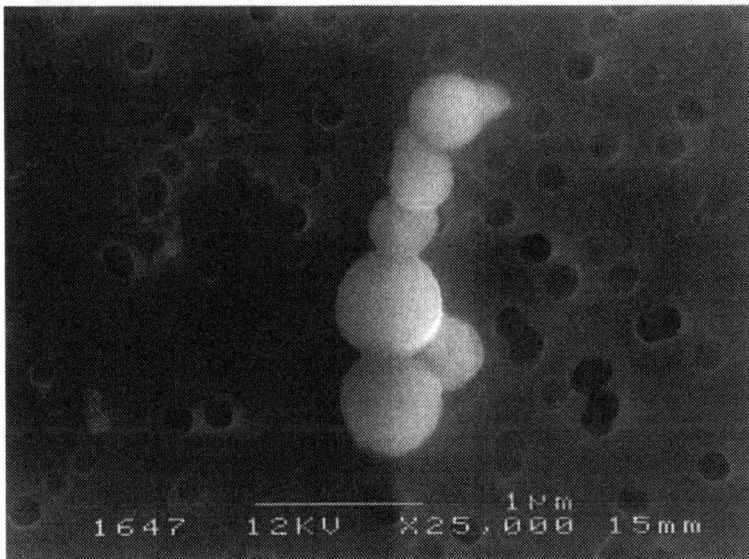

Figure 2 *Typical dust particle chain from the stainless steel melting shop*

Table 2 *Composition of the largest particle in the chain in Figure 2 (%)*

Mg	0.8	Mn	2.7
Al	0.3	Fe	44.0
Si	0.8	Ni	2.4
K	0.2	Zn	10.5
Ca	0.9	O	28.9
Cr	8.4		

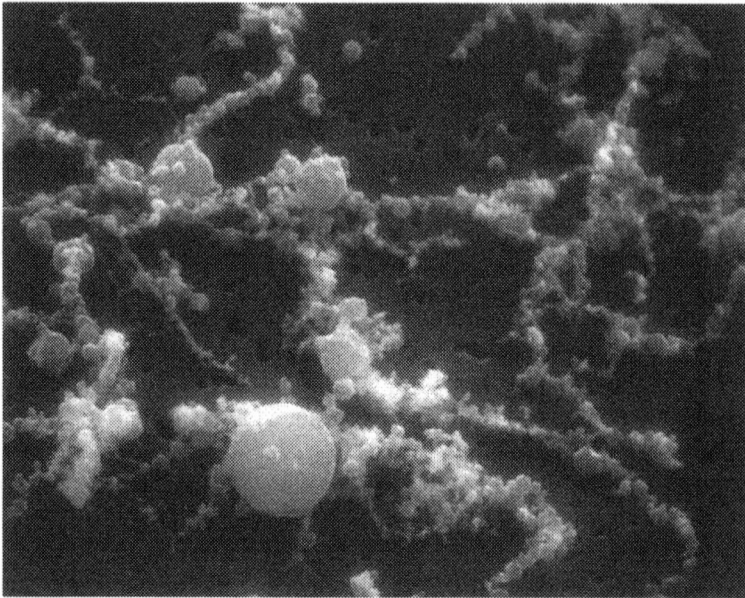

Figure 3   *Typical chains of particles in welding fume*

4   Health Effects

4.1   Chromium

Chromate dust and fumes of chromium trioxide have been reported to cause asthma. Chromates, along with other exposure agents in chromium plating, welding and ferrochromium production, have been connected with cases of occupational asthma or bronchitis. Obstructive effects on lung function have also been found among chromium workers. In one study, reduced forced vital capacity (FVC) and an increased prevalence of obstructive lung diseases were found among electro-furnace workers in a ferrochromium plant. Nodular pneumoconiosis is another finding amongst workers in chromate production. It has not been confirmed, however, in some studies. In a study on respiratory symptoms among 60 ferrochromium workers in Norway, pneumoconiosis was diagnosed from radiographical examinations. The differences in the production processes may well explain these findings.

An increased risk of lung cancer has been found among workers in the production of chromates and chromate pigments, as well as in chromium plating, although no conclusive data are available on lung cancer in ferrochromium production.[5]

4.2 Stainless Steel

Only one study has investigated long-term health effects in the production of stainless steel.[6] This study will be discussed in Section 5.

Three epidemiological studies[7-9] have investigated cancer mortality among workers in stainless steel production. These studies did not provide convincing evidence of elevated lung cancer risk. In one of the studies, however, an excess of lung cancer was reported among stainless steel foundry workers.

5   Health Effects in the Production Chain of Ferrochromium and Stainless Steel

The integrated production chain of Outokumpu Steel in Tornio, Finland is unique: the chromite mine, the ferrochromium smelter, the stainless steel melting shop and the hot and cold rolling mills are located in the same complex. The long-term respiratory health effects were investigated in a cross-sectional study.[6] 316 subjects were studied, who had worked in the same production department for at least eight years and been exposed to low levels of different chromium species. The average exposure time was 18 years.

The study included assessment of respiratory diseases and symptoms, profound lung function tests and chest X-ray-examination. The results indicated that an average exposure time of 18 years in modern ferrochromium and stainless steel production and low exposure to dusts containing Cr^{6+} or Cr^{3+} does not lead to any respiratory changes detectable by lung function tests or radiography, or to any increase in symptoms of respiratory diseases.

6   Conclusions and Research Needs

The research data on the health effects of chromium in the metallurgical industry are controversial. The quantity of data is, however, limited. Some studies indicate respiratory effects whilst others do not. This can in part be explained by the difference in exposure levels, but most probably also by the low bioavailability of potentially harmful chromium species. Despite verified exposure no accumulation or respiratory changes were observed in the Finnish stainless steel industry. The low bioavailability can be explained by the surface properties of the particles in this type of process.

There is a need to know more about the surface properties of the particles and the form of agglomerates in the metal aerosols encountered at the work places in metallurgical industry, in order to understand the harmful mechanisms in the target tissue and to identify those stages in the production processes which possess an elevated risk for health hazards.

7 References

1. M. Huvinen, M. Kiilunen, L. Oksanen, M. Koponen and A. Aitio, *J. Occup. Med. Toxicol.*, 1993, **2**, 205.
2. X.B. Cox, R.W. Linton and F.E. Butler, *Environ. Sci. Technol.* 1985, **19**, 345.
3. H.J. Cross, J. Beach, L.S. Levy, S. Sadhra, T. Sorahan and C. McRoy, *Manufacture, Processing and Use of Stainless Steel: A Review of the Health Effects*, EUROFER 1999.
4. M. Koponen, T. Gustafsson, P.L. Kalliomäki and L. Pyy, *Am. Ind. Hyg. Assoc. J.*, 1981, **42**, 596.
5. International Agency for Research on Cancer, IARC Monographs on the Evaluation of Carcinogenic Risks to Humans, Vol. 49, IARC, Lyon, 1990.
6. M. Huvinen, J. Uitti, A. Zitting, P. Roto, K. Virkola and P. Kuikka, *Occup. Environ. Health*, 1996, **53**, 741.
7. R.G. Cornell and J.R. Landis, *Mortality Patterns among Nickel/Chromium Alloy Foundry Workers*, in *Nickel in the Human Environment*, F.W. Sunderman *et al.* eds., IARC Scientific Publication, No 53:87-93, 1984.
8. J.J. Moulin, P. Portefaix, P. Wild, J.M. Mur, G. Smagghe and B. Mantout, *Br. J. Ind. Med.*, 1990, **47**, 537.
9. J.J. Moulin, P. Wild, B. Mantout, M. Fournier Betz, J.M. Mur and G. Smagghe, *Cancer Causes Control*, 1993, **4**, 75.

CHAPTER 22

The Importance of Chromium in Occupational Health

GRANT DARRIE

1 Introduction

Speciation can seldom be more significant to understanding the toxicological behaviour of a metal in the occupational or natural environment than it is in the case of chromium. In order to understand better the problems that this metal poses regarding speciation, a certain amount of background information has been included, regarding the chemistry and uses of chromium.

Chromium is encountered in several different valency states with characteristics that range from being an essential trace element which is thought to be important in glucose metabolism, through chemically inert/biologically inactive and ultimately to being a genotoxic carcinogen.

The purpose of this chapter is to review what is known, not known and speculated, with respect to the role of various chromium-containing species in human and environmental toxicology. This topic is examined against the background of the commercial importance of chromium in its various forms in meeting the perceived needs of society.

2 Occurrence, Production and Use

Chromium is a naturally occurring element, number 24 in the periodic table, found in trace levels in many rocks and soils and in substantial concentrations in some regions where chromite ore is found. Chromium is the 21st most abundant element in the Earth's crust with an average concentration of 185 ppm and is the 26th most common element in seawater at approximately 0.2 parts per billion. Apart from the rare mineral crocoite ($PbCrO_4$) all naturally occurring chromium is found in the trivalent state.

The most important and only commercial source of chromium is chromite ore with the general formula $(MgFe)(CrAlFe)_2O_4$ or $Mg(CrAl)_2O_4(FeCr)_2O_3$, in which the chromic oxide content falls typically within the range 15–65%. Approximately 12.25 million tonnes of chrome ore were mined and processed in 1998 (excluding Russia for which data are unavailable).[1]

2.1 The Redox Chemistry of Chromium

The essential details of the redox behaviour of chromium are shown in Figure 1, which is adapted from Pourbaix.[2,3] Although chromium exists in several valency states, only two, namely chromium $+3$ and $+6$ are of major significance both commercially and toxicologically and only the trivalent state is common in geochemistry.

2.2 Chromium Metallurgy

Although pure chromium metal finds fairly limited applications, the element has very marked and important alloying characteristics of which by far the most significant is in the steel industry. Although Figure 1 indicates that trivalent chromium, as the oxide itself or in chromite ore, could be reduced to metal by powerful aqueous reducing agents, such a reaction is of no commercial significance.

Ferrochromium made by the electrothermal reduction of chromite ore (route 1 in Figure 1) provides the source of chromium in the manufacture of stainless steels where the element imparts oxidation, chemical and wear resistance, as well

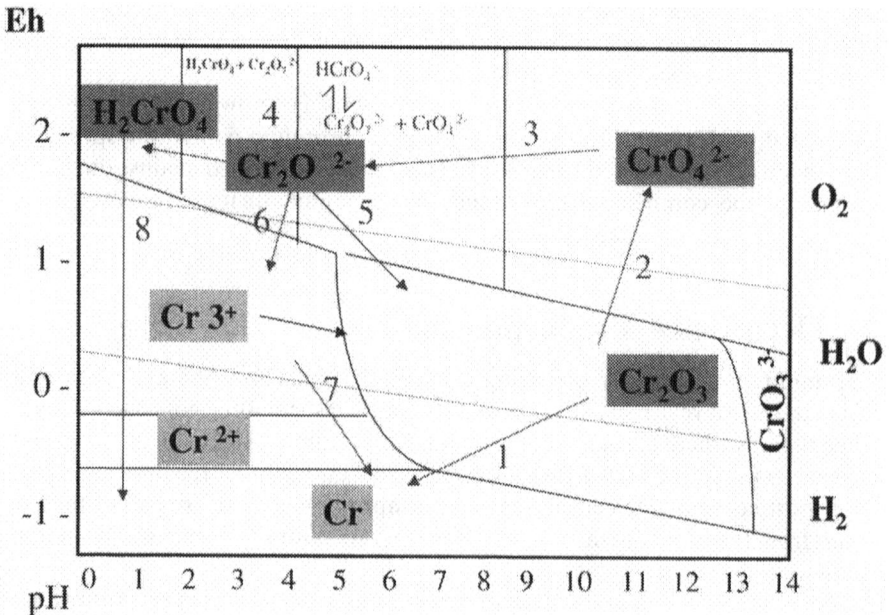

Figure 1 *Redox chemistry of chromium*

as hardness and strength retention at high temperatures. Stainless steels are mainly used in equipment for process engineering including chemical and nuclear plants, domestic applications and hospital equipment including surgical implants.

Pure chromium metal made either by the aluminothermic reduction of chromic oxide (route 1 in Figure 1) or by the electrolytic treatment of chromium alum made from ferrochromium (route 7 in Figure 1), is used in the manufacture of a range of non-ferrous alloys normally containing nickel and/or cobalt and used principally in the aerospace industry.

Electrolytic deposition of chromium metal on suitable substrates from either hexavalent or trivalent chromium solutions provides a surface engineering finish that has attractive and thus far unequalled properties for decorative (routes 7 and 8 in Figure 1) and functional (route 8 in Figure 1) requirements, primarily because of its corrosion resistance and low coefficient of friction. Many of the unique features of the metal and its various alloys arise from the formation of a surface layer of chromic oxide (reverse of route 1 in Figure 1).

Although the production of ferrochromium and of chromium metal from chromic oxide involves the reduction of trivalent chromium to the metallic state, the high temperatures involved lead to formation of fume which, in the presence of traces of volatile alkali metal salts and air, leads to the formation of hexavalent chromium in the form of chromate. Similarly during the manufacture of chromium-containing alloys, including stainless steels, and some applications, particularly welding, the presence of alkali metals encourages formation of chromates in the fumes evolved.

2.3 Chromium in Refractories

Because of its low reactivity and high thermal stability, chromite ore is used as a foundry sand for castings and mouldings and for industrial refractories placed in particularly harsh environments in the metallurgical and glass fibre industries. The use of chromium-containing refractories has declined in several areas of application and been completely replaced in some, *e.g.* cement and lime kilns, because of the known generation of hexavalent chromium compounds during use.

3 Chromium Chemicals

All chromium chemicals are manufactured from chromite ore by high temperature alkaline oxidation to form sodium chromate (route 2 in Figure 1), which is converted into the dichromate which in turn is the raw material for all commercial hexa- and tri-valent chromium products (routes 3–6 in Figure 1). This is summarised in Figure 2.

Chromite ore	Sodium carbonate	Filler/diluent

High temperature alkaline oxidation

Sodium chromate

Sodium dichromate

Solid or solution	Chromic acid	Other dichromates	Chromic oxide	Basic chromium sulfate

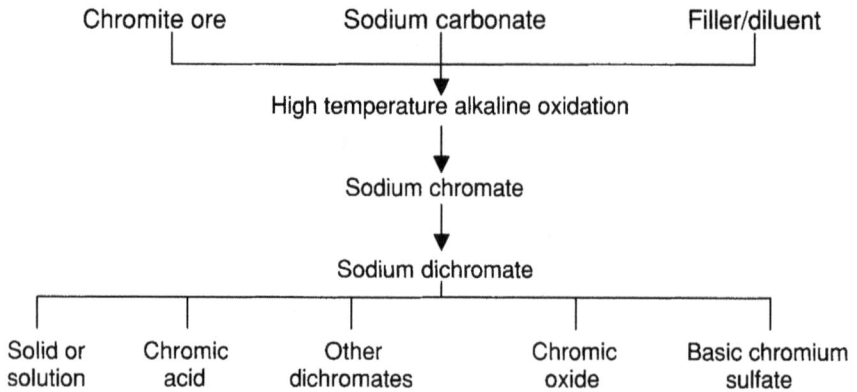

Figure 2   *Manufacture of chromium chemicals from chromite ore*

3.1   Hexavalent Chromium

Hexavalent chromium species range from chromates through dichromates to chromic acid as the pH falls – see Figure 1.

3.1.1   Chromates

Chromates are characterised by a wide range of solubility – see Table 1 for examples. Although sodium chromate is of major importance in chromium chemicals manufacture, relatively little is traded and used outside this environment. The low/intermediate-solubility chromates, *e.g.* strontium and zinc, are made from sodium dichromate or chromic acid and are used principally in anticorrosive primer systems on metal structures. The insoluble chromates, particularly lead chromate, are used as pigments, often in an encapsulated form.

3.1.2   Dichromates

All dichromates are readily/highly soluble in water.

Sodium dichromate represents the greatest quantity of hexavalent chromium traded and is used in a variety of processes:

- Manufacture of chromate pigments as above
- Manufacture of catalysts containing trivalent chromium
- As an oxidising agent in the manufacture of organic products including nicotinic acid, vitamin K and waxes
- As a raw material in the manufacture of chromium(III) sulfate tanning agents
- To introduce a very effective passivation/corrosion protection layer on to metal surfaces, particularly galvanised iron or aluminium. This layer contains not only hydrated chromic oxide, resulting from the redox reaction between the dichromate and the metal, but also the corresponding metal chromates formed *via* the oxides.

Table 1 *Solubility of chromates*

Product	Solubility g l^{-1}	Product	Solubility g l^{-1}
Ammonium chromate	~ 350	Potassium chromate	629
Barium chromate	0.004	Sodium chromate	873
Calcium chromate	0.11	Strontium chromate	1.2
Lead chromate	6×10^{-5}	Zinc chromate	2

After ageing, the surface layer of trivalent chromium provides the same protection as on the surface of stainless steel, without influencing the surface appearance of the host. The presence of traces of fixed chromate ensure that surface damage heals relatively quickly. For example on galvanised surfaces:

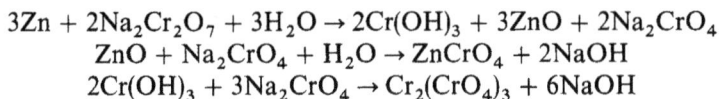

$$3Zn + 2Na_2Cr_2O_7 + 3H_2O \rightarrow 2Cr(OH)_3 + 3ZnO + 2Na_2CrO_4$$
$$ZnO + Na_2CrO_4 + H_2O \rightarrow ZnCrO_4 + 2NaOH$$
$$2Cr(OH)_3 + 3Na_2CrO_4 \rightarrow Cr_2(CrO_4)_3 + 6NaOH$$

The principal use of potassium dichromate is as a catalyst in the aluminothermic manufacture of chromium metal from chromic oxide, but it is also used in photographic chemicals, pyrotechnic products, engraving/lithography *etc.* Because of the nature of the aluminothermic reaction, the fumes given off contain hexavalent chromium as chromate. Ammonium dichromate is used principally in the manufacture of chromium dioxide – see Section 3.2.

3.1.3 Chromic Acid

The most acidic species containing hexavalent chromium is used principally in chromium plating, surface treatment and the timber treatment industries.

Chromium plating (route 8 in Figure 1) by electrolysis of chromic acid solutions is used either to provide a decorative, high lustre, tarnish resistant finish, *e.g.* for hospital equipment, or to provide a thicker functional finish to hard metal substances, providing a combination of high resistance to corrosion and wear plus a low coefficient of friction, for hydraulic equipment in industrial machinery.

The reactions involved in surface corrosion protection treatment of metals are similar to those listed above for sodium dichromate.

Chromic acid is reacted with copper oxide and either arsenic oxide (CCA) or boric oxide (CCB) principally to produce highly effective inorganic timber treatment agents, which are applied at professional treatment installations. In modern treatment plants, timber to be treated is firstly prepared under vacuum before the treatment solution is introduced under pressure thereby achieving preservative retention in the range 4–24%. Following the impregnation stage the treated wood is held at elevated temperature (typically 70 °C) to achieve fixation of the preservative ingredients before drying. Water-based timber preservatives containing either sodium or potassium dichromate first appeared early in the 20th century. Major developments took place in the second quarter of the same century, firstly with the appearance of preservatives containing mixtures of potassium dichromate and copper sulfate and subsequently including the addition of hydrated arsenic pentoxide. These salt-based agents have largely been superseded by oxide-based products based on chromic acid and copper oxide with any one of a variety of third components.

Chromium has no fungicidal or insecticidal role but rather acts as a fixative to prevent or minimise loss of the other ingredients by leaching, and as an anti-corrosion agent.

During the fixation stage, chemical reduction of the hexavalent chromium occurs accompanied by a rise in pH, leading to formation of a number of products with low solubility, particularly $CrASO_4$, $Cu(OH)CuASO_4$, $CuCrO_4$, $Cr2(OH)_4$ CrO_4, $Cr(OH)_3$ and a number of chromium and copper wood complexes.[4]

Chromic acid is also used in the manufacture of chromium dioxide – see below.

3.2   Tetravalent Chromium

Chromium dioxide (CrO_2), the only significant product containing tetravalent chromium, is made by reacting high specific surface ($\sim 50\,m^2\,g^{-1}$) chromic oxide, made by controlled decomposition of ammonium dichromate, with chromic acid under high pressure in autoclaves. The magnetic oxide is used extensively in high quality recording media in the audio, video and computer industries.

3.3   Trivalent Chromium

Trivalent chromium compounds show a very wide range of solubility, as shown in Table 2. The most commercially significant trivalent compounds are the basic sulfate and the oxide.

3.3.1   Chromic Sulfate

Basic chromium sulfate $Cr_2(OH)_{2x}(SO_4)_{3-x}$ is widely used in the manufacture of leather from animal hides or skins. Approximately 90% of all leather is produced using chromium salts.

Approximately 37% of all sodium dichromate manufactured world-wide is converted into basic chromium sulfate either as a prime product or as the by-product from an organic oxidation, *e.g.* the manufacture of vitamin K from

Table 2   *Solubility of trivalent chromium compounds*

Compound	Solubility
Chromic oxide	Insoluble
Chromic hydroxide	Solubility product $1.2 \times 10^{-28}\,mol\,l^{-1}$
Basic chromium sulfate	Approx. $500\,g\,l^{-1}$
Chromic chloride.$6H_2O$	$585\,g\,l^{-1}$

2-methylnaphthalene. Prime basic chromium sulfate is normally made by dissolving freshly precipitated chromium hydroxide in sulfuric acid or by reacting sodium dichromate solution with sulfuric acid and a sugar or starch, or with sulfur dioxide, *e.g.*

$$Na_2Cr_2O_7 + 3SO_2 + H_2O \rightarrow Cr_2(OH)_2(SO_4)_2 + Na_2SO_4$$

Chromium salts have been used in the tanning industry for over 150 years and whereas, historically, tanners tended to make their own tanning salts from sodium dichromate, those operating in the developed world are today supplied by producers who operate processes controlled to ensure the absence of hexavalent chromium in their products.

3.3.2 Chromic Oxide

Chromic oxide is made exclusively from hexavalent chromium compounds by processes designed to give the combination of physical and chemical features required by the user.

Although some chromic oxide is used in the manufacture of refractory bricks for the metallurgical industries, the more usual source of chromium for refractories is chromite ore.

For more sophisticated refractories containing up to 95% Cr_2O_3, material rheology is vital to enable the generation of stable, high solids content, aqueous suspensions necessary for the isostatic pressing processes used in brick manufacture for the fibreglass industry.

Physical features such as colour and particle size distribution are important to the effective use of chromic oxide in the surface coatings industry, where any or all of the following characteristics are required: light fastness; resistance to environmental or chemical corrosion; heat stability.

Manufacture of chromium metal from the oxide has been covered already. A variety of coloured ceramic glazes are manufactured by reacting chromic oxide with other metal oxides or mixtures at high temperature. Processes such as thermal decomposition of ammonium dichromate are used to give high specific surface products for the manufacture of chromium dioxide (see above) and catalysts for a wide range of applications.

3.3.3 Complexes

The bare trivalent chromium ion (Cr^{3+}) does not exist in solution because of very strong complexing characteristics, even in the case of the salts of strong mineral acids such as sulfate and chloride. Trivalent chromium readily forms stable complexes with many organic ligands. It is held to be an essential element necessary for normal metabolism of carbohydrates and lipids[5] and is available, as the picolinate or nicotinate, as a dietary supplement. Although trivalent chromium has little or no ability to enter human cells,[6] trivalent species, formed within cells following penetration by chromate ions, result in damage to the DNA, including formation of DNA–protein cross-links.[7,8] These are considered to be significant to chromium genotoxicity and carcinogenicity.

4 Toxicity

The toxicology of chromium and its compounds has probably been studied more widely than for any other metal but as yet some important aspects of its toxicological behaviour remain unresolved. Several comprehensive reviews have been published during the last ten years or so.[9-12]

4.1 Occupational

The acute and chronic toxicological data regarding the behaviour of chromium have been derived principally from animal studies and are highly species dependent. Table 3 summarises the available data for some hexa- and tri-valent chromium compounds.

4.2 Hexavalent Chromium

Hexavalent chromium compounds are heavily regulated through the Dangerous Substances Directive (67/548/EEC) *via* its progressive amendments and adaptations which identify hazards *via* standard risk (R) and safety (S) phrases on the basis of available data.[13] Notably, this is one of the few EC directives which are species specific and require regular speciation measurements. The regulation of workplace chromium exposure levels varies across European countries as shown in Table 4.

Occupational exposure in the USA is regulated by the Occupational Safety and Health Administration (OSHA) who set the current Permissible Exposure Limit (PEL) values (all as $mg\,Cr\,m^{-3}$): chromic acid and chromates 0.05 (ceiling values); chromium(III) compounds 0.5; chromium metal 1.0.

The International Agency for Research on Cancer (IARC) published a list

Table 3 *Toxicological data for chromium compounds*

Compound	Oral LD_{50} $mg\,kg^{-1}$	Inhalation $LC_{50}\,mg\,l^{-1}$	Dermal $LD_{50}\,mg\,kg^{-1}$	Irritancy/ corrosivity	Reference
Sodium chromate	52	0.104	1600	Not corrosive	14–17
Sodium dichromate	51	0.156	1170	Not corrosive	14–17
Chromic acid	52	0.217	57	Corrosive	14–17
Chromic oxide	> 5000	NA	NA	Not corrosive	18, 19
Basic chromium sulfate	> 3500	NA	NA	Not corrosive	18, 19

Table 4   *Summary of occupational exposure levels in European countries*

Species	Limit value (mg Cr m^{-3})	Countries
Cr metal/Cr(III) compounds	0.5	All
General Cr(VI)	0.005	Denmark
	0.02	Norway, Sweden
	0.05	All other (see notes)
Chromic acid	0.06	Sweden
Lead chromate	0.02	Norway
Zinc chromate	0.025	Germany
Strontium chromate	0.0005	Denmark

The 0.05 limit in Germany applies specifically to chemicals manufacture and welding; otherwise a limit of 0.025 mg Cr m^{-3} applies.
Italy and Spain use the American Conference of Governmental Industrial Hygienists (ACGIH) values – see Table 5.

Table 5   *Permissible exposure levels set by the American Conference of Governmental Industrial Hygienists*

Compound/group	Cancer class	Limit value mg Cr m^{-3} (TWA)
Zinc chromate I	A1	0.01
Water soluble Cr(VI) compounds (NOC)	A1	0.05
Insoluble Cr(VI) compounds (NOC)	A1	0.01
Chromite ore processing	A1	0.05
Lead chromate	A2	0.012
Strontium chromate	A2	0.0005
Calcium chromate	A2	0.001
Cr metal and Cr(III) compounds	A4	0.5

TWA = Time weighted average.
NOC = Not otherwise classified.
Cancer Class A1 = confirmed human carcinogen; A2 = suspected human carcinogen; A4 = not classified as human carcinogen.

of cancer evaluations for chromium and its compounds in 1990,[14] shown as Table 6.

The various acute toxicity effects of hexavalent chromium compounds are well known – see above. The behaviour of chromium metal and compounds towards biological systems is described by reference to Figure 3 which shows that the normal end product is chromium oxide or hydroxide (adapted from Pourbaix).[2,3]

Although most hexavalent chromium containing species are classified as carcinogenic, some, *e.g.* calcium, strontium and zinc, appear to be more potent and the interpretation of the evidence for lead chromate varies.

It is important to examine the accumulated evidence behind the above conclusions.

Table 6 *International Agency for Research on Cancer evaluation for chromium and its compounds*

Chromium and chromium compounds	Degree of evidence for carcinogenicity		Overall evaluation
	Human	Animal	
Chromium (VI) compounds			1
Chromium(VI) compounds as encountered in the chromate and chromate pigment production and chromate plating industries	Sufficient		
Barium chromate		Inadequate	
Calcium chromate		Sufficient	
Chromium trioxide		Limited	
Lead chromates		Sufficient	
Sodium dichromate		Limited	
Strontium chromate		Sufficient	
Zinc chromates		Sufficient	
Chromium(III) compounds	Inadequate	Inadequate	3
Metallic chromium	Inadequate	Inadequate	3
Welding fumes			2B
Welding fumes and gases	Limited		
Welding fumes		Inadequate	

Group 1: Carcinogenic to humans.
Group 2: Group 2A – probably carcinogenic to humans; Group 2B – possibly carcinogenic to humans.
Group 3: not classifiable as to its carcinogenicity to humans.
Group 4: probably not carcinogenic to humans.

4.3 Carcinogenicity

The first report of cancer in a worker involved in the manufacture of chromium pigments was reported over 100 years ago.[15] Since that time, several published epidemiological studies have consistently confirmed that exposure to hexavalent chromium compounds, encountered in chromate production, chromate pigments production and electroplating industries, has led to excess lung cancer with occasional reference to sino-nasal cancer. Most of these have been reviewed comprehensively.[9-12,21] Most recent updates for chromate production[16-25] indicate a significant reduction or elimination of the excess risk, due either to significant reduction in exposure levels and/or major changes in processing to either very low, or even zero, lime operations.

The most definitive chromate pigments study[26] indicated that exposure to lead chromate only resulted in no excess risk although exposure to zinc chromate

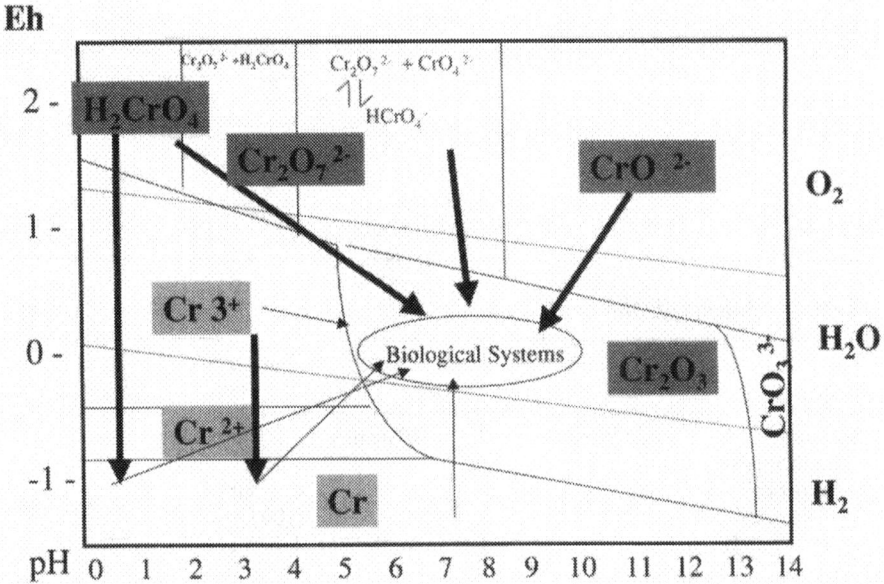

Figure 3 *Behaviour of chromium metal and compounds towards biological systems*

was associated with excess lung cancer. The most recent update on a plating industry cohort[27] shows a trend similar to that for chromate production, *i.e.* low/no excess risk for those employed since the early 1960s. Published studies for ferrochromium[28-30] and stainless steel production workers[31,32] show either a non-significant deficit or excess of lung cancer, but with highly likely exposure to other known carcinogens, *e.g.* polycyclic aromatic hydrocarbons. Stainless steel welders represent the largest population exposed to hexavalent chromium and, although several studies on welders have been published, few are specific to stainless steel. The most comprehensive multi-centre study[33] showed a higher lung cancer risk for mild steel welders than for those who worked with stainless steel.

Only the epidemiological studies for the pigments and plating industries can relate health effects to exposure to specific species. All studies in the chromate-producing industries include mixed exposure to chromates, dichromates and chromic acid. Few studies include occupational exposure data that would enable the derivation of a dose/response relationship. The 1975 Mancuso study[34] is still the main basis for USEPA risk assessments. Additional work by Crump,[35] which followed a group of 1000 workers occupationally exposed from the age of 20 until retirement at 65 concluded that $1 \mu g \, Cr^{6+} \, m^{-3}$ would result in 6–9 extra cases of lung cancer whilst $50 \mu g \, Cr^{6+} \, m^{-3}$ would result in an excess of 246–342 cases.

The conclusions of a more recent risk assessment study for the same cohort size,[36] employed for the same period but followed through to age 85, are shown in Table 7. These risk assessments will continue as new data are published.

Table 7 *Relationship between exposure levels and excess lung cancer deaths*[35]

Exposure level $\mu g\,Cr^{6+}\,m^{-3}$	Number of excess lung cancer deaths
50	5–28
25	2–14
10	1–6
5	0.5
1	0.1–0.6

Table 8 *Effects of inhalation exposure to chromium compounds on rats and mice*

Compound	Species	Dosing regime	Tumour incidence	Reference
Sodium dichromate aerosol	Male Wistar rat	25, 50, 10 $\mu g\,m^{-3}$	3/20 lung and 1 pharyngeal tumour ($100\,\mu g\,m^{-3}$)	38
Cr(III)/Cr(VI) oxide mixture	Male Wistar rat	100 μg total $Cr\,m^{-3}$ continuous/ 18 months	1/20 VS 0.40 for controls	38
Calcium chromate	C57BL/6 mouse	4.3 mg $Cr\,m^{-3}$– 5 h day^{-1}, 5 days/week for lifetime	45/1090 *versus* 24/1090 for controls	39

Some studies report cancer at sites other than the respiratory system, but the pattern does not have sufficient significance or consistency to justify a relationship to chromium exposure.[20]

Animal studies offer an opportunity to examine the chronic effects of exposure to specific substances under controlled conditions. Since the first studies on chromium compounds which were reported in 1959,[37] many others have been published in which different animal groups have been exposed to various chromium chemicals using a variety of protocols. These have been comprehensively reviewed elsewhere.[9-12] Only a few of these involve inhalation exposure and not all resulted in lung tumour formation rates above those in the controls. Results of the more important studies are summarised in Table 8.

Intratracheal instillation or intrabronchial implantation have also been used as alternatives to inhalation. Results of the more definitive studies are summarised in Table 9.

Levy *et al.*[38] reported on a major intrabronchial implantation study in which groups of 100 male and female Wistar rats were treated with a wide range of chromium containing chemicals, positive and negative controls. Significant tumour occurrence was observed only with calcium, strontium and zinc chromates.

Other study techniques such as intramuscular/intrapleural implantation[39,40] and direct injection[41] have shown tumour development only at the location of administration.

Table 9 *Effects of chromium compounds dosed into rats*

Compound	Species	Dosing regime	Tumour incidence	Reference
Sodium dichromate	Sprague Dawley rat	In the range 0.004–0.4 mg Cr kg^{-1}, 1–5 doses/week for 30 months	14/80 for 0.4 mg, 1 dose/week 1/80 for 0.09 mg, 1 dose/week 0/80–0/90 for all four other groups and controls	40
Calcium chromate	Sprague Dawley rat	0.08 mg Cr kg^{-1}, 5 doses/week 0.4 mg Cr kg^{-1}, 1 dose/week	6/90 13/80 0/80, 0/90	40

4.4 Genotoxicity

The many studies reporting the genotoxic effects of several hexavalent chromium compounds, principally *in vitro* with relatively few *in vivo*, have been comprehensively reviewed.[45] Positive effects are evident with all readily and sparingly soluble compounds but in the case of low-solubility materials, *e.g.* barium and lead chromates, prior solubilisation is necessary. Observed effects are DNA damage, reverse mutation, forward mutation, sister chromatid exchange, chromosomal aberrations, cell transformation and alterations in mitotic cell cycle. Evidence is gathering that redox processes might also be significant in the genotoxicity and carcinogenicity of chromium.[46–48]

4.5 Trivalent Chromium

Published acute toxicity data for insoluble trivalent compounds, *e.g.* chromic oxide, are typical of many other insoluble compounds. Some data are available for soluble salts such as sulfate and chloride (see Table 3), where their very acidic nature may be more significant than their chromium content. A case of suicide has been reported through drinking 0.4 l of a solution containing 48 g basic chromium sulfate.[49]

Apart from a cohort update by Mancuso,[50,51] there are no epidemiological studies that indicate that exposure to trivalent chromium may cause cancer, and the Mancuso conclusions have been challenged.[52]

Several animal studies using a variety of compounds and exposure regimes provide no evidence of carcinogenicity.[41,53–55] Whilst chromium(III) compounds appear to be able to produce genetic effects with purified nucleic acids or cell nuclei, there is generally no such activity in cellular systems.[45] These observations are almost certainly due to the relatively poor ability of trivalent chromium to cross cell membranes.[6,56]

5 Behaviour of Chromium in the Body

Trivalent chromium is poorly absorbed principally because, unless heavily complexed, it will precipitate under most physiological pH conditions. Even oral intake results in poor intestinal uptake.[57] This, coupled with the importance of trivalent chromium as an essential trace element in glucose metabolism has led to greater interest in increasing dietary uptake.[5] In contrast, hexavalent chromium species, predominantly chromate at physiological pH, behave very differently. Evidence indicates that the demonstrated ability of body fluids and long-lived non-target cells to reduce hexavalent chromium greatly reduces its potential toxicity including genotoxicity and carcinogenicity.[58]

Hexavalent chromium that survives reduction by body fluids is rapidly taken up by the erythrocytes penetrating the membrane *via* a general non-selective anion transport channel.[59,60] Possible mechanisms in chromium toxicity and carcinogenicity have been reviewed.[61-63]

It has been demonstrated that exposure to hexavalent chromium can cause DNA strand breaks in peripheral lymphocytes[64] and DNA–protein cross-links.[65] Despite many detailed studies, important details of chromium carcinogenicity remain unknown, particularly the role of other defence, cell repair and clearance mechanisms in determining why, on the basis of existing evidence, tumour generation appears to be exposure/administration site specific.

6 Conclusion

Chromium is an excellent case study of the importance of speciation measurements. Clearly, the toxicity of chromium is critically dependent on both the chemical form and concentration of the chromium in each exposure. The bioavail-ability and transport of chromium is clearly also species dependent. Whilst careful precautions have to be taken to avoid exposure to hexavalent chromium and its compounds, trivalent chromium is essential to life.

7 References

1. International Chromium Development Association data.
2. M. Pourbaix, *Atlas of Electrochemical Equilibria in Aqueous Solutions*, National Association of Corrosion Engineers, Houston, 1974.
3. J. Barnhart, *Reg. Toxicol. Pharmacol.*, 1997, **26**, 53.
4. P.A. Cooper, D.L. Alexander and T. Ung, *What is Chemical Fixation in Chromium-containing Waterborne Wood Preservatives: Fixation and Environmental Issues*, Forest Products Society, Madison, 1993, pp. 14–22.
5. R. A. Anderson, *J. Am. Coll. Nutr.*, 1998, **17**, 548.
6. A. Kortenkamp, D. Beyersman and P. O'Brien, *Toxicol. Environ. Chem.*, 1987, **14**, 23.
7. D. Beyersman, *Environ. Chem.*, 1989, **22**, 62.
8. M.D. Cohen, B. Kargacin, C.B. Klein and M. Costa, *Crit. Rev. Toxicol.*, 1993, **23**, 255.
9. S. Fairhurst and C.A. Minty, *Health and Safety Executive Toxicity Review 21*, HMSO, London, 1989.
10. ATSDR, *2000 Toxicological Profile for Chromium*, US Department of Health and

Human Services, Agency for Toxic Substances and Disease Registry, Atlanta, Ga. (update in print).

11. H.J. Cross *et al.*, *Criteria Document for Hexavalent Chromium*, commissioned and published by the International Chromium Development Association, Paris, 1997.
12. *Chromium and its Inorganic Compounds*, Report of the Dutch Expert Committee on Occupational Standards, 1998/01.
13. *Official Journal of the European Communities*, 1993, no. 258a; including 1996 revisions contained in 22nd ATP.
14. Allied Corp. Data, report no ma-285-84-5, Dec 3, 1984.
15. Allied Corp Data, report no ma-285-84-8, Jan 22, 1985.
16. Allied Corp Data report no ma-285-84-10, May 3, 1985.
17. Allied Corp Data report no ma-285-84-1, Dec 1, 1984.
18. Bayer ag data.
19. Bayer ag data.
20. *IARC Monographs on the Evaluation of Carcinogenic Risks to Humans*, Vol. 49, *Chromium, Nickel and Welding*, World Health Organisation, Lyon, France, 1990.
21. D. Newman, *Glasgow Med. J.*, 1890, **33**, 469.
22. S. Langard, *Am. J. Ind. Med.*, 1990, **17**, 179.
23. J.M. Davies, E.F. Easton and P.L. Bidstrup, *Br. J. Ind. Med.*, 1991, **48**, 299.
24. H. Pastides *et al.*, *Am. J. Ind. Med.*, 1994, **25**, 663.
25. U. Korallus *et al.*, *Int. Arch. Occup. Environ. Health*, 1993, **65**, 171.
26. J.M. Davies, *Br. J. Ind. Med.*, 1984, **41**, 158.
27. T. Sorahan *et al.*, *Occup. Environ. Med.*, 1998, **55**, 236.
28. L.V. Pokrovskaya and N.K. Shabynina, *Gig. Tr. Prof. Zabol.*, 1973, **10**, 23.
29. G. Axelsson *et al.*, *Br. J. Ind. Med.*, 1980, **37**, 121.
30. S. Langard *et al.*, *Br. J. Ind. Med.*, 1990, **47**, 14.
31. J.J. Moulin *et al.*, *Br. J. Ind. Med.*, 1990, **47**, 537.
32. J.J. Moulin *et al.*, *Cancer Causes Control*, 1993, **4**, 171.
33. I. Simonato *et al.*, *Br. J. Ind. Med.*, 1991, **48**, 145.
34. T.F. Mancuso, *Consideration of Chromium as an Industrial Carcinogen*, in *International Conference on Heavy Metals in the Environment*, ed., T.C. Hutchinson, Institute for Environmental Studies, 1975, pp. 343–356.
35. K.S. Crump, *Evaluation of Epidemiological Data and Risk Assessment for Hexavalent Chromium*, Occupational Health and Safety Administration, Washington DC, Contract no j-9-f-1-0066, 1995.
36. K.M. Steenland *et al.*, *Am. J. Ind. Med.*, 1996, **29**, 474.
37. A.M. Baetjer *et al.*, *Arch. Ind. Health*, 1959, **20**, 124.
38. U.M. Glaser *et al.*, *Toxicology*, 1986, **42**, 219.
39. P. Nettesheim *et al.*, *J. Natl. Cancer Inst.*, 1971, **5**, 1129.
40. D. Steinhoff *et al.*, *Exp. Pathol.*, 1986, **30**, 129.
41. L.S. Levy *et al.*, *Br. J. Ind. Med.*, 1986, **43**, 243
42. W.C. Hueper and W.W. Payne, *Ind. Hyg. J.*, 1959, 274.
43. W.C. Hueper and W.W. Payne, *Arch. Environ. Health*, 1962, **5**, 445.
44. A. Furst *et al.*, *Cancer Res.*, 1976, **36**, 1779.
45. S. De Flora *et al.*, *Mutat. Res.*, 1990, **238**, 99.
46. C.B. Klein *et al.*, *Chem. Res. Toxicol.*, 1991, **4**, 592.
47. A.M. Standeven and K.E. Wetterhahn, *Chem. Res. Toxicol.* 1991, **4**, 616.
48. X. Shi *et al.*, *J. Toxicol. Environ. Health*, B, 1999, **2**, 87.
49. P.V. van Heerden *et al.*, *Intensive Care Med.*, 1994,
50. T.F. Mancuso, *Am. J. Ind. Med.*, 1997, **31**, 129.

51. T.F. Mancuso, *Am. J. Ind. Med.*, 1997, **31**, 140

52. K.A. Mundt and L.D. Dell, OEM Report, *Carcinogenicity of Trivalent and Hexavalent Chromium*, 11,11, Nov. 1997.

53. W.C. Huepper, *J. Natl. Cancer Inst.*, 1955, **16**, 447.

54. W.C. Huepper and W.W. Payne, *Experimental Studies on Chromium Compounds, their Carcingenicity and their Importance for Industrial Medicine, in Proceedings of the 13th International Congress on Occupational Health*, Book Craftsmen Assoc., New York, 1961, pp. 473–486.

55. L.S. Levy and S. Venitt, *Carcinogenisis*, 1986, **7**, 831.

56. K.S. Kitagawa *et al.*, *Chem-Biol. Interact.*, 1982, **40**, 265.

57. R.M. Donaldson and R.F. Barreras, *J. Lab. Clin, Med.*, 1960, **68**, 484.

58. S. De Flora *et al.*, *Carcinogenisis*, 1997, **18**, 531.

59. K.W. Jennette, *Trace Elem. Res.*, 1979, **1**, 55.

60. Z.I. Cabantchik *et al.*, *Biochim. Biophys. Acta*, 1978, **515**, 239.

61. M.D. Cohen *et al.*, *Crit. Rev. Toxicol.*, 1993, **23**, 255.

62. S. De Flora *et al.*, *Chromium and Carcinogenisis*, in *Handbook of Metal–Ligand Interactions in Biological Fluids, Bioinorganic Medicine*, Vol. 2, G. Berthon ed., 1995, pp. 1020–1036.

63. C.B. Klein, *Carcinogenicity and Genotoxicity*, CRC Press Inc., 1996, pp. 205–219.

64. M. Gao *et al.*, *Human Exp. Toxicol.*, 1992, 1677.

65. M. Costa, *Scand. J. Work Environ. Health*, 1993, **19**, Suppl, 71.

CHAPTER 23

Speciation Related to Human Health

RIÁNSARES MUÑOZ-OLIVAS AND CARMEN CÁMARA

1 Introduction

The main aims of this contribution are (i) to stress the need for speciation in biological and clinical samples, (ii) to provide information on the main elements of interest and their species, (iii) to understand the role of these elements in organisms through knowledge of their distribution in various organs and (iv) to explain the main difficulties of speciation in these complex matrices. In order to achieve these aims the chapter includes a brief introduction dedicated to the actual importance of trace element determinations and speciation in human health.

The state of the art for the most relevant elements and their compounds is discussed. The last section will be dedicated to the need for a better understanding of the role of element speciation related to human health. It will conclude by highligthing the requirements for available CRMs (Certified Reference Materials) for speciation purposes in these type of samples.

There is not a generally accepted definition of the term 'trace element' as the elemental content may differ by orders of magnitude in the various body fluids and tissues. It is, however, more or less accepted that the term is used for elemental concentrations below $100\,\mathrm{mg\,l^{-1}}$. Another definition of a trace element, from a clinical point of view, is if its concentration does not exceed the Fe concentration in the same body fluid or tissue.[1]

A more useful classification of the elements, according to their importance to living organisms is into essential and non-essential categories. Essential trace elements are Co, Cu, Cr, Fe, I, Mn, Mo, Ni, Se, Si, V, and Zn, whereas Al, As, Cd, Sb, Sn, Pt, Hg, Pb, Bi, *etc.* are considered potentially toxic trace elements.[2]

It is recognised that such trace elements are involved in biogeochemical cycles, as well as in anthropogenic activities that cause their mobilisation, resulting in environmental pollution and human exposure. Since essential trace elements are clearly implicated in numerous biochemical processes, there is an increasing interest in the study of their health effects.

The toxicity, bioavailability and distribution of an element in any sample can only be determined if each chemical form can be identified and quantified. Since an element may exist in a sample in one or more different forms, including particulate, dissolved, complexed and ionic forms as well as in different oxidation states, it is imperative that the analytical technique used is able to provide accurate experimental data about these forms.[3]

The relationship between the amount of an element absorbed and the quantity utilised strongly depends on the chemical form. Other factors affecting bioavailability are: age, percentage intake, physico-chemical form of the element, multielement and antagonist interactions, functional and biochemical interactions.[4] Ideally, the bioavailability of the elements present in single or mixed foods should be determined before any attempt is made to interpret data on their contents in diets or foods. However, the characterisation of the different elemental species in complex samples is difficult, partly because of the low concentration present, and also because of the danger of changes in chemical forms during analysis.[5]

The improvement in the detection power of analytical techniques for trace elements has substantially increased our knowledge of the role of trace elements in human health. It is now recognised that trace elements can become limiting not only because of low environmental levels, but also because of imbalances in diets. Some of the more obvious examples include: iodine deficiency (goitre), with nearly 1600 million people at risk, selenium deficiency, resulting in such problems as Keshan disease, and zinc deficiency, giving rise to retardation in the growth of infants. It has also made it possible to study the interactions between trace elements and the activities of element-containing proteins, *e.g.* the glutathione peroxidase activity and the Se content in blood of children suffering from phenylketonuria.[6] Furthermore, several elements have been acknowledged to be essential in metalloproteins with specific biochemical functions.[2] For example, if Zn is linked to albumin, macroglobulins, enzymes or LMW (Low Molecular Weight) ligands like citrate, the characteristics and biological functions vary greatly. Likewise, Fe linked to hemoglobin in blood plays a completely different role to Fe linked to serum ferritin. In addition, the total concentration of Co in serum may be of interest in occupational exposure, whereas only the fraction present in vitamin B_{12} is specifically of biological importance.[7]

Several general reviews concerning speciation in clinical and biological samples have been reported[4,8] discussing sampling and sample preparation, methodology and speciation of a range of elements in blood, urine, milk, saliva and other fluids.

There are two important aspects of the concept of speciation concerning human health: (1) to establish which forms of a chemical element possess highest noxious effects and on the other hand, to clarify which species of an essential element are more available[9] and (2) to quantify the element in relation to the biological molecules to which it is bound. The latter helps to explain mobility, storage, retention and toxicity. A major problem is the hazard of contamination or losses of the element, due to the very low concentration encountered in biological materials.[10] Therefore, even more care must be taken than for environmental samples in order to maintain the integrity of the labile metal–ligand association and to check the mass balance of the protein and the trace element.

2 Usefulness of Samples

The criteria governing the acceptability of a clinical sample depends on the information needed, *e.g.* clinical diagnosis of deficiency or toxicity, biological control of environmental pollution, nutritional surveillance or forensic investigations. For these cases, not only the elemental content needs monitoring, but also the chemical form or oxidation state of the elements.[5,11]

Elemental analysis of human body fluids and tissues mainly concerns blood (plasma, serum) and urine. Other matrices of interest include human milk, saliva and hair. Analysis of these samples provides very different information.[12-15]

2.1 Blood

The levels of essential elements in blood have been considered the best indicator of the status of a nutrient in the body. However, this biomonitoring is not acceptable in all cases. There is a need to differentiate the moieties binding the element (speciation) and to determine the content of synergic or antagonistic elements as well as a combined analysis of other body samples.

2.2 Urine

Urine samples can indicate the dietary intake of trace elements, subnormal levels indicating a deficiency and abnormally high levels a possible contamination problem. Because of difficulties in interpreting the results of urine analysis, the data should only be used to supplement information obtained from blood, serum, hair *etc.*

2.3 Hair

External contamination is a major source of some trace elements in the hair and thus it is frequently analysed for trace elements, in order to characterise the exposure of an individual to contaminants. Hair can also be used to assess the intake of an ingested contaminant. For example, hair is a good indicator for Hg, and especially for CH_3Hg^+ because it has a greater rate of transfer to hair. After ingestion of organic Hg compounds has terminated, the level of Hg in hair decreases in parallel to that in blood.[5] At present, mercury in hair is considered a meaningful parameter for assessing the intake of CH_3Hg^+. Special care must be taken with the pre-treatment (cleaning step) in order to adequately differentiate the different methylmercury sources (environmental or internal).

2.4 Human Milk

In early childhood, milk is the most important food, since it is the only source of nutrition during the first months of a baby's life. It is therefore essential that all nutrients are present in milk in adequate quantities. There is an increasing interest in trace element speciation in this matrix.[2] Moreover, it is of primary importance to assess the concentration levels of the various species of toxic elements in milk, both to evaluate their safety and to understand their effect on humans.[7]

3 Main Analytes and Species

It is very difficult to clearly define elements as toxic and non-toxic. It has been well established that the toxicity or essentiality depends not only on the analyte concentration but also on the chemical form. In this chapter the analytes have been classified as (i) essential trace elements, (ii) probably essential trace elements and (iii) potentially toxic trace elements. Nevertheless, it must be kept in mind that this classification should not be considered too strictly. Figure 1 summarises the main species of the elements described, found in different human samples. In this chapter, chromium has not been considered because there is an entire chapter related to this element elsewhere in this book (Chapter 22).

3.1 Essential Trace Elements

3.1.1 Copper

A variety of symptoms have been associated with Cu deficiency in humans that include abnormal bone formation, osteoporosis, vascular troubles and anaemia.[2] On the other hand, copper poisoning in humans can result from contamination of foodstuffs or beverages by Cu containers. Massive accumulation of Cu can lead to hepatitis and liver cirrhosis. Cu bioavailability depends strongly on the chemical form: highly soluble species (sulfate or nitrate) are readily absorbed while carbonate is absorbed only after dissolving in the acids of the stomach. Other insoluble forms, such as CuO or CuS, are not metabolised at all.[5]

Copper speciation. Copper is widely distributed in biological tissues where it occurs largely in the form of organic complexes, many of which are metalloproteins and metalloenzymes (cytochrome oxidase, caeruloplasmin, superoxide dismutase, *etc.*). These are involved in a variety of metabolic reactions: (i) the utilisation of oxygen during cell respiration and energy utilisation; (ii) the synthesis of proteins in connective tissues of the skeleton and blood; (iii) neuroactive compounds concerned in nervous tissue function.[5] Copper in human blood is mainly distributed between the erythrocytes and the plasma. In erythocytes it occurs as the Cu–Zn metalloenzyme superoxide dismutase (60%), the remaining 40% being bound to other proteins.[2] In plasma, about 93% of Cu is firmly bound to the enzyme caeruloplasmin and the rest (7%) is bound to albumin and amino acids.

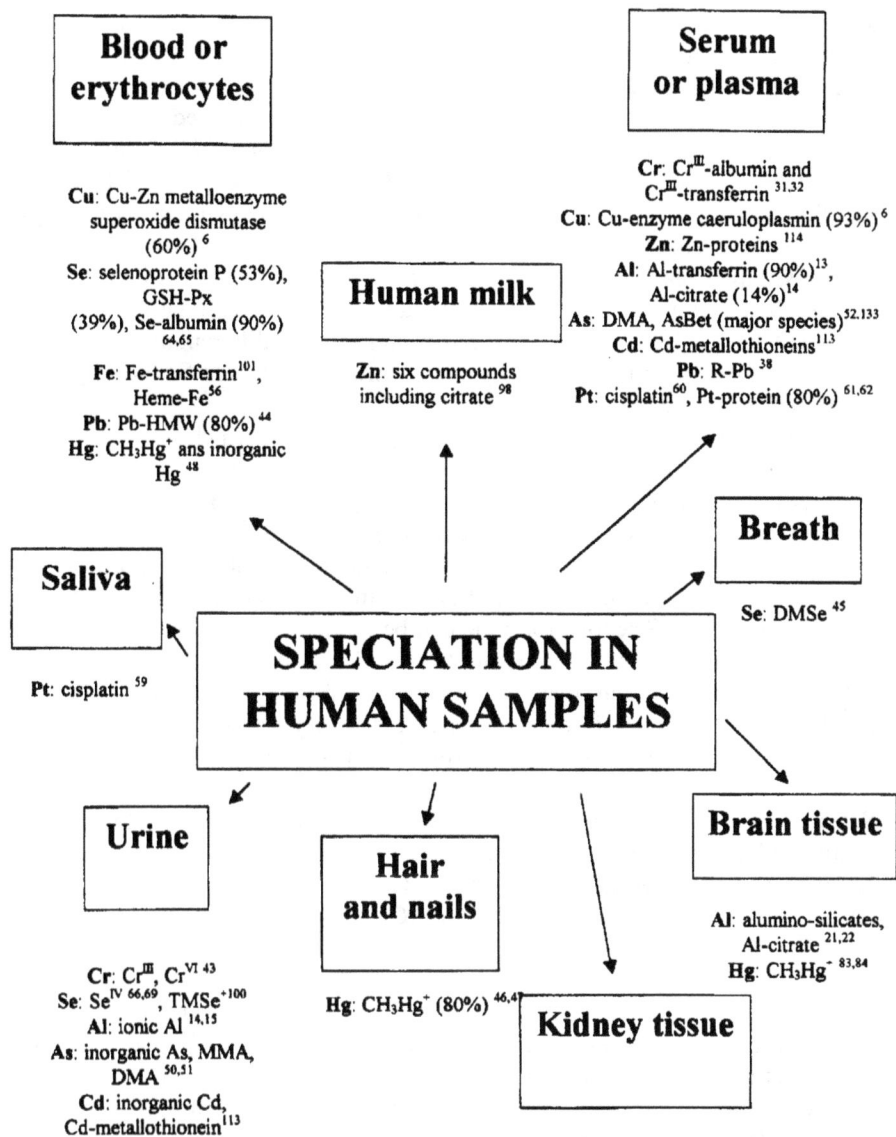

Blood or erythrocytes

Cu: Cu-Zn metalloenzyme superoxide dismutase (60%) [6]
Se: selenoprotein P (53%), GSH-Px (39%), Se-albumin (90%) [64,65]
Fe: Fe-transferrin [101], Heme-Fe [56]
Pb: Pb-HMW (80%) [44]
Hg: CH$_3$Hg$^+$ ans inorganic Hg [48]

Human milk

Zn: six compounds including citrate [98]

Serum or plasma

Cr: Cr^{III}-albumin and Cr^{III}-transferrin [31,32]
Cu: Cu-enzyme caeruloplasmin (93%) [6]
Zn: Zn-proteins [114]
Al: Al-transferrin (90%) [13], Al-citrate (14%) [14]
As: DMA, AsBet (major species) [52,133]
Cd: Cd-metallothioneins [113]
Pb: R-Pb [38]
Pt: cisplatin [60], Pt-protein (80%) [61,62]

Saliva

Pt: cisplatin [59]

SPECIATION IN HUMAN SAMPLES

Breath

Se: DMSe [45]

Urine

Cr: Cr^{III}, Cr^{VI} [43]
Se: Se^{IV} [66,69], TMSe$^+$ [100]
Al: ionic Al [14,15]
As: inorganic As, MMA, DMA [50,51]
Cd: inorganic Cd, Cd-metallothionein [113]

Hair and nails

Hg: CH$_3$Hg$^+$ (80%) [46,47]

Kidney tissue

Brain tissue

Al: alumino-silicates, Al-citrate [21,22]
Hg: CH$_3$Hg$^-$ [83,84]

Figure 1 *Main metallic species found in human samples*

3.1.2 Selenium

Until recently, the only known metabolic role of Se in mammals, including humans, was as a component of the enzyme glutathione peroxidase (GSH-Px) which, together with vitamin E, catalase and superoxide dismutase is one of the antioxidant defence systems of the body.[16] Actually, several other selenium-containing enzymes (*e.g.* glycine reductase, hydrogenase) have been described, and it is likely that selenoproteins other than GSH-Px remain to be discovered. There is, for example, growing evidence that an additional selenoenzyme protein is involved in thyroid activity.[17]. Keshan disease is the most famous illness caused by a deficit of Se in the human and animal diet, caused by the low Se content in soil and water in this area of China.[18] Selenium can be involved in some degenerative diseases: cancer, cardiovascular disease, cerebral thrombosis.[5] Some studies[19-21] with heart disease patients have shown that these people have low Se levels compared with controls, as well as lower GSH-Px activity. The possibility that increased supplementation of dietary Se might protect against the development of cancer in humans has generated great interest[22-24] (see also Chapter 18). Such an effect has been shown in different animal studies, together with small, but statistically significant, differences in Se blood plasma levels detected in people developing cancer.[16,25,26]

The biochemical mechanisms of Se toxicity have not been clearly established; *e.g.* selenite is a strong catalyst for the oxidation of SH^- groups, and this may be the basis of its inhibitory effect on protein synthesis. Inorganic Se must be reduced to selenide to become biologically useful. Selenide is either lost *via* urine or converted into selenocysteine and other proteins. On the other hand, selenomethionine can be incorporated into the protein chain by changing S and Se in a methionine base.[27] Selenomethionine can be retained at high levels, because it can be incorporated directly into muscle as well as into selenocysteine.

Selenium toxicity episodes in humans are more scarce than deficiency troubles. The signs of Se intoxication are: respiratory problems, skin infections, enzymatic systems inhibition, chromosomic aberrations *etc.*

Selenium speciation. As for the other trace elements, the samples most frequently examined have been blood, serum and urine. Human blood plasma and serum have been well investigated and methods such as ETAAS (Electrothermal Atomic Absorption Spectrometry),[28] and ICPMS (Inductively Coupled Plasma Mass Spectrometry) coupled to HPLC[29] have been developed in order to define the distribution of Se among blood proteins. In such studies, 53% of the Se has been found to be bound to selenoprotein P, 39% to GSH-Px and 9% to albumin. Whole blood Se values and the erythrocyte glutathione peroxidase activity have been determined in the diet of some adults.[30] Supplementation was performed with four different compounds: selenate, selenite, L-selenomethionine and high-Se yeast. Both the inorganic and the organic Se derivatives gave rise to steady-state levels of GSH-Px. However, whole blood selenium uptake was significantly higher with the organic derivatives, Se-Met being absorbed the most *via* amino acid routes.

Several methods combining HPLC and ICP/MS have also been reported for Se speciation in urine. In two papers,[31,32] selenite was the major Se species present in all samples, but also $TMSe^+$ was found in several samples. Only one study[33] revealed this compound as the most abundant in urine. In all cases, unknown species were detected that did not correspond to any known ionic or organic Se compound.

A method for the determination of organometallic species in breath was developed by Feldman *et al.*[34] by trapping at $-80\,^\circ$C and coupled GC–ICP-MS (Gas Chromatography–Inductively Coupled Plasma-Mass Spectrometry). From this study it was established that DMSe was detected in the breath of six persons.

Because of the evidence of cancer prevention by Se supplementation in the diet, different international programs have started during the past ten years, *e.g.* the PRECISE trial (Prevention of Cancer by Intervention with Selenium).[35]

3.1.3 Zinc

Most biochemical roles of Zn reflect its involvement in a large number of enzymes or as a stabiliser of the molecular structure of cellular constituents and membranes. For example, Zn participates in the synthesis and degradation of carbohydrates, lipids, proteins and nucleic acids. It plays an essential role in some processes of genetic expression.[5] Zinc is one of the metals bound by metallothioneins (MT), a class of cysteine-containing, low molecular weight (LMW), ubiquitous intracellular proteins with high affinity for metals. Mammalian MTs are characterised by a single chain protein containing between 61 and 62 amino acids, 20 being cysteine. Non-cysteine amino acids serve to connect the metal chelating cysteine residues.[36] Two isoforms, MT-I and MT-II, are expressed in all mammalian tissues, and only differ at neutral pH by a single negative charge. Each class is composed of different isoproteins; at least ten isoMT genes are expressed in humans. There is another MT isomer characterised as the brain-specific isomer MT-III, which could be related to Alzheimer's disease. It has been suggested that MT-III serves as a neuromodulator.

For all these reasons, Zn deficiency can provoke health disorders in humans, such as growth retardation, delay in sexual and skeletal maturation and an increased susceptibility to infections. Zn is also associated with other pathologies such as rheumatoid arthritis, chronic giardisis, tumours.[37]

Concerning toxicity, few instances of Zn poisoning have been reported. This element has a strong interaction with Cu and can be implicated in copper deficiency.[5] On the other hand, other elements with similar physicochemical properties, *e.g.* Cd, can restrict the absorption of Zn.

Zinc speciation. Investigation of the Zn-binding pattern in human milk was carried out by HPLC–ICP-AES.[38] Six Zn compounds including citrate were detected. It is also worthwhile underlining the negative correlation found between citrate and the maternal daily zinc intake. The potential involvement of the Zn compound MT-III in Alzheimer's disease has been studied revealing a decrease of up to 10-fold in the expression of this isomer in such patients.[36] Evidence for altered Zn metabolism in Alzheimer's disease includes the observation of some abnormalities in both the uptake and distribution of Zn, potentially as a consequence of MT-III down-regulation. Brätter *et al.*[38] studied more specifically the relationship of an alteration of Zn distribution in serum in an epileptic child with an associated hepatotoxicity. They also found significantly lower serum levels of total protein, albumin and selenium than in the controls.

3.1.4   *Iron*

Iron is involved in an extensive series of interactions which modify the utilisation or metabolism of both essential and potentially toxic trace elements. Synergistic interactions have been evidenced between Fe and Cu, for instance. Antagonistic behaviours are known for Fe and Cd or Fe and Pb. Transferrin is the primary transport protein for Fe in the bloodstream. Non-transferrin-bound Fe (NT Fe) represents a mobile fraction of Fe in blood. The nature of this last fraction is thought to be composed of a group of LMW Fe compounds associated with serum proteins as albumin or with ligands such as citrate. NT Fe may be responsible for liver and myocardial injury, as it is more avidly transferred to cells compared with transferrin Fe.[40] The study by Chyka *et al.*[41] with pigs and adult humans, aimed to elucidate the relationship between NT Fe and total serum Fe concentration. After subtoxic Fe doses, transferrin never reached saturation and consequently NT Fe did not increase with rising serum Fe concentrations.

Iron speciation. Smith and Harnly[42] described an ETAAS method for the simultaneous determination of heme and non-heme Fe species. Heme Fe is the name given to Fe bound with a porphyrin ligand, primarily hemoglobin or myoglobin. These compounds are relatively stable, survive the digestion process of the stomach and enter the intestine as intact Fe complexes. Heme Fe is absorbed at a higher rate than non-heme Fe.

3.2   Probably Essential Trace Elements

3.2.1   *Nickel*

Nickel has been identified in four typical enzymes of plants and micro-organisms. Evidence that Ni functions similarly in humans needs to be clarified. However, it is well known that skin allergy is provoked in some people exposed to nickel.[5] Several recent epidemiological and experimental studies have cast new light on the carcinogenic properties of nickel and its compounds. Nasal cancer among workers in Ni refineries was reported in the 1930s, and conclusive evidence of an increased risk of lung and nasal cancer in this group of people was presented in the 1960s.[43] The toxic effects of Ni compounds depend on the ability of Ni to enter target cells and to release nickel ions, which interact with cellular constituents and molecules. This implies different mechanisms of cellular uptake, biological transformation and cellular transport. Nickel sulfide (Ni_2S_3) is considered to be one of the most carcinogenic Ni compounds, as shown in experiments with animals and in human lymphocytes.[44] Other carcinogenic compounds are nickel sulfates and some oxides (see also Chapter 20).

Nickel speciation. In a paper of particular interest to toxicologists[45] the cytotoxicity and potential carcinogenicity of different Ni compounds was investigated. Biological uptake of insoluble Ni compounds was significantly greater than that of soluble Ni^{2+}. Consequently, the effects on cell transformation and growth were more potent.

3.2.2 Vanadium

Recently, the possibility that vanadium might play a role in the regulation of Na^+/K^+ exchange has been considered. The role of the vanadyl ion as an enzyme co-factor and its participation in metabolism have also been discussed. However, all these properties need to be investigated in humans.[46]

Vanadium speciation. Vanadium is a relatively toxic element for humans. A variety of signs of V toxicity exist because they vary both with species and with dosage. Thus, V^{5+} ion is absorbed 3–5 times more effectively than V^{4+}. Depending on the other substances present in the diet, V^{5+} can be transformed into V^{4+}, changing the percentage of V ingested. For example, ascorbic acid, Cr, Fe^{2+}, Cl^- and $Al(OH)_3$ can reduce V absorption and consequently its toxicity. Some of the disorders provoked by V are gastrointestinal disturbances, depressed growth, and cardiovascular disease. Toxicity usually occurs only as a result of high-level exposure to airbone vanadium at high concentrations.[5]

3.2.3 Cobalt

Cobalamins consist of a nucleus formed by four pyrrolidine rings and a cobalt atom coordinated with four nitrogen atoms. Four common and naturally occurring cobalamins are cyanocobalamin (vitamin B_{12}), which is a nutrient essential for cells and a deficiency of which produces delayed growth; hydroxocobalamin (vitamin B_{12a}); 5'-deoxyadenosylcobalamin (coenzyme B_{12}); and methyl-cobalamin (methylcoenzyme B_{12}), which is an important methylating agent with a high activity for transferring a methyl group to some metallic species, such as Hg, Se, Pb, and Sn.[47]

Cobalt speciation. Several papers have described the use of liquid chromatography coupled to different detectors for the separation and characterisation of cobalamins.[48-50] All these methods have been applied to the analysis of cobalamins in pharmaceutical products, but as far as we know, analysis of these Co-derivatives has not been performed in human samples.

3.3 Potentially Toxic Trace Elements

3.3.1 Aluminium

The risk of aluminium accumulation in the organism has been associated with several factors. These include a wide variety of uses, *e.g.* in packing and building materials, paints, treating patients with Al-containing antacids, long-term industrial exposure, hemodialysis treatment for renal failure patients.[51,52] The high bioavailability of 'free Al' in the physiological environment has led to the assumption that it may compete with other metals, causing changes in the activity of some enzymes.[53] Aluminium has been shown to interact with DNA, probably by cross-linking the protein chains.[54] Al is suspected of being involved in a number of neurological disorders, *e.g.* Alzheimer's disease and some liver diseases.[52,54] The involvement of Al in the pathogenesis of Alzheimer's disease (AD) was related to the identification of different target sites of accumulation in brain tissue in AD patients.[55,56] The same Al species (*e.g.* aluminosilicates, citrate complexes) were not detected in the same brain region of healthy individuals. In brief, apart from the analytical evidence, there is no direct experimental proof that demonstrates the direct implication of Al as an aetiological factor in AD.

Aluminium interacts with a number of other elements, including Ca, F, Fe and Mg, reducing their absorption. In some cases, Al has been used therapeutically to treat fluorosis and to reduce P absorption in uraemic patients.[5] Recent reports have suggested a protective role for Si against Al intoxication, resulting from the formation of highly insoluble aluminosilicates. These complexes may be formed in the extracellular environment under conditions of poor Al-transferrin binding and low citrate concentration (the highest binders of Al). The formation of such inactive compounds avoids Al moving into the intracellular compartments.[57,58]

Aluminium speciation. The monitoring of total Al in certain body tissues has been recognised to be a useful indicator of various diseases that have been associated with Al toxicity. Normal tissue levels of Al are very low (less than $10 \, \mu g \, l^{-1}$ in blood serum for healthy individuals). This fact makes Al speciation analysis difficult and requires highly sensitive analytical techniques.[51] The speciation techniques for Al in humans include a separation step that can be chromatographic or non-chromatographic. The latter are ultrafiltration and dialysis methods which separate the serum into two fractions (with a nominal cut-off of 10–30 kDa).[51] Based on presently available analytical data, the nature of the exact fraction separated by these methods is controversial, since Al contamination of samples can be a severe problem.[59] Assuming that an average value of 10% or less of Al in normal serum is ultrafilterable, it is reasonable to assume that the majority ($\sim 90\%$) should be bound to HMW (High Molecular Weight Proteins).[60] Speciation of these Al compounds needs a chromatographic separation method (gel filtration, electrophoresis, HPLC). By using these techniques, several authors have established transferrin as the only HMW serum Al-binding protein, in contrast with previous work that indicated Al also bound to albumin.[59,61,62] Some thermodynamic studies have confirmed this theory.[63] Concerning LMW proteins, Martin *et al.*[62] showed that citrate (although the strongest LMW Al binder in serum) would release Al^{3+} to transferrin. They calculated that 94.2% of the total Al in plasma would be bound to the transferrin and 4.9% to citrate. The complex, Al-citrate, is of great concern because the Al absorption in the gastrointestinal tract increases in the presence of citrate ligand, providing an effective pathway for the entry of Al into the circulatory system but, on the other hand, it induces Al urinary excretion.[63,64] Several approaches have been tried for speciation of LMW Al complexes in serum by using ion chromatography,[65] gel chromatography[66] or HPLC.[67]

There is still, however, a great necessity for further Al speciation studies to explain the role of Al-complexes in Al toxicity in humans.[64]

3.3.2 Arsenic

Arsenic is an element that raises much concern from both the environmental and human health points of view. Ingestion *via* food or water (generally as inorganic compounds) is the main pathway of this metalloid into the organism, where absorption takes place in the stomach and intestines, passing into the blood. It is then converted by the liver into less toxic forms and excreted mainly in urine.[68] Nowadays, it is acknowledged that different species of As produce diverse toxicological effects in human, with inorganic forms being more toxic than organic ones. For example, arsenobetaine (AsBet) and arsenocholine (AsChol) are well tolerated by living organisms.

Arsenic speciation. A review on the analytical methodology for As speciation in environmental and biological samples has been published by Burguera and Burguera.[69] The six arsenic species generally identified in biological samples are As(III), As(V), MMA, DMA, AsBet and AsChol. The matrices more usually employed for As monitoring in humans are blood and urine. Gas chromatography (GC) and liquid chromatography (HPLC) are widely used for the separation of As compounds. These chromatographic techniques have been coupled to powerful detectors such as AAS, ICP-AES (Inductively Coupled Plasma Atomic Emission Spectrometry) or ICP-MS. Capillary electrophoresis (CZE) is today one of the more promising separation techniques for As speciation. The hydride generation method has been extensively used in environmental and biological matrices. The conversion of species into forms able to generate the volatile arsine has been performed by using MW (Microwave energy) or by UV (photo-oxidation).[70] The selective reactivity of As species with acid and $NaBH_4$ was utilised by López-González *et al.*[71] to determine toxic (defined as As(III), As(V), MMA, and DMA, and non-toxic (As-proteins) arsenic species in urine by FI-HG-AAS (Flow Injection-Hydride Generation-Atomic Absorption Spectrometry). The same species have been differentiated by Bavazzano[72] in urine, by using a selective solvent extraction. Some workers have studied the relationship between blood As concentration and the degree of renal insufficiency.[72,73] These analyses revealed AsBet and DMA to be the major species present, but they also showed a three-fold increase in As in serum levels compared with healthy controls. The inorganic arsenic species were bound to proteins, mainly transferrin (about 5–6% of total As in serum). This binding may play an important role in arsenic detoxification. A recent speciation study of DMA and AsBet in three candidate lyophylised urine reference materials has been described.[74] The results from such a study offer the possibility of using these materials as CRMs, not only for total As but also for As species.

3.3.3   Cadmium

Cadmium has an extremely long residence time (over 20 years) in the human body and a significant proportion of the body burden is stored in liver and kidney, and bound to (MTs) metallothioneins (see also Section 3.1.3 on zinc). Metallothioneins are still the only biological compounds known to naturally contain this metal.[75] The synthesis of metallothioneins (MTs) is induced by the essential elements Cu and Zn in kidney and liver. Cd may replace these metals or share the protein with them.[5] The toxicity of Cd-thioneins then affects the kidney in particular, but can also produce skeletal damage (osteoporosis). The accumulation of Cd in the kidney increases with age, suggesting that Cd exposure will adversely affect health, particularly in older age groups. However, Cd absorption is higher in children (as in the case of Pb).[76] Cadmium, in contrast to Pb or Hg, is not transported into the foetus but is retained in the placenta.[76] Inhalation of Cd causes pulmonary infections, with cigarette smoke being a significant source of Cd.[77]

Cadmium speciation. There are diverse research groups working on Cd-MTs determination in order to characterise metal binding to MT.[78,79] Some attempts have been made to isolate quantitatively MTs from human fluids, in order to determine the level of environmental and industrial exposure to heavy metals. After MT isolation, covalent affinity chromatography (CAC) was applied, allowing specific isolation of the thiol-proteins.[80,81] This method is also valuable for Cu-MTs and Zn-MT. Recent studies of Cd speciation and its relevance in the human diet were discussed by Crews.[82] A review concerning the history of MT research with various methods of isolation, characterisation and quantification has been published.[39]

3.3.4   Lead

Lead is considered to be a toxic element as a consequence of a variety of biochemical effects. Among these are included neurological problems, haematological effects, renal dysfunction, hypertension and cancer (the IARC has classified Pb as a carcinogen), for which there is evidence for animals but not yet for humans.[5,76,83] Some studies with humans have suggested that increases in lead uptake have occurred when dietary Fe was low.[5]

The harmful effects of organolead compounds are considered to be much greater than those of inorganic lead. The toxicity of alkyllead species diminishes in the sequence $R_4Pb \rightarrow R_3Pb^+ \rightarrow R_2Pb^{2+} \rightarrow Pb^{2+}$ (where R is a methyl or ethyl group).[84] About 150 fatal cases of human intoxication with Et_4Pb have been reported in the literature. They were related to accidental exposures, but long term environmental exposure to low levels has been associated with a wide range of metabolic disorders and neurophysical deficits, especially in children.[85]

Lead speciation. In the past 10 years, speciation of organolead compounds has been the subject of various review papers.[86] In mammals, inhalation or adsorption of R_4Pb compounds results in the formation of trialkyllead in tissues and body fluids. With the exception of methylleads found in the blood of petrol workers, they are usually below the detection limits in blood and urine samples.[87,88] Some other papers have focused on Pb measurements in bone and other tissues from smelter workers.[89,90] The distribution of Pb among proteins in erythrocytes was studied by Bergdhal *et al.*,[91] using gel chromatography coupled to ICP-MS, who reported that 80% of lead was bound to HMW proteins (> 240 kDa).

3.3.5 Mercury

All forms of Hg are considered poisonous, but methylmercury is of particular concern since it is extremely toxic and it is frequently found in the environment. Some episodes of health damage by these compounds are very well known. In Minamata (Japan), methylmercury contamination caused severe brain damage in 22 infants whose mothers had ingested contaminated fish during pregnancy. A dramatic contamination of humans happened in Iraq because of the intake of wheat flour from seeds treated with organic mercury. In general, exposure to organic Hg can cause brain damage to a developing foetus since CH_3Hg^+ readily crosses the placenta.[92]

Organomercury compounds exhibit very high toxicity, can be formed naturally and bioaccumulate in both man and other living organisms. These factors have given rise to a great interest in understanding the distribution patterns, the interconversion between species and the noxious effects on the body, and to propose strategies for decontamination. The toxicological features of Hg depend on the form. Inorganic compounds may contain Hg in oxidation states $+1$ or $+2$. The absorption of these compounds in humans is about 7% since absorption of CH_3Hg^+ is of the order of 90–95%. Inorganic Hg is mainly retained in the kidneys, whereas organic Hg has a greater affinity for the brain. A genotoxic effect resulting in chromosomal aberrations has also been demonstrated.[93,94]

Diet also influences the tolerance of humans for CH_3Hg^+, and such variability must be taken into account when evaluating the toxic levels of organic mercurials. Some epidemiological studies suggest that high levels of Se may reduce the toxic effects of CH_3Hg^+ or produce a modification in the methylated activity with less conversion of inorganic Hg to CH_3Hg^+.[95] This antagonistic behaviour is well known, mainly by formation of mercury selenide.[26]

Mercury speciation. Many analytical methods are available for mercury speciation in environmental and biological samples.[96] However, data are still very scarce for humans. Good indicators of environmental and occupational Hg exposure are required. With this in mind, papers have been published on the determination of organic and inorganic Hg in hair and nail samples.[97,98] One of them[97] describes an extensive study on mothers and their infants, using CVAAS (Cold Vapour Atomic Absorption Spectrometry) to determine inorganic Hg and GC-AFS (Gas Chromatography–Atomic Fluorescence Spectrometry) for CH_3Hg^+. Methylmercury accounted for over 80% of the total Hg found in one mothers' hair. Tests on the neuro-development of the children showed very little evidence of any effect from exposure of the mothers to CH_3Hg^+. The second group of researchers[98] developed a new short extraction procedure coupled to HPLC–UV-CVAAS for determining both inorganic and CH_3Hg^+ compounds. A FI-CVAAS procedure was developed by Nixon[99] for the determination of inorganic and total Hg in human blood and urine. This technique has been adapted as a routine method for Hg monitoring.

3.3.6   Platinum

In the search for a cure for cancer and related diseases, compounds containing heavy metals such as Au and Pt have been found to have unique activities, useful for the treatment of these diseases.[100-102] A number of Pt compounds have been studied for this purpose, including cisplatin, carboplatin, iproplatin and tetraplatin. Essential measurements required for new materials of this kind are the determination of their purity as well as their stability. Some of them, *e.g.* cisplatin, have very low solubility in aqueous solution. However, inside the cell, formation of hydrated complexes can occur. It has been suggested these complexes can cause renal toxicity.[103] Other complexes, such as tetraplatin, decompose when placed in solution. In this case, tetraplatin undergoes a degradation process in solution leading to the formation of several fragments containing Pt. Because of this instability, it is important to quantify the Pt content of these moieties in order to evaluate the cytotoxicity and pharmacokinetics of these drugs.[3]

Nowadays, the addition of Pb to gasoline has been discontinued. Pt and other precious metals (Rh, Pd) are used as catalysts in exhaust systems to reduce harmful emissions. There is great interest in evaluating the removal of Pt compounds from the catalyser by exhaust gases, and their subsequent effect on humans. Nevertheless, studies on the distribution, speciation and environmental of Pt species impact are rare, due to the limited knowledge about microbial transformation products[104-106] (see also Chapter 9).

Platinum speciation. Platinum speciation in biological samples has become a great challenge for analytical chemists. The large number of possible transformations between the anti-tumour drugs and, *e.g.*, DNA, the instability of standard solutions and the great variety of equilibrium reactions are some problems that must be considered before attempting speciation studies.

De Waal *et al.*[107] have reviewed analytical methodologies, including sample preparation and stability studies for Pt. Bettmer *et al.*[108] have also summarised some analytical methods applicable to biological samples. Einhauser *et al.*[109] used both ICPAES and ETAAS to examine the protein binding of different Pt complexes, for the determination of the protein–Pt adducts in biological samples. ETAAS has also been used for determining cisplatin, transplatin and their hydrated complexes in saliva,[110] in human plasma[103] and also in studies to measure protein–Pt binding, formation of DNA adducts and optimal absorption and excretion time of the different drugs administered to cancer patients.[111,112]

In all cases, at least 80% of plasma Pt was protein bound. The two complexes studied (cisplatin and carboplatin) were found to form the same DNA adducts. However, the compounds differ in their kinetics for reaction with DNA, it being faster for cisplatin. Szuluha et al.[113] developed a method based on X-ray fluorescence to determine the distribution of Pt in cervical tumour tissue following administration of cisplatin. A two-dimensional electrophoretic method for the separation of Pt-proteins has been recently reported by Lustig et al.[114] A double-focusing high-resolution inductively coupled plasma mass spectrometrer was used.

3.3.7 Tin

Tin has no known biochemical function. Tin toxicity includes growth depression and anaemia, and it can modify the activity of several enzymes by interfering with the metabolism of Zn, Cu and Ca. When compared with inorganic tin, organotin compounds are highly toxic and attack the central nervous system.[115] All these studies have employed rats, but epidemiological studies suggest a similar behaviour in humans. When tin is fed at relatively high levels, it is poorly absorbed by humans. It appears that the lower the dietary intake the higher the percentage absorbed.[116] The major source of dietary tin is canned foods.

Tin speciation. Tin speciation studies have been focused on organotin species due to the important input of these compounds in the environment, especially from anti-fouling paints.[117,118] Some studies have also been applied with animals (rats) to determine the butyltin metabolites as TBT (tributyltin) is considered the most toxic tin species.

Neil et al.[32] investigated the capacity of Sn(II) and Sn(IV) ions (as $SnCl_2$ and $SnCl_4$) to activate heme oxygenase in cardiac tissue. The results from this investigation were that Sn(II) ions were more potent activators than Sn(IV). Their hypothesis was that citrate facilitates Sn transport across cell membranes. The relationship between heme oxygenase induction and oxidative stress is important in cardiac cells. The induction of heme oxygenase during oxidative stress may be a defence mechanism, acting to restore the antioxidant ratio inside the cell.

3.3.8 Antimony

The toxicity of Sb depends on its oxidation state (Sb(III) is normally more toxic than Sb(V)) and on its chemical form (normally inorganic oxyanions are more toxic than organic compounds), but even at very low concentrations Sb must always be considered very toxic.[119] Sb primarily attacks the liver and kidney, but inhalation of Sb compounds can produce fibrosis, cardiotoxicity, and high cancer risk (an increase in lung cancer has been observed among workers exposed to Sb).[120]

Antimony speciation. A method for the selective determination of Sb(III) and

Sb(V) in liver tissue using HGAAS was described by Rondon *et al.*.[121]

4   Conclusion: Needs for Elemental Speciation in Human Studies

The different steps required for speciation analysis of metallic compounds need monitoring procedures. These steps are summarised in Figure 2. The problems encountered at each step are described in this section.

4.1   Sample Collection

Obtaining a representative sample from any kind of matrix is one of the most difficult tasks in analytical chemistry. This problem is more acute in the case of clinical samples because of the varying degrees of invasiveness depending upon the type of matrix required. There are important parameters that must be evaluated in advance of collection. Generally, these samples suffer from temporal variations, or other external factors. Some samples must be taken in the morning without any alcohol or medicine having been taken for several days before

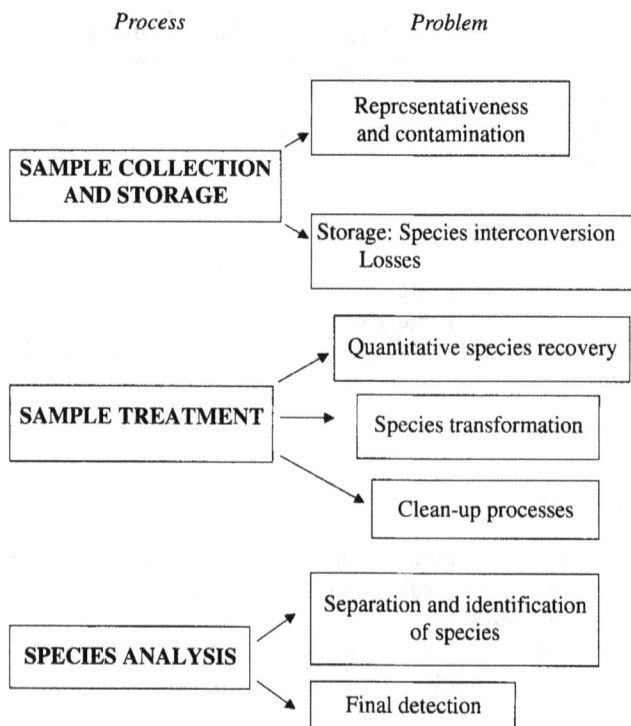

Figure 2   *Main problems encountered for elemental speciation in biological and clinical samples*

sample collection. Urine and blood are heavily influenced by such factors. Furthermore, a suitable quantity of sample is not always available, and the lack of representativeness can sometimes be remarkable. For example, the lipophilic character of organolead compounds favours their accumulation in particular tissues of living organisms and stresses the need for a dissection of the parts of interest and homogenisation prior to analysis.[85] Finally, the low concentration of most analytes and their species of interest make the risk of contamination a very important parameter to consider. Special care with these samples must therefore be taken.

4.2   Storage and Stability of the Species

The normal way to store samples prior to speciation analysis is by freezing them. This is of special relevance for biological tissues because the enzymatic activity and natural proteolysis and autolysis continue after sampling and could alter the speciation.

The materials employed for trace elements storage need to be safe, clean, uncontaminated, non-degradable and, occasionally, sterilised containers. Polyethylene (PE) vials and bottles are acceptable for dry powders, providing very low temperatures are not required. If such temperatures are desired, Teflon containers are recommended. Even so, it is prudent to verify that the containers have not been treated with chemical agents, *e.g.* lubricants, pigments or antifungal or antibacterial agents, which can introduce contaminants.[5] For example the adsorption of organolead compounds on to the walls of sampling bottles is likely to cause losses of more than 90% within a few hours. It is therefore recommended to perform the extraction immediately after sampling.[85]

The stability of the different species of one element is strongly related to the sample storage procedure adopted. If the physico-chemical conditions (temperature, container material, pH, light exposure *etc.*) are unsuitable, species interconversion or species degradation can occur. There is a lack of knowledge as to how the stability of the species depends on the sample matrix. A stability study on As species in urine[11] has shown that As(III) and AsC are unstable, and oxidation to As(V) and AsB, respectively, is observed. A clean-up procedure was applied, consisting of the addition of ethanol, and the mixture was chilled in a solid CO_2/acetone bath.[122] The residue obtained showed a better stabilisation of arsenic species during 60 days at room temperature. These results are very encouraging as such procedures may enable preservation of clinical samples before analysis.

4.3   Pre-treatment of the Sample

Problems with sample treatment prior to speciation measurements vary with the clinical material being investigated. For those cases in which the analytes are bound to proteins, the application of acid, basic or enzymatic hydrolysis to the preparation of tissues for elemental speciation analysis should be further investigated, so as to avoid interconversion of species. Supercritical fluid extraction (SFE) is a promising extraction technique because of its efficiency and selectivity. Extractions by SFE can be performed at low temperatures and in a very short time which helps to prevent the oxidation of labile compounds.[10] With both treatments, two problems may occur:

- The ability to achieve a quantitative extraction of the species. It has been demonstrated that the yield of extractable Se from fish-tissue significantly increases when two enzymatic extractions are carried out.[123]
- How to avoid species transformation during sample treatment. For instance, when a basic hydrolysis is performed the species found depend upon the pH conditions. Thus, considerably more research in this area is required.

4.4 Clean-up Processes

Speciation analysis usually demands the use of chromatographic techniques. A preliminary clean-up step is frequently advisable. In this step, it is vital to avoid losses or species transformation. In addition, after the sample clean-up procedure, some species have been shown to be more stable. Stabilisation of As compounds in urine has already been mentioned in this chapter.[11]

4.5 Analysis

Two main problems have been detected during the determination of the different species in biological substrates. Some chromatograms, from specific detectors such as ICP-MS, contain unidentified peaks. This problem could be reduced by having more standards available, eventually to include all of the unknown species. This problem is very common with some arsenic and selenium species identification in, for example, fish samples.[31,124]

A second important aspect to be considered during the analysis is how to achieve good instrumental resolution, and to get as much information as possible in only one chromatographic run.

4.6 Methods Validation

The EC Standards, Measurements and Testing Programme (SMT) recognised the need for collaborative projects to establish and to improve the state of the art of speciation analysis in Europe. Several programmes have been undertaken over the past 10 years in various environmental matrices. In some cases, certification campaigns have been carried out with subsequent production of CRMs.[125] However, there is a need for similar exercises for clinical samples. Thus, validation of analytical methods is still difficult and clinical CRMs are a real challenge for the future.

4.7 Regulations

The European Community participates in the organisation of international meetings concerning the environment, foodstuff and occupational health and hygiene and their implementation. The purpose of such conventions is to adopt directives, implemented by the laws or regulations of the Member States. A series of principles regarding environmental protection has been set out within the last 20 years, which include a range of elements and their compounds. However, to our knowledge, no data are available on regulation of the different chemical forms of trace elements related to human fluids and tissues. This represents another need which must be addressed.

5 References

1. W. Mertz, R. Brätter and P. Schramel, in *Trace Element Analytical Chemistry in Medicine and Biology*, de Gryter, Berlin, 1980, pp. 727.
2. E.J. Underwood, in *Trace Elements in Human and Animal Nutrition*, 4th Edn., Academic Press, New York, 1977.
3. I.T. Urasa, in *Element Speciation in Bioinorganic Chemistry*, S. Caroli, ed., John Wiley and Sons, New York, 1996, Volume 135, pp. 121–154.
4. A. Taylor, S. Branch, H.M. Crews, D.J. Halls, L.M. Owen and M. White, *J. Anal. At. Spectrom.*, 1997, **12**, 119R.
5. World Health Organization (WHO), in *Trace Elements in Human Nutrition and Health*, WHO, Geneva, 1996.
6. K. Kasperek, E. Land and L.E. Feinendegen, in *Trace Element Analytical Chemistry in Medicine and Biology*, de Gryter, Berlin, 1980, p. 75.
7. E. Coni, A. Alimonti, A. Bocca, F. De la Torre, D. Pizzuti and S. Caroli, in *Element Speciation in Bioinorganic Chemistry*, S. Caroli, ed., John Wiley and Sons, 1996, Volume 135, pp. 255–285.
8. A.K. Das, R. Chakraborty, M.L. Cervera and M. de la Guardia, *Mikrochim. Acta*, 1996, **122**(3–4), 209.
9. S. Caroli, in *Element Speciation in Bioinorganic Chemistry*, S. Caroli, ed., John Wiley and Sons, New York, 1996, Volume 135, pp. 1–18.
10. K. Vercoutere and R. Cornelis, in *Quality Assurance for Environmental Analysis*, Ph. Quevauviller, E. Maier and B. Griepink, eds., Elsevier Science Publ., Amsterdam, 1995, pp. 195–213.
11. M.A. Palacios, M. Gómez, C. Cámara and M.-A. López, *Anal. Chim. Acta*, 1997, **340**, 209.
12. R. Cornelis, F. Borguet and J. De Kimpe, *Anal. Chim. Acta*, 1993, **283**, 183.
13. B. Wallaeys and R. Cornelis, *Sci. Total Environ.*, 1988, **71**, 401.
14. P. Borella, A. Bargellini, S. Salvioli, C.I. Medici and A. Cossarizza, *J. Immunol. Methods*, 1995, **186**, 101.
15. P. Borella, A. Bargellini, S. Salvioli and A. Cossarizza, *Clin. Chem.*, 1996, **42**, 319.
16. M. Simonoff and G. Simonoff, in *Le Sélénium et la Vie*, ed. Masson, Paris, 1991.
17. J.R. Arthur, F. Nicol and G.J. Beckett, *Biochem. J.*, 1990, **272**, 537.
18. B.Q. Gu, *Chin. Chim. Med. J.*, 1983, **96**, 251.
19. M.B. Mihailovic, D.M. Avramovic, I.B. Jovanovic, O.J. Pesut, D.P. Matic and V.J. Stojanov, *J. Environ. Pathol. Toxicol. Oncol.*, 1998, **17**, 285.
20. H. Koehler, H.J. Peters, H. Pankau and H.J. Dick, *Biol. Trace Elem. Res.*, 1988, **15**,

157.

21. R.W. Marcus, *Maryland Med. J.*, 1993, **42**, 669.
22. L.C. Clark, B. Dalkin, A. Krougrad, G.F. Combs and B.W. Turnbull, *J. Am. Med. Assoc.*, 1996, **276**, 1957.
23. L.C. Clark, B. Dalkin, A. Krougrad, G.F. Combs and B.W. Turnbull, *Br. J. Urol.*, 1998, **81**, 730.
24. P.R. Taylor and D. Albanes, *J. Natl. Cancer Inst.*, 1998, **90**, 1184.
25. G.F. Combs and L.C. Clark, *Nutrition Rev.*, 1985, **43**, 325.
26. H.D. Foster, *Med. Hypoth.*, 1997, **48**, 355.
27. L. Fishbein, in *Metals and Their Compounds in the Environment*, 1991, Fishbein eds., Washington, DC, p. 1153.
28. Y. Harrison, D. Littlejohn and G.S. Fell, *Analyst*, 1996, **121**, 189.
29. T.M. Bricker and R.S. Houk, in *Frontiers in Analytical Spectroscopy*, Royal Society of Chemistry, Cambrdige, 1995, p. 109.
30. J. Clansen and S.A. Nielsen, *Biol. Trace Elem. Res.*, 1988, **15**, 125.
31. M.A. Quijano, A.M. Gutiérrez, C. Pérez-Conde and C. Cámara, European Winter Conference, 1999, Pau, France.
32. T.K. Neil, N.G. Abraham, R.D. Levere and A. Kappas, *J. Cell. Biochem.*, 1995, **57**, 409.
33. R. Muñoz-Olivas, N. Gilon, M. Potin-Gautier and O.F.X. Donard, *J. Anal. At. Spectrom.*, 1996, **11**, 1171.
34. J. Feldmann, T. Riechmann and A.V. Hirner, *Fresenius' J. Anal. Chem.*, 1996, **354**, 620.
35. S. Moesgaard and R. Morrill, *The Need for Speciation to Realise the Potential of Selenium in Disease Prevention*, Chapter 18 in this book.
36. M. Aschner, *FASEB J.*, 1996, **10**, 1129.
37. K.M. Hambidge, in *Trace Elements in Human and Animal Nutrition*, 5th Edn., Academic Press, San Diego, 1987, p. 1.
38. P. Brätter, V.E. Negretti de Brätter, S. Recknägel and R. Brunetto, *J. Trace Elem. Med. Bio.*, 1997, **11**, 203.
39. M. Nodberg, *Talanta*, 1998, **46**, 243.
40. S. Singh, R.C. Hider and J.B. Porter, *Anal. Biochem.*, 1990, **186**, 320.
41. P.A. Chyka, T.D. Mandrell, J.E. Holley and B.E. Beegle, *Vet. Human Toxicol.*, 1996, **38**, 24.
42. C.M.M. Smith and J.M. Harnly, *J. Anal. At. Spectrom.*, 1996, **11**, 1055.
43. A. Aitio and L. Tomatis, in *Trace Elements in Health and Disease*, Royal Society of Chemistry, Cambridge, 1991, p. 159.
44. H.F. Hildebrand, P. Shilari, F.Z. Arrouijal, A.M. Decaestecker and R. Matínez, in *Trace Elements in Health and Disease*, Royal Society of Chemistry, Cambridge, 1991, p. 179.
45. Y.C. Hong and C.Y. Park, *K'atollik Taehak Wihakpu Nonmunjip*, 1994, **47**, 273.
46. F.H. Nielsen, in *Trace Elements in Human and Animal Nutrition*, 5th Edn., Academic Press, San Diego, 1987, p. 275.
47. W.P. Ridley, L. Dizikes and J.M. Wood, *Science*, 1977, **197**, 329.
48. P. Viñas, N. Campillo, I. López-García and M. Hernández-Córdoba, *Anal. Chim. Acta*, 1996, **318**, 319.
49. K. Akatsuka and I. Atsuya, *Fresenius' J. Anal. Chem.*, 1989, **335**, 200.
50. M. Morita, T. Uehiro and K. Fuwa, *Anal. Chem.*, 1980, **52**, 349.
51. A. Sanz-Medel, B. Fairman and K. Wrobel, in *Element Speciation in Bioinorganic Chemistry*, S. Caroli, ed., John Wiley and Sons, New York, 1996, Volume 135, pp. 223–254.

52. R.C. Massey and D. Taylor, in *Aluminium in Food and the Environment*, Royal Society of Chemistry, London, 1988.
53. R.E. Viola, J.F. Morrison and W.W. Cleland, *Biochemistry*, 1980, **19**, 3131.
54. M.R. Willis and J. Savory, *Lancet*, 1983, **2**, 29.
55. D.P. Perl and A.R. Brody, *Science*, 1980, **208**, 297.
56. J.M. Candy, A.E. Oakley, J. Klinowsky, T.A. Carpenter, R.H. Perry, J.R. Atack, E.K. Perry, G. Blesset, A. Fairbairn and J.A. Edwardson, *Lancet*, 1986, **2**, 354.
57. J.S. Chapell and J.D. Birchall, *Inorg. Chim. Acta*, 1988, **153**, 1.
58. E.M. Carlisle and J.M. Curran, *Alzheimer Dis. Assoc. Disord.*, 1987, **1**, 83.
59. E. Blanco-González, J. Pérez-Parajón, J.I. García-Alonso and A. Sanz-Medel, *J. Anal. At. Spectrom.*, 1989, **4**, 175.
60. M.R. Pereiro-García, M.E. Díaz-García and A. Sanz-Medel, *J. Anal. At. Spectrom.*, 1987, **2**, 699.
61. A.B. Soldado-Cabezuelo, M. Montes-Bayón, E. Blanco-González, J.I. García-Alonso and A. Sanz-Medel, *Analyst*, 1998, **123**, 865.
62. C.S. Múñiz, J.M. Marchante-Gayón, J.I. García-Alonso and A. Sanz-Medel, *J. Anal. At. Spectrom.*, 1998, **13**, 283.
63. R.B. Martin, J. Savory, S. Brown, R.L. Bertholfe and M.R. Willis, *Clin. Chem.*, 1987, **33**, 405.
64. D. Sandrine, M. Filella and G.J. Berthon, *J. Inorg. Biochem.*, 1990, **38**, 241.
65. P.M. Bertsch and M.A. Anderson, *Anal. Chem.*, 1989, **61**, 535.
66. F.Y. Leung, A.E. Nibloch, L. Bradley and A.R. Henderson, *Sci. Total Environ.*, 1988, **71**, 49.
67. A.K. Datta, P.J. Wedlund and R.A. Yokel, *J. Trace Elem. Electrolytes Health Dis.*, 1990, **4**, 107.
68. S. Caroli, F. La Torre, F. Petrucci and N. Violante, in *Element Speciation in Bioinorganic Chemistry*, S. Caroli, ed., John Wiley and Sons, New York, 1996, Volume 135, p. 445.
69. M. Burguera and J.L. Burguera, *Talanta*, 1997, **44**, 1581.
70. B. Amran, F. Lagarde, M.J.F. Leroy, A. Lamotte, C. Demesmay, M. Ollé, M. Albert, G. Rauret and J.F. López-Sánchez, in *Quality Assurance for Environmental Analysis*, Ph. Quevauviller, E. Maier and B. Griepink, eds., Elsevier Science Publ., Amsterdam, 1995, pp. 285–304.
71. M.A. López-González, M. Gómez, C. Cámara and M.A. Palacios, *Mikrochim. Acta*, 1995, **120**(1–4), 301.
72. P. Bavazzano, A. Perico, K. Rosendahl and P. Apostoli, *J. Anal. At. Spectrom.*, 1996, **11**, 521.
73. X. Zhang, R. Cornelis, L. Mees, R. Vanholder and N. Lameire, *Analyst*, 1998, **123**, 13.
74. R. Cornelis, X. Zhang, L. Mees, J.M. Christensen, K. Byrialsen and C. Dyrschel, *Analyst*, 1998, **123**, 2883.
75. J.H.R. Kägi and A. Schäffer, *Biochemistry*, 1988, **27**, 8509.
76. G.K. Davis and W. Mertz, *Trace Elements in Human and Animal Nutrition*, 5th Edn., Academic Press, San Diego, 1987, p. 301.
77. G. Oberdörster, *J. Am. College Toxicol.*, 1989, **8**, 1251.
78. J. Wang and W.D. Marshall, *Analyst*, 1995, **120**, 623.
79. A.K.M. Kabzinski and T. Takagi, *Biomed. Chromatogr.*, 1995, **9**, 123.
80. A.K.M. Kabzinski and T. Paryjczak, *Chem. Anal.*, 1995, **40**, 831.
81. A.K.M. Kabzinski, *Talanta*, 1998, **46**, 335.
82. H.M. Crews, *Spectrochim. Acta, Part B*, 1998, **53**, 213.
83. EPA, *Air Quality Criteria for Lead*, US EPA, Research Triangle Park, NC, 1986.

84. R.J.C. Van Cleuvenbergen and F.C. Adams, in *Handbook of Environmental Chemistry*, O. Hutzinger, ed., Springer, Berlin, 1990, p. 97.

85. R. Lobinski, W.M.R. Dirkx, J. Spuznar-Lobinska and F.C. Adams, in *Quality Assurance for Environmental Analysis*, Ph. Quevauviller, E. Maier and B. Griepink, eds., Elsevier Science Publ., Amsterdam, 1995, pp. 319–356.

86. R. Lobinski, W. Dirkx, J. Spuznar-Lobinska and F.C. Adams, *Anal. Chim. Acta*, 1994, **286**, 381.

87. B. Gercken and R.M. Barnes, *Anal. Chem.*, 1991, **63**, 283.

88. M. Blaszkiewicz, G. Baumhoer and B. Neidhart, *Fresenius' J. Anal. Chem.*, 1986, **325**, 129.

89. L. Gerhardsson, V. Englyst, N.G. Lundström, G. Nordberg, S. Sandberg and F. Steinvall, *J. Trace Elem. Med. Biol.*, 1995, **9**, 136.

90. M.L. Bleeker, F.E. McNeill, K.N. Lindgren, V.L. Mastn and D.P. Ford, *Toxicol. Lett.*, 1995, **77**, 241.

91. I.A. Bergdhal, A. Schutz and A. Grubb, *J. Anal. At. Spectrom.*, 1996, **11**, 735.

92. D.O. Marsh, G.J. Myers and T.W. Clarkson, *Clin. Toxicol.*, 1981, **10**, 1311.

93. F. Bakir, *Science*, 1973, **181**, 230.

94. C. Cox, *Environ. Res.*, 1989, **49**, 318.

95. I.R. Rowland, in *Reproductive and Developmental Toxicity of Metals*, Clarkson and Nordberg, eds., Plenum Press, New York, 1983, p. 745.

96. Y. Draebaek and A. Iverfeldt, in *Quality Assurance for Environmental Analysis*, Ph. Quevauviller, E. Maier and B. Griepink, eds., Elsevier Science Publ., Amsterdam, 1995, pp. 305–318.

97. E. Cernichiari, T.Y. Toribara, L. Liang, D.O. Marsh, M.W. Berlin, G.J. Myers, C. Cox, C.F. Shamlye and O. Choisy, *Neurotoxicology*, 1995, **16**, 613.

98. R. Falter and H.F. Schoeler, *Fresenius' J. Anal. Chem.*, 1996, **354**, 492.

99. D.E. Nixon, G.V. Mussmanand T.P. Moyer, *J. Anal. Toxicol.*, 1996, **21**, 17.

100. B. Rosenberg, L.Van Camp, J.E. Trosko and V.H. Mansour, *Nature*, 1969, **222**, 385.

101. S.J. Lippard, *Pure Appl. Chem.*, 1987, **59**, 731.

102. J. Christodoulou, M. Kashani, B.M. Keohane and P.J. Sadler, *J. Anal. At. Spectrom.*, 1996, **11**, 1031.

103. H.C. Ehrsson, I.B. Wallin, A.S. Anderson and P.O. Edlund, *Anal. Chem.*, 1995, **67**, 3608.

104. F. Alt, A. Bambauer, K. Hoppstock, B. Mergler and G. Tölg, *Fresenius' J. Anal. Chem.*, 1993, **346**, 693.

105. C. Wei and G.M. Morrison, *Anal. Chim. Acta*, 1994, **284**, 587.

106. M. Moldovan, M. Gómez and M.A. Palacios, *J. Anal. At. Spectrom.*, 1999, **14**, 1163.

107. W.A.J. De Waal, F.J.M.J. Maessen and J.C. Kraak, *J. Pharm. Biomed. Anal.*, 1990, **8**, 1.

108. J. Bettmer, W. Buscher and K. Cammann, *Fresenius' J. Anal. Chem.*, 1996, **354**, 521.

109. T.J. Einhauser, M. Glanski and B.K. Keppler, *J. Anal. At. Spectrom.*, 1996, **11**, 747.

110. L.J.C. van Warmerdam, O. van Tellingen, R.A.A. Maes and J.H. Beijnen, *Fresenius' J. Anal. Chem.*, 1995, **351**, 777.

111. H.J.M. Groen, A.H.D. van der Leest, E.G.E. de Vries, D.R.A. Uges, B.G. Szabo and N.H. Mulder, *Br. J. Cancer*, 1995, **72**, 992.

112. F.A. Blommaert, H.C.M. van Dijk-Knijnenburg, F.J. Dijjt, L. den Engelse, R.A. Baan, F. Berends and A.M.J. Fichtinger-Schepman, *Biochemistry*, 1995, **34**, 8474.

113. K.L. Szluha, I. Uzonyi, J. Bacso, L. Lampe, I. Czifra, M. Peter, C. Villena and W. Schmidt, *Microchem. J.*, 1995, **51**, 238.

114. S. Lustig, J. De Kimpe, R. Cornelis and P. Schramel, *Fresenius' J. Anal. Chem.*, 1999,

363, 484.

115. P. Fritsch, G. de Saint Blanquat and R. Derache, *Toxicology*, 1977, **8**, 165.

116. M.A. Johnson and J.L. Greger, *Am. J. Clin. Nutr.*, 1982, **35**, 655.

117. R. Morabito, S. Chiavarini and C. Cremisini, in *Quality Assurance for Environmental Analysis*, Ph. Quevauviller, E.A. Maier and B. Griepink, eds., Elsevier, Amsterdam, 1995, p. 437.

118. R. Ritsema, F.M. Martin and Ph. Quevauviller, in *Quality Assurance for Environmental Analysis*, Ph. Quevauviller, E.A. Maier and B. Griepink, eds., Elsevier, Amsterdam, 1995, p. 490.

119. C. Cámara and M.B. de la Calle, *Encyclopedia of Analytical Science*, 1995, p. 136.

120. P.J. Craig, in *Organometallic Compounds in the Environment. Principles and Reactions*, Longman, Harlow, 1986.

121. C. Rondon, J.L. Burguera, M. Burguera, M.R. Brunetto, M. Gallignani and Y. Petit de Pena, *Fresenius' J. Anal. Chem.*, 1995, **353**, 133.

122. M. Gómez, M.A. López, M.A. Palacios and C. Cámara, *Chromatographia*, 1996, **43**, 507.

123. P. Moreno, M.A. Quijano, A.M. Gutiérrez, M.C. Pérez-Conde and C. Cámara, European Winter Conference, 1999, Pau, France.

124. E.H. Larsen, C.R. Quétel, R. Muñoz-Olivas, A. Fiala-Medioni and O.F.X. Donard, *Mar. Chem.*, 1997, **57**, 341.

125. Ph. Quevauviller, *Method Performance Studies for Speciation Analysis*, Royal Society of Chemistry, Cambridge, 1998.

CHAPTER 24

Risk Assessment and Trace Element Speciation

JOHN H. DUFFUS

1 Introduction

The need for risk assessment springs from the requirement of government regulators and industrial managers to optimize the safe use of the chemicals on which our human welfare is increasingly dependent. This is the area covered by the term 'risk management'. It is absolutely dependent upon accurate risk assessment. The generally accepted model for risk assessment of potentially toxic chemicals is shown in Figure 1. The key steps are hazard identification, dose–response assessment and exposure assessment. All these components contribute to risk characterization, an estimate of the probability of harm to any individual in a population at risk. As the figure shows, risk assessment is dependent on continuing research and, indeed drives research because of the need to improve the input data and its subsequent interpretation.

Risk assessment of substances in the environment[1] illustrates how the above approach is applied in practice. Once a substance has been identified as a potential hazard, the predicted no effect concentration (PNEC) or, in the USA, the toxicological benchmark concentration (TBC) is calculated from fundamental toxicological data by the application of modifying factors appropriate to the uncertainty of the data (see next paragraph). Then, the predicted or estimated environmental concentration (PEC or EEC) is determined from the known chemical properties of the substance and relevant characteristics of the environment (discharge patterns, physico-chemical properties, possible metabolism, particularly by micro-organisms, and other potential environmental transformations) using worst case assumptions according to the precautionary principle. The risk is expressed as the ratio of PEC or EEC to PNEC or TBC. If the risk index is well below 1, the risk may be regarded as negligible. If the risk index is above 1, the situation must be further evaluated, as the precautionary assumptions may not be realistic in the particular situation for which the risk assessment has been required.

Research

Laboratory and field observations of adverse heath effects and exposure to particular agents

Information on extrapolation methods for high to low dose and one species to another

Field measurements, estimated exposures, & characterization of populations

Risk assessment

Hazard identification

(Does the agent cause the adverse effect?)

Dose-response assessment (What is the relation between dose and incidence of adverse effect in the population at risk?)

Risk characterization

(What is likely to be the estimated incidence of the adverse effect in the population at risk

Exposure assessment (What exposures are currently experienced or expected under prevailing conditions?)

Risk management

Development of regulatory options

Evaluation of public health, economic, social, and political consequences of regulatory options

Regulatory agency, industrial and local management decisions and actions

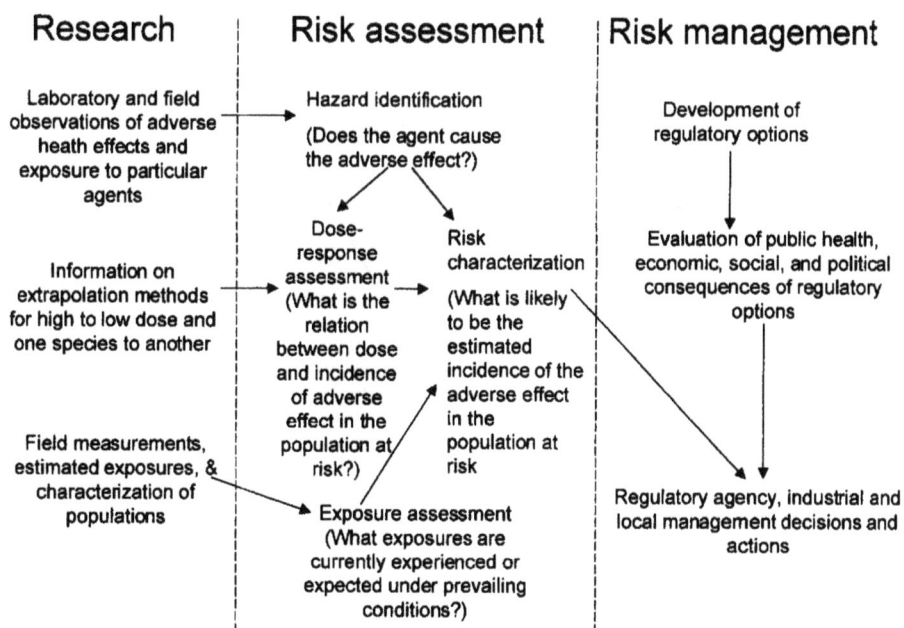

Figure 1 *Risk assessment in relation to research and risk management*

The PNEC or TBC calculation is based on the result of toxicity tests that were originally devised for organic pollutants. In principle, the threshold concentration for the appearance of an adverse effect in the most sensitive stage of the most sensitive species is determined and divided by an uncertainty factor of at least 10 to establish the PNEC. In practice, the information available is deficient in various ways and greater uncertainty factors must be applied to data that may only be available in terms of LC_{50}s. Being based on assumptions appropriate to organic chemicals, these tests fail to take into account the changes in chemical speciation and hence of bioavailability of trace elements in waters which are chemically different from those used in the test systems. Most concentrations of metals in published tests have been calculated and not measured, and chemical speciation has been ignored. For example, the pH in many test systems, including those recommended by the OECD,[2] may change by 1 pH unit or more during a test and thus chemical speciation also varies in the course of a single test. Figure 2 shows how aluminium speciation, and hence bioavailability and toxicity, can change with pH and concentration.[3] Thus, any extrapolation of values obtained in such tests to the natural aquatic environment for risk assessment must be very cautious. Other important factors affecting speciation, such as redox potential, are also uncontrolled in most current test systems, adding further doubts about their validity as a basis for risk assessment. Such tests give us little clue as to the exact nature of the toxic species for which risk must be assessed.

A general problem with environmental risk assessment is that in practice it has

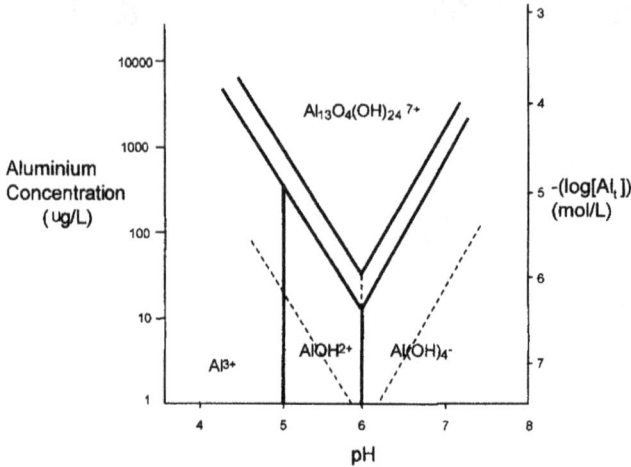

Figure 2 *Chemical speciation of aluminium in solution as a function of concentration and pH (after Nieboer et al., 1995[3])*

been related to single substances, assuming no interactions with other substances that may be present. The reality is that there are multiple complex exposures in the natural environment, and this is particularly the case for trace elements. The problem of interactions in complex exposures will be discussed below.

2 Trace Elements as Hazards

Conventionally, trace elements have been divided into those that are regarded as essential and those that are regarded as toxic,[4] as shown in the Periodic Table in Figure 3.

This is a convention that is another over-simplification which can lead to incorrect risk assessment and hence to poor risk management. The fundamental rule stated by Paracelsus[5] still applies. All substances are potentially toxic and only the dose determines whether they are essential or cause harm. Thus, the distinction between 'essential' elements and 'toxic' elements is entirely artificial. It is also chemically inaccurate, since biological essentiality or toxicity depends not simply on the element but also on its chemical speciation. In general, it appears that the biologically available form of any element is the free ion in solution. Consideration of cadmium compounds generally supports this view but indicates an important area where the reverse may be true. Cadmium acetate and cadmium chloride have been thoroughly tested and, as soluble salts, shown to be bioavailable and to produce adverse effects.[6] Cadmium oxide, which is minimally soluble, does not appear to have systematic effects but it appears to have direct effects on lung tissue following inhalation and deposition. Arguably, the soluble salts are not bioavailable to the lung tissue because they are quickly removed by the blood circulation. The particulate insoluble cadmium oxide remains and causes chronic damage. In addition to its solubility, the oxidation

1	2	3	4	5	6	7	8	9	10	11	12	13	14	15	16	17	18
H (E)																	He
Li	Be (T)											B (t)	C (E)	N (E)	O (E)	F	Ne
Na (E)	Mg (E)											Al	Si (t)	P (E)	S (E)	Cl (E)	Ar
K (E)	Ca (E)	Sc	Ti	V (t)	Cr	Mn (t)	Fe (t)	Co (t)	Ni (t)	Cu (t)	Zn (t)	Ga	Ge (t)	As (T)	Se (t)	Br	Kr
Rb	Sr	Y	Zr	Nb	Mo (t)	Tc	Ru	Rh	Pd	Ag	Cd (T)	In	Sn (t)	Sb (T)	Te	I (t)	Xe
Cs	Ba	* Lu	Hf	Ta	W	Re	Os	Ir	Pt	Au	Hg (T)	Tl	Pb (T)	Bi (T)	Po	At	Rn
Fr (T)	Ra	** Lr	Rf	Db	Sg	Bh	Hs	Mt	Uun	Uuu	Uub						

*lanthanides	La	Ce	Pr	Nd	Pm	Sm	Eu	Gd	Tb	Dy	Ho	Er	Tm	Yb
**actinides	Ac	Th	Pa (T)	U (T)	Np (T)	Pu (T)	Am (T)	Cm (T)	Bk (T)	Cf (T)	Es (T)	Fm (T)	Md (T)	No (T)

Figure 3 *The Periodic Table, marked to show essential major elements (E), essential trace elements (t), and so-called toxic elements (T)*

state of an ion may be crucial. Where chromium is concerned, only the Cr(VI) form as chromate is readily taken up by living cells. Even so, epidemiological studies suggest that almost insoluble chromates are the ones associated with lung cancer and not the soluble salts.[7] The same may be said of nickel compounds associated with lung cancers.[8] The soluble nickel sulfate gave negative results in thorough US National Toxicology Program tests of carcinogenicity.[9]

It is worth emphasizing that biological effects depend upon both chemical and physical speciation. This explains why the risk to human health from processing ore varies with the stage in ore processing. While breathing in hard metal dusts may lead to localized respiratory effects, including asthma, pulmonary irritation and edema,[10] metal roasting or welding leads to the production of metal fumes (oxides of lead, arsenic, cobalt, copper, and zinc), which can have systemic effects such as metal fume fever.[11-13]

Chemical speciation is a problem for the risk assessor because legally binding regulations about exposure levels for trace elements usually ignore bioavailability and therefore may not be very helpful. Elements, other than carbon, are typically regulated as classes of compounds containing a named element rather than as specific salts, oxides, or other compounds. The problem for risk assessment may be illustrated in relation to lead. Investigations into lead bioavailability using rodent bioassays demonstrated that the bioavailability of the lead in ore from Skagway, Alaska was low relative to that of other ores.[14] This information helped explain the relatively low blood lead levels ($< 15\,\mu g\,dl^{-1}$) observed in Skagway residents despite high residential soil lead levels ($> 500\,ppm$) and demonstrated a lower adverse health risk from environmental lead than would have been predicted based on environmental monitoring and the use of default assumptions of total lead bioavailability.[15]

3   Risk Assessment of Trace Elements – General Considerations

Even if the biologically available form of an element is its free ion in solution, the free ion may be derived from other chemical species, and the thermodynamic equilibrium for production of the ion or the rate of its release may be the limiting factor for its uptake by living cells. Knowing which chemical species determine the rate and amount of trace element uptake by living organisms is essential for risk assessment. Chemical speciation includes possible redox states, complexes and the presence of elements in particulate forms. In order to determine the relevant chemical species for risk assessment, three questions must be answered.[16,17]

1   What is the mechanism of uptake of the element of concern?
2   How do trace elements interact in the uptake process? Interactions may occur outside the organism, where elements may interact directly or compete for transport sites, or inside, where elements may compete for binding sites that regulate trace element transport systems.
3   Which chemical species controls the rate of trace element uptake and excretion by the cell?

Many trace metals are found as cations that may be complexed by inorganic and organic ligands, or adsorbed on to or bound within particles. Many metals can cycle between different oxidation states. Complexation and redox cycling determine biological effects of elements because of the large differences in reactivity, kinetic lability, solubility and volatility between different chemical species. Most trace elements are absorbed by living organisms from aqueous solution, whether it be from the sea, freshwater, soil water, aerosols or dietary intake to the gut. Complexation of trace elements by inorganic ligands in water has been well studied. For seawater, most Ni, Mn(II), Zn, Co(II) and Fe(II) are present as free aquo ions. Some metals such as Cd, Cu(I), Ag(I), and Hg(II) are complexed by chloride ions. Others such as Cu(II) and Pb(II) are complexed with carbonate, while Fe(III) and Al complex with hydroxide ions.[18] Surface seawater has a fairly constant pH and major ion composition, and inorganic speciation of trace elements generally varies little throughout the surface water of the world's oceans. On the other hand, there are large variations in chloride concentration, alkalinity, pH and redox potential in fresh and estuarine waters, soil water *etc.* These variations lead to changes in inorganic complexation. In fresh waters, pH can vary from <5 to >9 and this alters the hydroxide and carbonate complexation of Fe(III), Cr(III), Al, Cu(II) and Pb. In estuaries, large salinity gradients strongly affect the extent of chloride complexation to Cd, Hg, Cu(I) and Ag.

Naturally occurring organic complexation of trace elements has not been studied as thoroughly as inorganic complexation. Organic complexation of Cu(II) has been the most extensively studied. More than 99% of Cu(II) is complexed to organic ligands in almost all aquatic systems except deep oceanic water[19-25] Copper(II) is bound by unidentified organic ligands present in low concentrations and having extremely high conditional stability constants (log $K \sim 13$ in seawater and 14–15 in lakes at pH 8).[22-24,26,27]

Some studies indicate that Fe(III) is also >99% complexed in near surface seawater by unidentified organic ligands.[28-30] Zinc, lead and cadmium are also complexed organically in surface seawater. Electrochemical measurements indicate that 98–99% of zinc in surface waters is complexed to organic ligands,[31,32] while 50–99% is organically bound in estuarine and fresh waters.[20,33,34] On the other hand, manganese(II) forms only weak co-ordination complexes and there is no evidence for much organic chelation of this element.[35]

Redox transformations alter the speciation and bioavailability of at least eight trace metals, Fe, Mn, Cu, Co, Ag, Hg, Cr and Sn. The oxidation states usually differ markedly in regard to acid–base chemistry, ionic charge, solubility, ligand exchange kinetics and stability of co-ordination complexes. The stable or metastable forms of manganese, Mn(III) and Mn(IV) oxides, are insoluble and, hence, not conventionally regarded as bioavailable. However, it may be that in particulate form they pose problems for aquatic filter feeders similar to those caused by particulates in the lungs of air breathing mammals. Mn(II) is highly soluble and readily taken up by cells but is liable to oxidation by molecular oxygen to form Mn(III) and Mn(IV). Manganese(II) is released into surface waters from photochemical and chemical reduction of Mn oxides by organic matter, and can persist for days to months because of its slow oxidation kinetics.[36,37]

Iron is quantitatively the most important micronutrient metal. It occurs in oxygenated water mainly in the thermodynamically stable state, Fe(III), which is only sparingly soluble in the absence of organic chelation. Photochemical or biological reduction of Fe(III) can increase the biological availability of iron since the resulting Fe(II) is more soluble, has much more rapid ligand exchange kinetics and forms much weaker complexes than Fe(III). The Fe(II) formed is unstable and rapidly re-oxidizes to Fe(III), especially at high pH.[38]

4 Cationic Metal Uptake Mechanisms and Their Relation to Chemical Speciation

Trace elements are usually transported into cells by specialized proteins. Appropriate chemical species bind to receptor sites on the proteins and then either dissociate back into the medium or are transported across the membrane and released into the cytoplasm (Figure 4). The rate of uptake equals the concentration of the elemental species bound to the transport protein multiplied by the kinetic rate constant for transport across the membrane and release into the cytoplasm.

Binding to receptor sites is related to the 'hardness' or 'softness' of the cations[39] (Table 1). 'Hard' cations are of small size and low polarizability owing to poor deformability of the electron sheath. 'Hard' cations, if free in aqueous solution, tend to bind to the 'hard anions', hydroxide and phosphate which are common on the outer surface of living cells. Hydroxyl derivatives of the 'hard' cations lack this affinity and therefore have poor bioavailability. 'Soft' cations

E-particle ⟶ E-water-solution ⟶ Ligand

E-precipitate

E-ligand-solution

E-cell surface ligand

E-carrier

E-phospholipid

E-ligand-phospholipid-solution

E-ligand1

E-ligand2

E-water-solution

E-functional site

E-precipitate (or in organelle)

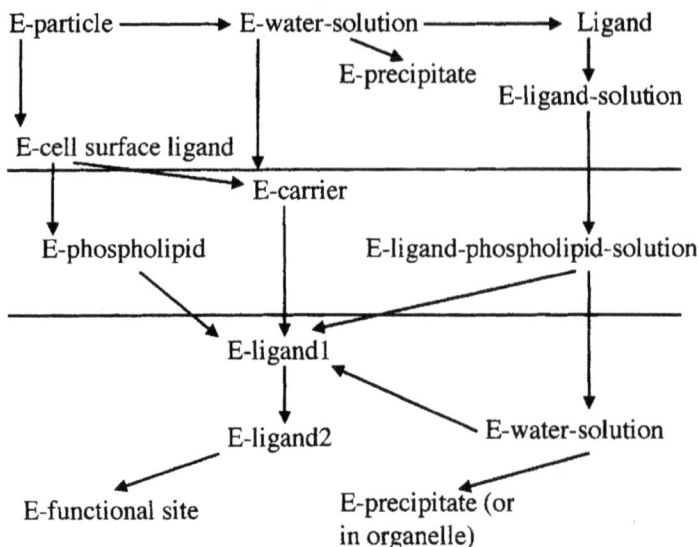

Figure 4 *Schematic representation of the relationships between the extracellular chemical species of an element and the uptake of that element by living cells*

Table 1 *Hard and soft cations*

Hard (class a) cations
Lewis acids (electron acceptors) of small size and low polarizability (deformability of the electron sheath) (hardness)

Li(I), Na(I), K(I), Rb(I), Cs(I), Be(II), Mg(II), Ca(II), Sr(II), Ba(II), Al(III), Sc(III), La(III), Cr(III), Mn(III), Fe(III), Co(III), Si(IV), Ti(IV) Zr(IV), Hf(IV)

Soft (class b) cations
Lewis acids (electron acceptors) of large size and high polarizability (softness)

Cu(I), Ag(I), Au(I), Tl(I), Ga(I), Cd(II), Hg(II), Sn(II), Tl(III), Au(III)

Intermediate (borderline) cations
Fe(II), Co(II), Ni(II), Cu(II), Zn(II), Pb(II), Sb(III), Bi(III)

---

are of large size and high polarizability. 'Soft' cations, such as lead, cadmium and mercury, bind to nitrogen or sulfur centres from which they dissociate slowly. This is the basis of their toxicity. Further, the 'soft' cations compete for binding sites with 'essential' cations. Thallium competes with potassium, while lead and cadmium compete for zinc and calcium sites.

The kinetics of uptake will vary with changes in inorganic metal speciation (*e.g.* the presence of aquo ions or hydroxide or chloride complexes). At equilibrium, the concentration of bound element and, therefore, the uptake rate is related to the external free ion concentration. Here the uptake system is effectively under thermodynamic control[17] (Figure 5). Such binding equilibration might be expected for elements with rapid ligand exchange kinetics, such as copper(II)

Equilibrium control   Kinetic control

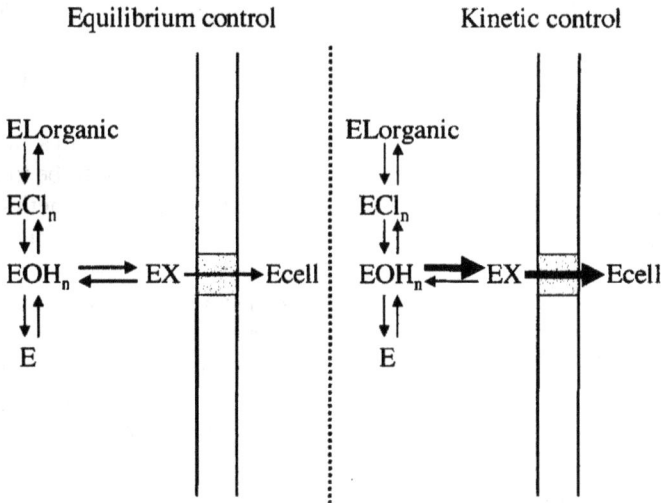

Figure 5 *Schematic diagram to show the extremes of equilibrium (thermodynamic) control and kinetic control of element binding to cell membrane transport molecules*

and cadmium. Iron(III), whose ligand exchange rates in seawater are only 1/500th of those for copper[40] show uptake under kinetic control by the rate of metal binding to membrane transport sites. The rate of binding is related to the concentration of the labile dissolved inorganic iron(III) species [$Fe(OH)_2{}^+$, $Fe(OH)_3$ and $Fe(OH)_4{}^-$)] whose exchange kinetics are rapid enough to permit appreciable rates of iron coordination to transport sites. Other elements with slow exchange kinetics, such as nickel, aluminium and chromium(III), may also be under kinetic control, and their uptake determined by the concentration of kinetically labile inorganic species.[40] A number of these chemical species, as in the case of iron cited above, will be the result of hydrolysis. Hydrolysis of metal ions is the process by which trivalent and more highly charged cations react with water to form hydroxo or oxo complexes which vary in ionic charge and may be anionic and/or multinuclear. Hydrolysis is the cause of the speciation changes with pH shown for aluminium in Figure 2. The tendency to hydrolyse increases with dilution and the hydrolysis products vary with pH.

Hydrolysis decreases the availability of simple ions as can be seen above for iron(III). This is why aluminium and beryllium compounds in aqueous solution are poorly absorbed by cells at pHs near neutrality, where they occur as a mixture of hydroxide complexes. Clearly, any risk assessment for exposure to aluminium or beryllium should incorporate a quantitative consideration of the relative concentrations of these complexes.

Many elements, such as zinc, are taken up by more than one transport system with differing binding affinities. Each transport system may be the main one depending upon the chemical conditions (*e.g.* low *vs.* high metal ion concentrations[41]). Each system may be under differing controls (kinetic or thermodynamic), and thus it may be important to know which system predominates in order to predict effects of chemical speciation on uptake rates. Once the uptake limiting species (one or more) are identified, risk assessment must concentrate on determining the species concentration and the effective exposure for the element of concern.

In cases of kinetic control, trace elements in organic chelates and bound particulates (*e.g.* colloidal metal oxides) are generally assumed to be unavailable for direct uptake since their dissociation and ligand exchange kinetics (or their diffusion rates in the case of colloids and particulates) are too slow to permit rapid donation of free ion to membrane transport sites.[42] However, there are certain chelates, such as the crown ethers, which can bind metal ions, dissolve in the phospholipid bilayer of the cell membrane and pass through by simple diffusion (see below). This appears to be true also for certain hydrophobic copper complexes.[43] Moreover, we have very little knowledge of what happens to particulates bound to membrane receptors, for example in lung tissue or gills. Such interactions must depend critically on chemical speciation in the particulates and on their surface chemistry.[44]

Thus, in some cases, the uptake of metals is related to the free ion concentration, while in others it is related to the concentration of kinetically labile inorganic species (free ions plus labile inorganic complexes) and, in some cases, to the concentration of particulates or bioavailable complexes. In constant ionic media such as near-surface seawater equilibrated with the atmosphere (pH $\sim$ 8.2), the inorganic speciation is constant and concentrations of free ions and inorganic complexes are related to one another by constant ratios[45] (Table 2). Under those conditions, it probably does not matter whether one defines trace element availability in terms of concentrations of free metal ions or of dissolved inorganic metal species, although it is important to know the concentration of the truly bioavailable species if complex interactions are suspected.

The inorganic complexation of many trace metals such as manganese, zinc,

Table 2   *Inorganic speciation of some trace metals in marine systems at 25 °C, 1 atm pressure, 3.5% salinity, and pH 8 (computer model calculation from unknown total concentrations and available stability constant data)*[45]

Metal	Main chemical species
Al^{3+}	$Al(OH)_3$ 100%
Cr^{3+}	$Cr(OH)_3$ 100%
Mn^{2+}	Mn^{2+} 58%, $MnCl^+$ 37%, $MnSO_4$ 4%, $MnCO_3$ 1%
Fe^{2+}	Fe^{2+} 69%, $FeCl^+$ 20%, $FeCO_3$ 5%, $FeSO_4$ 4%, $Fe(OH)^+$ 2%
Fe^{3+}	$Fe(OH)_3$ 100%
Co^{2+}	Co^{2+} 58%, $CoCl^+$ 30%, $CoCO_3$ 6%, $CoSO_4$ 5%, $Co(OH)^+$ 1%
Ni^{2+}	Ni^{2+} 47%, $NiCl^+$ 34%, $NiCO_3$ 14%, $NiSO_4$ 4%, $Ni(OH)^+$ 1%
Cu^{2+}	Cu^{2+} 9%, $CuCO_3$ 79%, $Cu(OH)^+$ 8%, $CuCl^+$ 3%, $CoSO_4$ 1%
Zn^{2+}	Zn^{2+} 46%, $ZnCl^+$ 35%, $Zn(OH)^+$ 12%, $ZnSO_4$ 4%, $ZnCO_3$ 3%
Mo(VI)	MoO_4^{2-} (100%)
Cd^{2+}	Cd^{2+} 3%, $CdCl^+$ 97%
Pb^{2+}	Pb^{2+} 3%, $PbCl^+$ 47%, $PbCO_3$ 41%, $Pb(OH)^+$ 9%, $Pb(SO_4)$ 1%
Si(IV)	H_4SiO_4 92%, $H_3SiO_4^{-1}$ 5%

cobalt and nickel is negligible in most saline and fresh waters and the inorganic speciation is dominated by the free aquo ions. For these elements, the distinction between thermodynamic equilibrium and kinetic control may be of little practical importance. However, for metals that are strongly complexed by chloride ions, such as cadmium, mercury, silver and copper(I), whose inorganic speciation varies widely with salinity, knowledge as to the type of transport control may be critical in predicting how expected changes in metal uptake rates may accompany variations in major ion composition of the water. Similarly, for metals which are highly complexed to hydroxide, such as iron(III), aluminium and chromium(III), and for metals complexed to carbonate ions, such as copper(II) and lead, changes in pH or alkalinity in freshwater systems can lead to large differences in inorganic speciation, and therefore metal transport behaviour, and related risk assessment.

As already indicated (Figure 4), there are exceptions to the membrane protein carrier transport mechanism for trace metal uptake. Some neutrally charged, non-polar complexes, such as $HgCl_2$ and CH_3HgCl, can diffuse directly across phospholipid membranes owing to their lipid solubility.[46] In experiments with a coastal diatom, the uptake and toxicity of Hg(II) within an Hg/salinity matrix was related to the computed concentration of $HgCl_2$ and not that of Hg^{2+} or of total inorganic Hg species, implying that uptake occurred *via* diffusion of the neutral $HgCl_2$ complex through the cell membrane.[47] Once inside the cell, the $HgCl_2$ takes part in ligand exchange reactions with biological ligands, such as sulfhydryls, providing an intracellular sink for the diffusing Hg. Other less polar neutrally charged chloro-complexes, such as AgCl and Cu(I)Cl, also may be taken up by the same diffusion/ligand exchange mechanism. Similarly, lipophilic chelates, such as those with 8-hydroxyquinoline and dithiocarbamate, are taken up by the same process.[48,49] Although uptake of such chelates has not yet been shown to be important in natural unperturbed systems, their uptake and exchange may provide a significant uptake pathway in aquatic environments receiving pollutant inputs of synthetic organic ligands that form such lipophilic chelates.

Chelates with certain biological ligands can be transported into cells by specific membrane transport proteins. Many microorganisms release strong iron-binding ligands (siderophores) into their surrounding medium under iron limiting conditions. Siderophores complex and solubilize iron. The siderophore-iron chelates are then transported into the cell by specific membrane transport proteins, after which the iron is released for assimilation by metal reduction or degradation of the siderophore.[50] The production of siderophores is widespread in eubacteria and fungi, and also occurs in many, but not all, cyanobacteria.[51]

4.1 Bioavailability of Metals in Sediments

Determining the bioavailability of metals sorbed to sediments is a key to understanding their potential to accumulate in aquatic organisms and to induce toxic effects. It is clear from the published data that total metal concentrations in sediments are poorly related to the bioavailable fraction.[52-54] Total metal concentrations in sediments which produce defined toxic effects can differ by a factor of 10–100 for different sediments. A number of approaches to determining metal bioavailability associated with sediments have been suggested, including carbon normalization and sorption of metals in oxic freshwater sediments to particulate carbon and the oxides of iron and manganese.[55]

Recently, the dominant role of the sediment sulfides in controlling metal bioavailability has been demonstrated.[56–58] Sulfides are common in many freshwater and marine sediments and are the predominant form of sulfur in anaerobic sediments (usually as iron sulfide). The ability of sulfide and metal ions to form insoluble precipitates with water solubilities well below the toxic concentrations in solution is well established.[56] This accounts for the lack of toxicity from sediments and sediment pore waters even when high metal concentrations are present.[58] It has been shown that the solid-phase sediment sulfides that are soluble in weak cold acid, termed acid volatile sulfides (AVS), are a key factor in controlling the toxicity of metals such as copper, cadmium, nickel, lead and zinc. Toxicity due to these metals is not observed when they are bound to sediment and when, on a molar basis, the concentration of AVS is greater than the sum of the molar concentrations of metals. When the ratio of the sum of the simultaneously extracted metals to AVS concentration exceeds 1.0 on a molar basis, toxic effects due to metals may be expressed, if the metal(s) are not complexed by other ligands. Thus, the metal:AVS ratio can be used to predict the fraction of the total metal concentration present in sediment that is bioavailable and hence provides the basis for risk assessment.

Limitations to the metal:AVS ratio approach occur when the AVS concentration is low, for example in fully oxidized sediments. Most sediments have at least a small zone where the sediments are oxic near the sediment–water interface. The importance of this zone has been demonstrated for copper relative to AVS and accumulation of copper in the midge (*Chironomus tentans*).[59] In these situations, other phases (*i.e.* iron and manganese oxides, dissolved organic carbon and particulate organic carbon) can play an important role in determining the bioavailability of metals.

5   Uptake of Anions and Non-metals[39]

Like the uptake of metal cations, the uptake of anions and non-metals by biological systems depends on membrane restrictions and thermodynamic and kinetically energized binding traps and steps. Neutral molecules can diffuse into cells. Thus, $Si(OH)_4$ and $B(OH)_3$ can carry silicon and boron to all parts of a cell by free diffusion across membranes. Compounds such as $B(OH)_3$ may be trapped by binding to *cis*-diols, probably to polysaccharides, giving moderately labile ring condensation products. $Si(OH)_4$ or any weak dibasic acid may react in the same manner. Molybdenum, in the form $Mo(OH)_6$, behaves like $Si(OH)_4$.

Most non-metals are available to biological systems as simple anions, *e.g.* F^-, Cl^-, Br^-, I^-, or oxyanions such as $H_3PO_4^-$, HPO_4^{2-}, $H_2ASO_4^-$, AsO_2^-, $B(OH)_4^-$, SO_4^{2-} and SeO_4^{2-}. Molybdenum and vanadium can also be included here as MoO_4^{2-} and VO_4^{3-} and the principles of uptake for all these elements are similar. These ions cannot cross membranes without carriers and, once captured, they remain inside the cell unless there is a specific outward pumping system.

In the same way as for cations, there are two major selection mechanisms based on thermodynamic equilibrium binding properties:

1 Selection by size and charge, taking into account the huge influence of hydration of anions on binding to proteins
2 Selection by differences in binding affinity for different types of cationic centres such as H^+, $-NH_3^+$, or M^{n+}.

5.1 Selection by Charge and Size; the Hydration of Anions and Binding to Proteins

The various anions of Group VII and oxyanions of Groups IV–VI show considerable variation in size, permitting separation on that basis alone. The associated thermodynamic selectivity may be much more significant than the great differences in abundance and availability of these elements. On the basis of similar sizes, only the anions F^- and OH^-, or Br^- and SH^-, can show competition.

There are some differences in the nature of the binding sites for anions compared with those for cations. Anions are negatively charged and they have a relatively lower charge density because they are large. This means that anions must bind to clustered, positively charged groups, such as the ammonium groups $-NH_3^+$ the guanidinium group, and metal ions, and/or through very extensive hydrogen bonding. Rarely, binding may occur at hydrophobic centres or regions containing $-OCH_3$, $-SCH_3$, $-CH_2-$, phenyl, *etc.* These centres will take up anions only if they are surrounded by a 'buried' positive charge generated by the fold of the protein.

Hydration opposes binding if the binding sites are hydrophobic and the less hydrated anions will be bound preferentially to such sites. The so-called 'lyotropic' or Hofmeister series is obtained; this is the reverse of the hydration free energy. Thus, in binding strength:

$$I^- > Br^- > Cl^- > F^- \text{ and } ClO_3^- > BrO_3^- > IO_3^- > HPO_4^-$$

The non-polar sites in proteins which give such series often contain (buried) charged histidine residues or, sometimes, arginine and lysine residues.

5.2 Selection of Anions by Differences in Binding Affinity for Different Cationic Centres

Two different types of binding behaviour to cations or cationic centres can be recognized. For class 'a' (hard) cations the order of binding by anions is

$F^- > Cl^- > Br^- > I^-$, and $O^{2-} > S^{2-}$; for class 'b' (soft) cations the order is the reverse. The order of binding of anions to class 'a' metal ions is overwhelmingly due to electrostatic effects, but binding to class 'b' metal ions is driven mainly by covalence.

The main cations concerned in biological systems are Mg^{2+}, Ca^{2+}, Cu^{2+}, Cu^+, Zn^{2+}, Co^{2+}, Co^{3+}, Fe^{2+}, Fe^{3+}, Mn^{2+} and Mn^{3+}, but only Cu^+ is a class 'b' cation. Na^+ and K^+ cannot normally be considered as anion binding sites since they form only very weak complexes and are highly mobile. However, they may have a role in a few specific cases.

Thus, of the free aqueous ions, Cu^+ will amplify the lyotropic effect, while all other hydrated cations should oppose it to different degrees.

$$Mg^{2+} > Ca^{2+}$$

and

$$Mg^{2+} > Mn^{2+} > Zn^{2+} > Cu^{2+}$$

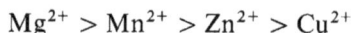

Simultaneous uptake of an anion and a cation is also a possibility. The most frequently met example is that of the binding of polyphosphates such as ATP and ADP with Mg^{2+}. The binding appears to be to very hydrophilic protein regions containing additional positive charges on basic side-chains of the proteins. The binding is very selective since the combined requirements of the anion and the cation have to be satisfied simultaneously.

5.3   Kinetic Binding Traps for Anions, with or without Accompanying Redox Reactions

Some neutral molecules can be trapped by condensation reactions with various organic compounds forming kinetically stable covalent bonds, *e.g.* $B(OH)_3$ which condenses with *cis*-diols. Other neutral molecules must be activated, reduced, or oxidized before they can bind to other compounds through kinetically stable covalent bonds.

Phosphate anions can only be retained weakly by ionic interaction with positively charged groups, *e.g.* $-NH_3^+$, but they can be retained in a kinetic trap by condensing with –OH groups. In these forms phosphate is transported, *e.g.* as sugar phosphate, stored, *e.g.* as polyphosphate and ATP, and transferred to other molecules, simple or polymerized, to give a variety of substances, *e.g.* RNA and DNA.

Inside cells, SO_4^{2-} is reduced and sulfur is retained in sulfide, S^{2-} the thiol RSH, or the disulfide –S–S– form. The same reactions occur with SeO_4^{2-}, which is bound as selenothiol, and with NO_3^-, which is incorporated as an amino group or other nitrogen-containing molecule. In the halide group, F^- and Cl^- are usually handled as such, but I^- and Br^-, once taken into a cell, can be selectively absorbed by enzymes, and the fact that they can be much more easily oxidized than Cl^- permits the reactions $Br^- \rightarrow Br$ and $I^- \rightarrow I$ in the presence of

Cl^-, *e.g.* by the action of many peroxidases. The reactive radicals so produced attack organic compounds readily, *e.g.* phenols, and thus Br and I are inserted, covalently linked, into organic molecules. This is well seen in the biosynthesis of the thyroid hormone. In some cases, the free halogen can be formed and liberated, *e.g.* I_2 in some seaweeds.

There are situations in which reduction can change an anion into a cation. The uptake of molybdenum and of vanadium initially as MoO_4^{2-} and HVO_4^{2-}, respectively, *i.e.* in a nonmetal form, is based on this principle. They are reduced to oxocations or simple cations and bound into proteins or cofactors. Oxovanadium(IV), VO^{2+}, is like nickel in its complexes with polyaminocarboxylic acids and binds quite strongly to N/O donors at pH ~ 7, but with certain strong complexing agents it can lose the oxo-group to give V(IV) complexes. The vanadium-containing fungus *Amanita muscaria* contains this species, V(IV), bound to an *N*-hydroxy derivative of iminodiacetic acid forming a very stable octa-coordinated complex in which the *N*-hydroxy group is ionized and binds the metal. The unusual accumulation of vanadium in tunicates as V(III) or VO^{2+} from the low concentrations of HVO_4^{2-} found in the environment is probably due to this change in oxidation state, so that there is no 'saturation' inside cells relative to the ionic species available outside, even when no internal ligand appears to be present and the retention of the ion (V^{3+} or VO^{2+}) is ensured by storage in vesicles.

The reactions of molybdenum are unusual, since as a metal element it is the closest to the non-metals. On entering cellular systems, the hydroxide form of this ion $MoO_2(OH)_4^{2-}$ can react not just with OH groups to form MoO_2^{2-} bound species but, since it is a soft 'b' class metal ion, it can give

$$MoO_2(OH)_4^{2-} + 2RSH \rightarrow [MoO_2(OH)_2(RS)_2]^{2-} + 2H_2O$$

The more stable forms of this anion are probably $[MoO_3(RS)_2]^{2-}$.

6 Competition and the Uptake and Toxicity of Non-nutrient Elements

Coordination sites are never entirely specific for any chemical species of an element. Biological ligands that have evolved to bind a particular species will also bind competing species with similar ionic radii and coordination geometry. Such competitive binding can occur at transport sites, active sites of metalloproteins, or feedback control sites such as those regulating the number or activity of specific membrane transport proteins. Competition often occurs for binding to trace element uptake sites, such as those for manganese ions. Competing metal ions such as those of copper, zinc and cadmium inhibit uptake of manganese ions by their transport system. This effect has been seen in several diatoms and chlorophytes.[60-62] The inhibition of manganese ion uptake by competing metal ions results both from direct binding of these metal ions to membrane transport sites and probably also from binding to control sites regulating the V_{max} of the transport system. The control site binding results in a reduction in the cell's capability for feedback regulation of intracellular manganese ion concentration. The inhibition of manganese ion uptake by other metal ions, such as cadmium, causes manganese deficiencies at low $[Mn^{2+}]$. Such deficiencies can be the fundamental cause of inhibition of growth by potentially toxic metals, and emphasize the inherent linkages between speciation, metal toxicity and metal nutrition.

Although in algal cells studied the transport system for manganese is the only system responsible for uptake in the environmental $[Mn^{2+}]$ range, it has a higher affinity for binding competing metal ions such as cadmium ions. This is because of the low reactivity of manganese ions toward complexation by organic ligands compared with reactivity of zinc and cadmium ions, the so-called Irving–Williams order of affinity.[39] Because of these inherent differences in reactivity, it may be impossible for a manganese transporter to have a higher affinity for Mn^{2+} than it does for more reactive divalent metal ions, such as Zn^{2+}, Cd^{2+} and Cu^{2+}. In most unpolluted coastal waters, the much higher concentrations of Mn^{2+} relative to those of competing metal ions (Zn^{2+}, Cd^{2+} and Cu^{2+}) compensate for the transport site's lower affinity for manganese. In polluted environments, free ion levels of the other potentially toxic metals may become high enough to reduce manganese uptake and cause manganese deficiencies. Thus, risk assessment for any of these metal ions must consider the concentrations of the others as well as of manganese.

Elements are often taken up by more than one transport system, each system being important under different sets of conditions. Although cadmium ions are taken up by the manganese system at high $[Zn^{2+}]$, uptake at low $[Zn^{2+}]$ is dominated by a high affinity cadmium ion transport system that is under negative feedback control by intracellular zinc ions. In *T. pseudonana*, the rate of Cd^{2+} uptake by this system increases by over a 1000-fold as $[Zn^{2+}]$ is decreased from 10^{-10} to 10^{-12} M. Cobalt also appears to be taken up by this system, and its uptake also increases substantially with decreasing $[Zn^{2+}]$. Because of its uptake by manganese and zinc related systems, cadmium uptake by cells can be as heavily influenced by external manganese and zinc ion concentrations as it is by free cadmium ion levels. Again, risk assessment must consider these ions as a group and not as acting independently.

Once inside the cell, a competing element may bind to nutrient element sites such as the active sites on metalloproteins. The bound competing element may have coordination geometry, Lewis acidity, redox behaviour or ligand exchange kinetics which do not permit reaction of the nutrient element with the site. There may be a loss of metabolic function and inhibition of metabolism. Damage will be made worse by high internal ratios of potentially toxic to nutrient element.

Cells have evolved detoxification mechanisms for some potentially toxic elements. In eukaryotic algae, potentially toxic elements, such as cadmium, copper, zinc and mercury, are detoxified intracellularly by binding to phytochelatin, an inducible polypeptide containing two or more cysteine residues.[63,64] Such binding, however, can also enhance metal uptake by providing a non-toxic intracellular sink for accumulated elements.

Cells may also have efflux systems for potentially toxic elements, as observed for cadmium in diatoms. For at least one species, *Thalassiosira weissflogii*, there is a system for the export of cadmium-phytochelatin chelates.[65] Cadmium and zinc ion efflux systems involving plasmid encoded membrane proteins have also been identified in bacteria.[66] Efflux systems are induced by high intracellular element concentrations and decrease toxicity by minimizing toxic element accumulation. However, even when these systems are completely successful in detoxifying competing intracellular elements, the cells may be left with an induced deficiency of a nutrient element.

Competitive interactions between nutrient and inhibitory elements are common. Much information comes from studies on phytoplankton, where there is evidence of competition between manganese and copper, manganese and zinc, manganese and cadmium (see above), zinc and copper[67,68] zinc and cadmium[69,70] cobalt and zinc,[71] iron and cadmium[72] and iron and copper.[73] In addition, uptake of the thermodynamically stable oxyanions, molybdate[74] and chromate[75] is competitively inhibited by sulfate, which is stereochemically very similar to both anions. The chromate/sulfate antagonism appears to be related to uptake of chromate by the sulfate transport system.[79] As a consequence of these interactions, variations in salinity and accompanying changes in sulfate influence both Cr toxicity and Mo nutrition. Chromate and molybdate should also interact antagonistically but there appear to be no published data on this.

7 Conclusions

A number of complex interacting factors influence trace element accumulation by living cells. Trace elements are generally taken up into cells by membrane transport proteins which have selective binding sites for different chemical species. The amount of an element bound to external receptor sites on these proteins and, therefore, the intracellular transport rate is usually related to the external concentration of either free ions with appropriate properties or kinetically labile inorganic species (free ions plus inorganic complexes). Organic complexation and particulate binding often decrease element uptake rates by decreasing the concentrations of free ions and labile inorganic complexes. However, in certain circumstances, organic complexes such as the crown ethers may facilitate trace element uptake. In addition, particulates which bind to tissues such as lung or the gill epithelia may provide a focus of prolonged trace element exposure and toxicity. Transport systems are never entirely specific for a single element, and these systems often show competition between similar chemical species, resulting in inhibition of uptake of essential elements and uptake of competing potentially toxic elements. Because of these competitive interactions, ion ratios often control the cellular uptake of both toxic and nutrient elements. Such interactions also result in inherent inter-relationships between toxicity and nutrition. It is important to define trace element species interactions clearly before carrying out risk assessment because of such profound effects on trace element availability and toxicity.

8 References

1. European Commission, *Technical Guidance Document in Support of Commission Directive 93/67/EC on Risk Assessment for New Substances* and *Commission Regulation (EC) No 1488/94 on Risk Assessment for Existing Substances*, 4 parts, Office for Official Publications of the European Communities, Luxembourg, 1996.
2. *Guidelines for Testing of Chemicals*, Organization for Economic Cooperation and Development, Paris, 1982 onward.
3. E. Nieboer, B.L. Gibson, A.D. Oxman and J.R. Kramer, *Environ. Rev.*, 1995, **3**, 27.
4. R.J.P. Williams and J.J.R. Frausto da Silva, *The Natural Selection of the Chemical Elements*, Clarendon Press, Oxford, 1996.
5. Paracelsus (Theophrastus ex Hohenheim Eremita), *Von der Besucht*, Dillingen, 1567.
6. US National Academy of Science, *Toxicological Assessment of the Army's Zinc Cadmium Sulfide Dispersion Tests*, National Academy of Sciences, Washington, DC.
7. J.H. Duffus, *Sci. Progr.*, 1996, **79**, 311
8. M.H. Draper, in *Carcinogenicity of Inorganic Substances – Risks from Occupational Exposure*, ed. J.H. Duffus, Royal Society of Chemistry, Cambridge, 1997, p. 181.
9. US National Toxicology Program (NTP), *NTP Technical Report on the Toxicology and Carcinogenesis Studies of Nickel Sulfate Hexahydrate (CAS No. 10101-97-0) in F3441N Rats and 86C3F Mice (Inhalation Studies)*, NIH PubL No. 94-3370, US Department of Health and Human Services, Public Health Service, National Institutes of Health, Bethesda, MA, 1994.
10. G. Chiappino, *Sci. Total Environ.*, 1994, **150**, 65.
11. W.A. Burgess, *Recognition of Health Hazards in Industry*, Wiley, New York, 1995.
12. M.W. Roife, R. Paine, R.B. Davenport and R.M. Stricter, *Am. Rev. Respir. Dis.*, 1992, **146**, 1600.
13. T. Gordon and J.M. Fine, *Occup. Med.*, 1993, **8**, 504.
14. M.P. Dieter, H.B. Matthews, R.A. Jeffcoat and R.F. Moseman, *J.Toxicol. Environ. Health*, 1993, **39**, 79.
15. J.P. Middaugh, C. Li and S.A. Jenkerson, *Health Hazard and Risk Assessment from Exposure to Heavy Metal Ore in Skagway, AK (Final Report)*, Alaska Department of Health and Social Services, Anchorage, 1989, cited in R.A. Ponce and E.M. Faustman, in *Reproductive and Developmental Toxicology*, ed. K.S. Korach, Marcel Dekker, New York, p. 449.
16. R.J.M. Hudson, *Sci. Total Environ.*, 1998, **219**, 95.
17. W.G. Sunda and S.A. Huntsman, *Sci. Total Environ.*, 1998, **219**, 165.
18. R.H. Byrne, L.R. Kump and K.J. Cantrell, *Mar. Chem.*, 1988, **25**, 163.
19. W.G. Sunda and PJ. Hanson, in *Chemical Modeling in Aqueous Systems: Speciation, Sorption, Solubility and Kinetics*, ed. E.A. Jenne, ACS Symposium Series 93, American Chemical Society, Washington, DC, 1979.
20. C.M.G. Van den Berg, A.G.A. Merks and E.K. Dursma, *Estuar. Coast. Shelf Sci.*, 1987, **24**, 785.
21. K.F.I. Coale and K.W. Bruland, *Limnol. Oceangr.*, 1988, **33**, 1084.
22. K.F.I. Coale and K.W. Bniland, *Deep-Sea Res.*, 1990, **34**, 317.
23. J.W. Moffett, L.E. Brand and R.G. Zika, *Deep-Sea Res.*, 1990, **37**, 27.
24. W.G. Sunda and S.A. Huntsman, *Mar. Chem.*, 1991. **36**, 137.
25. H. Xue, A. Oestreich, D. Kistler and L. Sigg, *Aquat. Sci.*, 1996, **58**, 69.
26. J.W. Moffett, *Deep-Sea Res.*, 1995, **41**, 1273.
27. X. Hue and W.G. Sunda, *Environ. Sci. Technol.*, 1997, **31**, 1902.
28. M. Gledhill and C.M.G. van den Berg, *Mar. Chem.*, 1994, **47**, 4129.

29. E.L. Rue and K.W. Bruland, *Mar. Chem.*, 1995, **50**, 117.
30. J. Wu and G.W. Luther, *Mar. Chem.*, 1995, **50**, 159.
31. K.W. Bruland, *Limnol. Oceanogr.*, 1989, **34**, 269.
32. J.R. Donat and K.W. Bruland, *Mar. Chem.*, 1990, **28**, 301.
33. B.L. Lewis, G.W. Luther and T.M. Church, *Electroanalysis*, 1995, **7**, 166.
34. H. Xue, D. Kistler and L. Sigg, *Limnol. Oceanogr.*, 1995, **40**, 1142.
35. J.S. Roitz and K.W. Bruland, *Anal. Chim. Acta*, 1997, **344**, 175.
36. W.G. Sunda and S.A. Huntsman, *Limnol. Oceanogr.*, 1987, **32**, 552.
37. W.G. Sunda and S.A. Huntsman, *Deep-Sea Res.*, 1988, **35**, 1297.
38. W.L. Miller, K. Lin, D.W. King and D.R. Kester, *Mar. Chem.*, 1995, **50**, 63.
39. J.J.R. Frausto da Silva and R.J.P. Williams, *The Biological Chemistry of the Elements*, Oxford University Press, Oxford, 1991.
40. R.J.M. Hudson and F.M.M. Morel, *Deep-Sea Res.*, 1993, **40**, 129.
41. W.G. Sunda and S.A. Huntsman, *Limnol. Oceanogr.*, 1992, **37**, 25.
42. F.M.M. Morel, R.J.M. Hudson and N.M. Price, *Limnol. Oceanogr.*, 1991, **36**, 1742.
43. M. Ahsanullah and T.M. Florence, *Mar. Biol.*, 1984, **84**, 41.
44. E. Nieboer, G.G. Fletcher and Y. Thomassen, *J. Environ. Monit.*, 1999, **1**, 1.
45. D. Turner, M. Whitfield and A.G. Dickson, *Geochim. Cosmochim. Acta*, 1981, **45**, 855.
46. J. Gutknecht, *J. Membrane Biol.*, 1981, **61**, 61.
47. R.P. Mason, R.J. Reinfelder and F.M.M. Morel, *Environ. Sci. Technol.*, 1996, **30**, 1835.
48. J.L. Stauber and T.M. Florence, *Mar. Biol.*, 1987, **94**, 511.
49. J.T. Phinney and K.W. Bruland, *Environ. Sci. Technol.*, 1994, **28**, 1781.
50. J.B. Nielands, *Annu. Rev. Nutr.*, 1981, **1**, 27.
51. S.W. Wilhelm and C.G. Trick, *Limnol. Oceanogr.*, 1994, **39**, 1979.
52. R. Ruiz, F. Romero and G. Besga, *Toxicol. Environ. Chem.*, 1991, **33**, 1.
53. E. DeVevey, G. Bitton, D. Rossel, L.D. Ramos and S.M. Guerrero, *Bull. Environ. Contam. Toxicol.*, 1993, **50**, 253.
54. H.E. Alien and D.J. Hansen, *Water Environ. Res.*, 1996, **68**, 42.
55. E.A. Jenne, *Report to the US Environmental Protection Agency*, Office of Water Regulations and Standards, Criteria and Standards Division, Washington, DC, 1987.
56. D.M. Di Toro, J.D. Mahoney, D.J. Hansen, K.J. Scott, M.B. Hicks, S.M. Mayr and M.S. Redmond, *Environ. Toxicol. Chem.*, 1990, **9**, 1487.
57. D.M. Di Toro, C.S. Zarba, D.J. Hansen, W.J. Berry, R.C. Swarz, C.E. Cowan, S.P. Pavlou, H.E. Alien, N.A. Thomas and P.R. Paquin, *Environ. Toxicol. Chem.* 1991, **10**, 1541.
58. G.T. Ankley, G.L. Phipps, E.N. Leonard, D.A. Benoit, V.R. Mattson, P.A. Kosian, A.M. Cotter, J.R. Dierkkes, D.J. Hansen and J.D. Mahoney, *Environ. Toxicol. Chem.*, 1991, **10**, 1299.
59. J.M. Besser, C.G. Ingersol and J.P. Giesy, *Environ. Toxicol. Chem.*, 1996, **15**, 286.
60. B.A. Hart, P.E. Bertram and B.D. Scaife, *Environ. Res.*, 1979, **18**, 327.
61. W.G. Sunda and S.A. Huntsman, *Limnol. Oceanogr.*, 1983, **28**, 924.
62. W.G. Sunda and S.A. Huntsman, *Limnol. Oceanogr.*, 1996, **41**, 373.
63. B.A. Ahner, S. Kong and F.M.M. Morel, *Limnol. Oceanogr.*, 1995, **40**, 649.
64. B.A. Ahner and F.M.M. Morel, *Limnol. Oceanogr.*, 1995, **40**, 658.
65. J.G. Lee, B.A. Ahner and F.M.M. Morel, *Environ. Sci. Technol.*, 1996, **30**, 1814.
66. S. Silver, B.O. Lee, N.L. Brown and D.A. Cooksey, in *The Chemistry of Copper and Zinc Triads*, Royal Society of Chemistry, Cambridge, 1993, pp. 38–53.
67. J.G. Rueter and F.M.M. Morel, *Limnol. Oceanogr.*, 1982, **26**, 67.
68. W.G. Sunda and S.A. Huntsman, *Limnol. Oceanogr.*, 1998, **43**, 1055.
69. J.G. Lee, S.B. Roberts and F.M.M. Morel, *Limnol. Oceanogr.*, 1995, **40**, 1056.

70. W.G. Sunda and S.A. Huntsman, *Environ. Sci. Technol.*, 1998, **32**, 2961.
71. W.G. Sunda and S.A. Huntsman, *Limnol. Oceanogr.*, 1995, **40**, 404.
72. G.I. Harrison and F.M.M. Morel, *J. Phvcol.*, 1983, **19**, 495.
73. L.S. Murphy, R.R.L. Guillard and J.F. Brown, *Biol. Oceanogr.*, 1984, **3**, 187.
74. R.W. Howarth, R. Marino and J.J. Cole, *Limnol. Oceanogr.*, 1988, **33**, 688.
75. G.E. Riedel, *Aquat. Toxicol.*, 1985, **7**, 191.

Overview

CHAPTER 25

An Overview

LES EBDON AND LES PITTS

1 Introduction

The purpose of producing this book was not to provide another analytical methods manual, of which there are now several detailing speciation methodologies, but rather to give an overall picture, of use to anybody who needs to know about speciation, from the researcher to the legislator, from the informed layman to the food producer.

The preceding chapters have outlined the current 'state-of-the-art' in the field of speciation, in the areas of the environment, industrial health and hygiene, and food and nutrition. The progress made in recent years is clear but so also are some of the needs, both in terms of basic research, extraction technologies, separation techniques, sensitive detection, customised instrumentation, provision of more CRMs and in the sensible application of the knowledge gained.

Advances in instrumentation have been crucial in enlightening the researcher. Such advances need to continue, since every answer seems to provoke more questions. In this concluding chapter, some thoughts will be given as to where we go from our present state of knowledge.

2 Analytical Needs and Instrumentation

Over the last half-century, many major advances in the field of analytical chemistry have led to the lowering of detection limits. This alone has enabled chemists to study mechanisms and interactions, and has made possible the measurement of speciation at levels encountered in environmental samples. Problems, such as the damage done to the marine ecosystem by organotin compounds, have provided the impetus for instrument manufacturers to develop new instruments, and for analytical chemists to develop methodologies to detect vanishingly small quantities of pollutant. A number of disciplines have combined to give the analytical chemist ever more powerful (and costly) instrumentation with which to track down the last picogram of analyte, and automation has enabled these low limits to become available for many routine determinations.

During this period, probably the three most important instrumental developments having a bearing on speciation have been those in atomic spectrometry, mass spectrometry and separation science rather than in electroanalysis, as might have been expected fifty years ago. Special mention must also be made of the computer. Virtually all instruments sold these days contain a microcomputer, or are run by an external computer. Computer controlled automation has made possible the analysis of large batches of samples, with little or no human intervention, hence reducing the overall cost of, and time for, analysis. It has permitted instrumental self-diagnosis and optimisation, thus contributing to an improvement in the quality of results. Computerisation has also changed the role of the analytical chemist, so that a greater emphasis on the quality of results is both possible and essential, given the quantities of data produced.

Successive developments in atomic spectrometry, from atomic absorption spectrometry (AAS) to inductively coupled plasma atomic emission spectrometry (ICP-AES) to inductively coupled plasma mass spectrometry (ICP-MS) have led to advances in the understanding of biogeochemical cycles and processes within the environment, drug and food metabolism within the body, all at lower and lower levels. The coupling of separation techniques such as gas or liquid chromatography to these methodologies has pushed the study of speciation forward. As more has been discovered about the roles of the compounds of elements, it has become more apparent that speciation is vital to understanding the impact of trace elements in environmental, clinical and nutritional systems.

Presently, research is underway in the area of high cost instrumentation – development is taking place to further refine the instruments already in use, to further lower detection limits and reduce interference effects. Investigations are taking place into the reduction of background noise within the mass spectrometer and the removal of dimer and other interferences, by employing a hexapole system in place of the more normal quadrupole, collision cells, shielded torches and higher resolution magnetic sector instruments, but this work is unlikely to revolutionise the field of analytical chemistry in the same way that the ICP-MS itself has done. Advances in semiconductor technology have resulted in the lowering of noise levels and the ability to extract signals from noisy backgrounds, the advances in computer software also helping in this task. More sophisticated computer software is also being developed to extract information from transient signals, and this is an area of intense interest to a number of manufacturers. Mass spectrometry offers the capability, with less harsh ionisation required than an atmospheric plasma, to provide information on the nature and identity of the species in a given sample. As sensitivity continuously improves, more and more 'unknown species' are observed and identification becomes an increasing challenge. Hence the interest in electrospray, low pressure plasmas, and other 'softer' ionisation approaches. The challenge is to obtain fragment containing mass spectra, with a sensitivity approaching that offered by the ICP-MS.

During the meetings of the Speciation 21 Network, there has been one overriding plea – for simple, low-cost, portable instrumentation to be developed for specific elements, which will enable screening to take place. In many industrial processes and in most environmental monitoring studies, the knowledge that is required is not the exact concentration of a particular material, but is it above or below a particular threshold level. Is this a simple task? The fact that such instruments are not currently available suggests that either they are not easy to design and manufacture, or that the instrument companies do not consider that they can make a profit from them. The answer, in fact, is probably a mixture of both – it is possible to visualise how such an instrument may be made to screen levels in drinking water for example, but how it could be modified to accept wheat, soil or even seawater samples is another matter. The difficulty in making such an instrument lies far more in the area of sample preparation and presentation – for such an instrument to return its development and marketing costs, and for it to be available at a low price, it must be very versatile in the types of sample that it can handle.

3 Sample Preparation

The area of sample collection, preservation, preparation and presentation is one that requires considerable study. The importance of sampling itself is often overlooked and yet it is vital to obtaining meaningful results. It is obvious that different types of sample require diverse collection and storage methods, but these vital areas are frequently disregarded by scientists and left to untrained staff, who may have little understanding of sampling theory on the many chemical processes. For speciation studies, it is imperative that, in the handling and extraction of the samples, the species are preserved, and this is often the major problem in the whole analytical exercise. Frequently, the chemist performing the analysis cannot be certain that these procedures have left the species unaltered. One new technology, which has helped in the area of sample preparation, has been the introduction of focused microwave systems. This technique can reduce the time taken to perform an extraction, and the temperature at which that extraction takes place, both of these often benefiting preservation of species. It is, however, a laboratory based technology, and the development of a portable microwave extraction system poses challenges. As a new technology, much has still to be learnt about adventitious changes in speciation caused by microwave energy. Automated solvent extraction systems are also proving their value in certain areas and could be made portable – they would appear to offer some potential benefits for speciation studies. The use of immobilised enzymes and bacteria for separating analytes of interest from complex matrices is also an area which appears to hold great promise, but one which will require much co-operation between biotechnologists and analytical chemists. Such interdisciplinary co-operation is often difficult to achieve and requires funding from organisations such as the European Union if it is to take place.

Solid phase extraction is another relatively new technique, which greatly assists the analytical chemist, and a huge growth in the use of the method has occurred over the past few years. Unlike the very high costs involved in developing major new instrumentation, the costs involved in developing new sorbents is much lower, and manufacturers have reacted very rapidly in producing new materials with an ever increasing list of applications.

There is clearly considerable scope for the development of the portable screening instrument, but whether such devices will ever appear on the market is open to question. We are much more likely to see purpose built analytical equipment capable of laboratory screening for speciation. Some systems, developed under European Union funding, such as an integrated separation, microwave induced plasma spectrometer[1] or the use of chemically modified sample introduction capillaries for ICP-MS, which separate species as they are introduced to the plasma,[2] are examples of such approaches.

4 Certified Reference Materials

Another topic that attracts continued interest, from all involved in speciation studies is the availability of Certified Reference Materials (CRMs). CRMs are materials as similar as possible to real samples, *e.g.* soils, plants, metals, foodstuffs, analysed by several laboratories using a variety of analytical techniques, who obtain sufficient independent agreement to allow a statement of the believed concentration of trace elements in the sample. They offer the best way in which to ensure that the methodologies employed and developed by various analytical laboratories are able to provide acceptable answers. It is essential to employ a CRM with a similar matrix to that of the samples. They have been in use for many years for total elemental composition, supplied by a number of sources world-wide, including the BCR within the European Community. CRMs are prepared under rigorous conditions, and are analysed by a panel of 'specialist laboratories'. It would be nice, but untrue, to say that there is always agreement in the values obtained by the 'expert laboratories', even when analysing for total concentrations, and outliers have to be removed. The situation becomes more challenging when speciated CRMs are being considered, and hence the production costs of speciated CRMs are very high. Speciated CRMs are going to assume a greater significance as legislators take greater account of the importance of the subject of speciation when framing legislation. There is, after all, no point in specifying that a foodstuff must contain no more than x ng g^{-1} of arsenic(III) and y ng g^{-1} of arsenic(v) if the results cannot be checked against an acceptable standard. A major difficulty is predicting which new CRMs are going to be required. The process of producing a new CRM may take up to five years. It is therefore necessary that future requirements are anticipated well in advance, so that the speciated CRMs are available prior to changes in legislation being introduced. This requires that there is excellent communication between the scientific community concerned, legislators and CRM producers, and that finance be available for the level of CRM production which is likely to be needed.

Another change which is worthy of consideration is for 'one shot' CRMs. At present, CRMs are normally sold in containers holding between 10 and 100 g of the material. When supplied, the CRM has a known and specified water content, and is frequently packed under an inert gas such as nitrogen. As soon as the CRM reaches the end user, the seal is broken, the gaseous protection is removed, water vapour may be absorbed and the material composition changes. How much better it would be if instead of one bottle containing 50 g of material, it was possible to purchase a number of sealed vials containing say 1 g of CRM. This has the added advantages that contamination of the CRM within the end users laboratory would be reduced. The reduced sample size does pose problems for homogeneity. One challenge which has not yet been met is that the dried, finely powdered CRMs currently available are different from the bulky, wet samples received in most laboratories. Speciation may change upon drying, powdering and sterilising such samples. There is a need to develop speciated CRM materials, more typical of those encountered, and to preserve them, so that the speciation is unaltered over a reasonable period of time.

As will have been seen, there is an urgent and overriding requirement for the range of speciated CRMs to be dramatically increased.

5 Spiking

Spiking is the addition of a known quantity of analyte to the sample and its subsequent extraction using the selected analytical. The degree to which the spike is extracted is then frequently assumed to be the recovery efficiency of the extraction process employed. The spiking process may be the best method currently available for this determination, but may not yield the correct recoveries. Consider an estuarine sediment – it is formed by the deposition over many years of all manner of humic, alluvial and other substances, a very complex matrix indeed, along with the analyte of interest. This analyte becomes bound over time, by complex interactions with this matter in a partially saline medium. This material is then removed, dried, and has a spike added to it, frequently contained in an organic solvent due to solubility problems of the analyte in water, the spike and sample are then left for say 24 hours to equilibrate, and then the spike is extracted. It is unlikely that the spike has become distributed fully throughout the sample, and become bound in the same way as the original analyte. This is an area which may well account for large variations in reported results. This technique may be the best/only one available at present, but much more research is required in this area to obtain a more meaningful method of measuring the effectiveness of the extraction processes used. This is even more important when speciation is being considered, since the interactions between the analyte species and that of the substrate material requires special attention.

6 Speciation at the Cellular Level

One branch of speciation which also requires considerably more research to be undertaken involves speciation at the cellular level. It is only within the past few years that the fate of many elemental species within the organs of the body has been fully explored, progress having been previously hindered by the lack of sufficiently sensitive analytical techniques. Work is now taking place to determine which cells within each organ are involved in the various processes involving the elements, but it is what takes place within the cell itself which presents the next major challenge to the analyst. The problem itself is quite simple, but the solution is not. Metals can be detected within the cell by various methods – at the most simple level by a total determination, having killed the cell. This provides the answer as to how much was there, but not in what form. It is generally true to say that science has little knowledge of what ligands are attached to metal atoms when they are inside the cell. The difficulty is that once the cell membrane is ruptured so that the contents of the cell are accessible, the speciation of the element concerned may well be altered. There is a need to determine the element within the cell itself, without causing the cell to alter its normal functions. Potentially, nuclear magnetic resonance (NMR) spectrometry offers a way forward but, with the exception of iron, most elements of interest are found in the body at concentrations well below those measurable by NMR. Without more fundamental information about interactions at the cellular level, better models for toxicity and pharmacology cannot be devised. Without such models, both of these sciences will be hampered and forced to focus on increasingly complex animal and human experiments.

7   Communication

Although this book sets out to cover the subject of speciation, comments made here apply to all areas of science. The Speciation 21 network has illustrated the great and overriding need that exists for scientists to communicate more effectively, both with each other and, perhaps more significantly, with non-specialists and with the general public. The network brought together representatives from industry, academia, regulatory authorities, medicine and instrument manufacturers, under the broad headings of food and nutrition, occupational health and the environment. At each of the expert meetings, much was learnt by all the delegates – the position and viewpoints of 'polluters' of the environment (and this clearly includes every one of us) has been set against those charged with the responsibility of detecting and monitoring the anthropogenic inputs, and suggestions as to ways in which it may be ameliorated have been propounded. Explanations have been given by the manufacturers of raw materials through to finished products as to the problems they face, and their genuine desire to learn how they may improve their manufacturing processes and so minimise any impact upon the environment and their workplace. With the increased legislative emphasis along the lines of 'the polluter pays' and with potentially large settlements to victims of industrial illness and accident, this attitude makes sound economic sense. Complaints have been heard from manufacturers about the ill informed and reactive regulations placed upon them by ill informed but well intentioned regulators, who often demand lower emissions than background levels – indeed mention was made of one discharge consent which apparently made it illegal for the company concerned to pour a glass of drinking water into their drains!

The network has allowed, possibly for the first time, truly multidisciplinary discussions of these problems to take place and has illustrated the dire need for scientists to improve the way in which they communicate. The broader topic of how scientific advances are reported must therefore be considered.

The long established measure of the success of a research scientist is the number of their papers which are published in peer reviewed scientific journals. Unfortunately, these journals are rarely read by anybody other than research scientists working in the same field. A few scientists are prepared to write for the 'popular' scientific press, where the results may be read by journalists and, provided the subject is considered sensational enough, may be reported to the general public. It may even be reported on radio or television, provided it can be covered by twenty second sound bites. In the field of speciation and analytical chemistry, however, it is difficult to imagine how the improvement in a detection limit of one order of magnitude may be packaged to interest the 'man in the street', without suffering misleading sensationalism. Indeed, in most countries, distrust of science and authority as a whole has reached a level at which scientific results and regulatory decisions are generally not discussed rationally.

One particular problem is that the media prefer the paradigm of the American law court to the paradigm of scientific debate. For every proposition, there is a proponent and an opponent. Neither must exhibit any doubt about their case, and must express 100% confidence. The truth then should emerge from debate between the protagonists. Science is not like this. Scientists rarely have 100% confidence levels in analytical measurements, for example, and deal in the balance of probabilities. They prefer to explain these to a sceptical public who persist in the belief that they are being paid to put one side of an argument. Reputation counts for less than voice quality or quick wittedness. The validity of measurement is an unknown concept to the general public.

One requirement is a way of explaining recent advances and how these affect everybody, whether for good or ill, and indicating in simple terms why what was good last week has become problematic this week. Non-scientists are understandably very confused by the stream of advice emanating from the scientific and 'para-scientific' community regarding health and life style, which totally contradicts what they have been told by scientists in the past. Unfortunately, it is difficult to explain how the continuous lowering of detection limits enables the determination of elements and compounds in environments and foods, where they have previously not been detected. This may, or may not, be significant for health.

The public cynicism of science is enhanced when the recent past is remembered. DDT, CFCs, PCBs and organotins were all heralded as wonderful products which would make the planet a better place. We know more about the problems now. Why should the public trust scientists any more? Why should they believe that genetically modified organisms (GMOs), for example, offer any advantages, other than making massive amounts of money for multinational biotechnology companies? The argument that they are subject to regulation and testing is no longer sufficient for many people, since these were in place when the aforementioned chemicals were first introduced. Many believe that the only way in which the public's trust in science will be restored is for far greater openness and easier access to information. One development, which has done more than any other since the invention of the printing press to influence how information is distributed, is the Internet. It can be expected to revolutionise the way in which knowledge is transferred and transmitted.

It has been said that the amazing thing about the Internet is its anarchy. Nobody can police it, and anybody can publish on it. Serious scientists must use the Internet to publish the results of their research and also explain, in layman's language, why their results make a difference. The speed at which information can be transmitted from site to site is now determined only by the speed of data entry. This is a challenge to the traditional scientific publication, with their three to twelve month lag time. Already, electronic versions of peer reviewed journals are well established. One problem for the publishers is how to meet the costs of publication, since once on the Internet they become public property. Whatever the solution to the problem, publishing the results of scientific research is bound to move from the printed journal. It is absolutely essential that scientists embrace the changes in information technology as they occur, but remain mindful of the needs to sustain the quality of information, as well as enhancing the availability of results.

As well as offering the opportunities for publication, the Internet also provides hitherto unequalled opportunities for people, anywhere in the world, to learn new skills and develop greater knowledge. It is in this area in which good, simple communication skills will be of great value.

The Speciation 21 Network has shown the value of multidisciplinary gatherings when considering matters affecting food, the environment and occupational health. The increasing compartmentalisation of research means that, in many cases, scientists continue with their research in isolation. The ability of the scientists to understand the needs of all stakeholders is of growing importance. To be competitive, industrial scientists need almost instant answers to problems. Next week, those problems may have faded into the background and new dilemmas require attention, and here researchers must provide the answers quickly. Researchers must be prepared to summarise their results in a suitable format, so that industrialists may gain rapid access to the information that they require. This is especially true in the area of speciation since it is a rapidly evolving science, requiring an interdisciplinary approach, and the results produced are vital to producers, environmentalists and legislators.

8   Legislation

While all are concerned to limit pollution, the rules and regulations governing industry and industrial pollution must be based on sound science, and must be realistic and achievable.

Complaints have been voiced from industrial delegates to the Speciation 21 Network meetings regarding the constraints placed upon them by governments and legislators. Examples have been given such as setting discharge consents below background levels. What is required is that legislators become rapidly familiar with the advances that have been made in the area of speciation, and then that they are sufficiently well informed in current scientific thinking that they are able to apply this knowledge sensibly, for the good of all, both public and industry alike.

Since speciation is a relatively new concept, many legislators have had little or no training in it. Those that have may have difficulty in keeping fully up to date with the rapid advances that have taken place within this field. In addition, there must always be a reluctance to add to the costs placed on industry and governments. This position is not improved when conflicting reports regarding the potential or actual damage caused by pollutants are published. Industrial concerns will obviously apply pressure for legislation to place as little cost on shareholders as possible, and globalisation allows industries to move to lower cost areas. Governments therefore face a formidable challenge. They have to impose restrictions which protect the workers within the industry, those living near that industry and the environment itself, whilst at the same time not imposing such constraints that will cause the industry concerned to lose competitiveness. National interests have to be considered, since multinational companies cannot be expected to have a loyalty to any particular country, and may close an operation in one country and move to another, if it is financially advantageous so to do. Regulations are therefore seen as a necessary evil, often to be implemented by agreement. It is therefore understandable that the legislators would resist new complications such as speciation.

Since one of the original aims of the European Economic Community was to provide a common market, it is imperative that environmental and occupational protection legislation be applied equally throughout the member countries. Legislators will always be in the unenviable position of having to catch up with events, unless innovation stifling measures are introduced. New products and processes appear at ever increasing rates, so the need for our legislators to be well informed has never been greater. While common legislation throughout the European Union has been challenging, the need for the common application of regulations is even more demanding. Disputes over analytical measurements are a frequent barrier to trade. If regulations specifying speciation are to be introduced, these need to be backed by a greater commitment to valid analytical measurements. Hence the importance of greater understanding of the underlying science, of more dialogue and for more CRMs to allow measurement disputes to be resolved.

Another legislative area in which speciation must be urgently considered concerns food. Under present regulations, manufacturers and processors of foodstuff are permitted to quote the recommended daily allowance (RDA) for a particular element and then the total concentration in their product. These figures represent the presumed requirements of the body, but absorption and utilisation depend upon speciation. In many cases, the element is present in the food in a non-bioavailable form, and the foodstuff concerned is therefore not contributing at all to the intake for the consumer. Thus a food label, whilst being factually correct, may not be helpful to the consumer. The same applies to the rapidly growing dietary supplement industry. New legislation is required which will lead to better labelling and the display of not only the total concentration of an element in a product, but the proportion which is available to the body. It follows again, that new CRMs are required to be produced as a matter of urgency, so that they are in place ahead of the introduction of such legislation.

9   Conclusion

The Speciation 21 Network has illustrated the growing maturity of the science of speciation. It has brought together those with interests in instrumental and methodological development with a wide range of users. Regulators have been impressed by the measurement advances made. Toxicologists, food scientists, environmental scientists and clinicians have understood more clearly the current state of the art in speciation and emphasised those areas of crucial value. As a result, it should be possible in the next decade to direct instrumental advances and new methodology in a more targeted way, to identify the key CRMs required and to establish robust regulations for protecting the environment, consumers and workers, but without placing excessive burdens on producers or regulators.

Subject Index

www.ingramcontent.com/pod-product-compliance
Lightning Source LLC
Chambersburg PA
CBHW050520190326
41458CB00005B/1610